Ultrafine Particles in the Atmosphere

THE ROYAL SOCIETY

Ultrafine Particles in the Atmosphere

Editors

L. M. Brown
University of Cambridge, UK

N. Collings
University of Cambridge, UK

R. M. Harrison
University of Birmingham, UK

A. D. Maynard
National Institute for Occupational Safety and Health,
Centers for Disease Control and Prevention, USA

R. L. Maynard
Department of Health, London, UK

ICP

Imperial College Press

Published by

Imperial College Press
57 Shelton Street
Covent Garden
London WC2H 9HE

Distributed by

World Scientific Publishing Co. Pte. Ltd.
5 Toh Tuck Link, Singapore 596224
USA office: Suite 202, 1060 Main Street, River Edge, NJ 07661
UK office: 57 Shelton Street, Covent Garden, London WC2H 9HE

British Library Cataloguing-in-Publication Data
A catalogue record for this book is available from the British Library.

First published by The Royal Society in 2000
Published in 2003 by Imperial College Press

ULTRAFINE PARTICLES IN THE ATMOSPHERE

ISBN 1-86094-358-6

Printed in Singapore by World Scientific Printers (S) Pte Ltd

PREFACE

The past decade has seen mounting evidence that atmospheric particles are more damaging to health than previously thought. Epidemiological studies relating population health to airborne particle concentrations provided some of the first indications of a hitherto unsuspected toxicity. There is a remarkable consistency of effect between different geographic locations, showing a correlation with the incidence of respiratory and cardiovascular diseases. Initially, these findings were received with some scepticism, because there appeared to be no plausible biological mechanism. However toxicological studies have since indicated a range of mechanisms that underpin the epidemiology, whereby airborne particles may damage human health. Many studies have now shown that some particles become more toxic per unit mass as their size decreases. Thus attention is focused upon particulate surface area or number per unit mass, rather than mass fraction, and one is led to consider 'ultrafine particles': those of effective diameter less than one-tenth of a micrometer.

In March 2000 The Royal Society (London) hosted a multi-disciplinary discussion meeting to address issues surrounding ultrafine particles in the atmosphere. Eminent researchers from a wide range of disciplines met together over two days to consider the current state of our knowledge and understanding of ultrafine aerosol generation, characterization, transportation, exposure and toxicity. The result was a comprehensive overview of what we know, and what information is yet needed, about ultrafine particles and how they potentially impact our health. This collection of papers is based on the lectures given at the meeting, and likewise comprehensively documents the current state of affairs regarding ultrafine particles in the atmosphere. Chapters 1 to 3 consider the characterization of ultrafine particles. Chapters 4 to 8 follow on by considering the sources of ultrafine aerosols. The remaining chapters deal with the health effects associated with ultrafine particulate exposure: chapters 9 to 13 focus on the toxicology of ultrafines, while chapters 14 to 16 conclude by considering the epidemiology of ultrafine aerosol exposure. Sadly, Professor Glen Cass, author of chapter 2 on the chemical composition of atmospheric ultrafine particles, passed away prematurely in July 2001.

It is only relatively recently that particles in the ultrafine size range have received serious attention. The first problem is to characterize them. Chapters 1 to 3 address this issue. Not only is it necessary to look at overall particle shapes and sizes, but the chemical composition also needs to be known. It seems that quite firm conclusions can be drawn, from studies in quite different locations. Ultrafines account for a very large proportion of the number of particles in the atmosphere, although only a modest proportion of the surface area, and a minute proportion of the mass. Although transport vehicles represent the main source of them in urban areas, interesting new results show that even in remote marine atmospheres ultrafine particles are formed, which grow to become nuclei for cloud condensation. In urban environments, the particles appear to be predominantly organic compounds in nature, with further contributions from elemental carbon, trace metals, sulphate, nitrate, and ammonium. It is noteworthy that modern physical techniques are now able to look at particles of nanometre size, and analyze them, and one can expect further technical developments to enhance this capability. It appears that ultrafine particles in engine exhaust comprise two families: the larger ones from the formation of solid carbonaceous particles during the combustion process, and the smaller ones from gas-to-particle conversion processes during dilution of exhaust gases. Control of dilution can influence the relative proportions of the two families of particles.

What are the major sources of ultrafines? Chapters 4 to 8 explore the answers to this question. Gas-to-particle conversion that is, nucleation in a supersaturated atmosphere — is one source. The nucleation of sulphuric acid requires a third species such as ammonia to explain observed aerosol generation. Detailed models can be constructed using the chemistry once it is elucidated, from which it is possible to predict the formation of sulphates from sulphur dioxide oxidation as well as of secondary organic particles, produced by the oxidation of both natural and anthropogenic organic compounds. When combined with predictions of long-range transport of air masses, such models display an impressive capability to predict local sulphate concentrations deriving from distant sources in the UK and in continental Europe. On the other hand, emissions from internal combustion engines caused by exhaust dilution processes can be explained in detail by nucleation processes involving sulphuric acid and subsequent condensation of organic matter largely coming from lubricating oils. By control of the sulphur content, it appears possible to suppress the formation of such particles and to manipulate particle size distributions through vehicle design and fuel consumption. However, from open flames fed by hydrocarbon combustion one sees solid carbonaceous particles a few nanometres in diameter, comprised of elemental carbon. These coagulate, growing ultimately into diffusion-limited aggregates of smaller clusters. Do these processes occur in the workplace, and if so, how can they be controlled? How might it be possible to set out an ultrafine particle convention for use alongside other particle size conventions to use in regulating exposure within the workplace? The answer to this question depends upon the potential health risks associated with the various types of ultrafines, and leads therefore to the question of particle toxicology and its mechanisms, the subject of chapters 9 to 13.

Examination of the patterns of deposition of very small particles in the respiratory system has led to an appreciation that, while ultrafine particle deposit well in the distal parts of the lung, even smaller particles, less than 10 nm in diameter, do not: most deposit in the upper airways. The interaction between such particles and the material that lines the respiratory tract has been studied, and one can see that the low surface tension produced by a surfactant film aids particle transfer through the liquid lining layer. Perhaps most remarkably, ultrafine particles have been shown to have unusual toxicological potency. Particles of titanium dioxide, aluminium oxide and carbon black of less than 50 nm in diameter have all been shown to be much more toxic per unit mass than similar particles ten times larger. The reasons for this property are being explored. The smallest of such particles may display extraordinary surface chemistry which arises from geometrical constraints on the packing of ions in what amounts to a 'particulate molecule' than a section of bulk material. Thus there is not only high specific area, but surfaces of special composition which may be rich in metallic species and play an important part in free radical generation; an effect displayed by oil fly-ash, but not in that produced by small particles of carbon black or latex, which are metal-free.

When it comes to studying the effects of ambient aerosol upon the health of populations, one can show that variations in daily counts of events such as deaths and hospital admissions are related to day-to-day changes in the mass concentration of particles. Chapters 14 to 16 are concerned with detailed epidemiological studies which produce robust confirmation of these effects. The view that there is a causal association between mass concentrations of particles and ill-health is now generally accepted. However there is evidence suggesting that the association is stronger with smaller-diameter particles. In some studies, number concentration of particles is found to be more strongly related to ill-health than mass concentration. Other studies confirm this by showing that mass concentrations of smaller particles may be better correlated to health effects than those of larger particles. A limited number of cohort studies suggest that long-term exposure to even low-concentrations of fine particles ($PM_{2.5}$, or sulphate) may be associated with reduced life expectancy.

Thus, although ultrafine particles contribute very little to the mass concentration of the ambient aerosol, they may contribute disproportionately to its toxicity. The 'ultrafine hypothesis' seeks to explain at least some of the reported associations between mass concentration and indices of ill-health by suggesting that mass concentration is a surrogate for the number or surface area of ultrafine particles. If true, this view will have far-reaching implications: control of mass concentration without control of the ultrafine component will have little effect in reducing damage to health. The study of ultrafine particles in the atmosphere, and their pathways into the human body, is a new and vital multi-disciplinary subject. With increased awareness of ultrafine aerosol exposure in the environment, workplace and in the home, and in particular with the emergence of aerosol-based nanotechnologies, this is an area of research which will no doubt attract intense study over the next few years.

L. M Brown
N. Collings
R. M. Harrison
A. D. Maynard
R. L. Maynard

Contributions originating from the proceedings of a Royal Society Discussion Meeting first published in Philosophical Transactions of the Royal Society, Series A, Vol. 358 No. 1775, pp. 2561–2797.

A Discussion Organized and Edited by L. M. Brown, N. Collings, R. M. Harrison, A. D. Maynard and R. L. Maynard

Discussion held 15 and 16 March 2000

CONTENTS

CHAPTER 1

MEASUREMENT OF NUMBER, MASS AND SIZE DISTRIBUTION OF PARTICLES IN THE ATMOSPHERE

Roy M. Harrison, Ji Ping Shi, Shuhua Xi, Aftab Khan, David Mark, Rob Kinnersley and Jianxin Yin

Division of Environmental Health and Risk Management, University of Birmingham, Edgbaston, Birmingham B15 2TT, UK

Typical size distributions for airborne particles are described and the significance of the ultrafine fraction highlighted. Size distributions may be expressed in terms of either mass (volume), surface area or number, and the interpretation of each is discussed together with appropriate measurement methods. The sources of ultrafine particles in the atmosphere include both primary emissions and secondary particles formed through homogeneous nucleation processes within the atmosphere. Examples of measurements of atmospheric ultrafine particles are given, highlighting situations with high concentrations of primary ultrafine particles and also situations where gas-to-particle conversion through homogeneous nucleation gives rise to bursts of new particle production. Finally, the relationship between particle mass and number within the atmosphere at a polluted site is examined.

Keywords: ultrafine particles; particle size distribution; particle nucleation processes; road-traffic emissions

1. Introduction

Atmospheric particulate matter is inherently more difficult to study than gas-phase components of the atmosphere. It is highly variable in size and in chemical composition, and, indeed, individual particles may have a very complex make-up (Harrison & van Grieken 1998). Simple spectroscopic techniques which can be applied to qualitative and quantitative analysis of gas-phase species in the atmosphere are not applicable in useful ways to the determination of aerosol composition. The subject therefore progressed rather slowly as a topic of research until it received three major stimuli in the 1980s. These were as follows.

(a) Recognition of the importance of airborne particles through both direct and indirect mechanisms as regulators of global climate (Horvath 1998). Direct mechanisms relate to the absorption and scattering of solar and terrestrial radiation, while the indirect mechanisms, which may ultimately prove more important, operate through the role of airborne particles as cloud condensation nuclei (Charlson *et al.* 1987).
(b) Discovery of the Antarctic ozone hole and a recognition that reactions on polar stratospheric cloud particles were key to the chemistry leading to dramatic ozone depletion.
(c) The discovery that both acute and chronic exposure to airborne particles is associated statistically with a range of adverse health outcomes and a growing acceptance that these relationships are causal (COMEAP 1995).

In both the first and last of these areas, ultrafine particles are a source of especial interest. In the case of climate regulation, one of the key areas of interest is the atmosphere over the oceans, where cloud condensation nuclei develop from the growth of new initially ultrafine particles formed by gas-to-particle conversion processes (Charlson *et al.* 1987). In the area of human health, toxicological studies using rat models have shown that ultrafine particles are considerably more toxic per unit mass than coarser particles of the same material (Donaldson & MacNee 1998). Additionally, one of the hypotheses explains the unexpected link between particulate matter exposure and cardiovascular disease outcomes in terms of the capability of ultrafine particles to penetrate the pulmonary interstitium (Seaton *et al.* 1995).

This paper will set the background to many of the subsequent papers on research on airborne particulate matter by describing the origins and measurement methods for airborne particles, the interrelationship between methods and measurements, and by giving some examples of particle measurements illustrative of specific phenomena in the atmosphere.

2. Size Distribution of Particles in the Atmosphere

There are three distinct modes into which airborne particles can typically be divided. These may be described as follows.

Transient nuclei mode. These are particles typically less than *ca.* 100 nm in diameter, which are relatively newly formed, having arisen from the condensation of involatile materials to form new particles which subsequently grow by condensation processes. The formation can occur both in hot combustion gases and in metallurgical processes, involving, for example, the condensation of lead atoms from the vapour to form particles, or within the atmosphere itself from chemical reactions of gases to form involatile species which condense to form particles. In order for new particles to form through a process known as homogeneous nucleation, a very substantial supersaturation of vapour needs to occur. The only clearly recognized example of this process in the atmosphere is the oxidation of sulphur dioxide to sulphuric acid, which is able to undergo binary nucleation with water or ternary nucleation with water and ammonia (Korhonen *et al.* 1999). It is likely that most new particles formed in the atmosphere arise from this oxidation process, although subsequent particle growth may be enhanced through condensation of semi-volatile organic compounds (Marti *et al.* 1997). Newly formed nucleation mode particles are typically of the order of 1–2 nm in diameter, but rapid growth generally ensues. The mode in the number distribution of nucleation mode particles is typically *ca.* 20–30 nm in diameter.

Accumulation mode. Particles in the transient nuclei mode can grow both by condensation of low volatility materials and through coagulation. In doing so they are likely eventually to enter the accumulation mode which describes particles between *ca.* 100 nm and 2 μm in diameter. Accumulation mode particles are subject to rather inefficient loss from the atmosphere by wet and dry deposition processes and, because of their low number concentration, are not subject to significant further growth through coagulation. They have an atmospheric lifetime of several days and can therefore travel over very long distances within the atmosphere.

Coarse particle mode. After a minimum in abundance at *ca.* 1–2 μm, there is a subsequent growth in particle abundance (in terms of mass but not number) for particles which extend in size up to *ca.* 100 µm, although above 10 µm diameter their atmospheric lifetime becomes rather short. These coarse mode particles are formed through mechanical attrition and disintegration processes such as the formation of sea spray from bubble bursting in the ocean and the wind-blown suspension of land surface dusts and soil. They therefore arise quite differently from transient nuclei and accumulation mode particles and can be quite distinct in their chemical composition (QUARG 1996).

The term *ultrafine particles* does not have a universally agreed definition but is widely accepted as describing particles of less than 100 nm in diameter. The further widely used term *nanoparticles* again has no universally agreed definition, but is widely used to describe particles of less than 50 nm in diameter.

Table 1. Influence of particle size on particle number and surface area for a given particle mass.

particle diameter (µm)	relative number of particles	relative surface area
10	1	1
1	10^3	10^2
0.1	10^6	10^4
0.01	10^9	10^6

Airborne particles are most frequently measured as either number-weighted or mass-weighted distributions. Because of the cube dependence of volume and, thus, mass on diameter, the two kinds of distribution look extremely different. Figure 1 illustrates this through plotting the number-weighted, surface-area-weighted and mass-weighted distributions of atmospheric particles measured in Birmingham, UK. Measurements were made using a scanning mobility particle sizer (SMPS) and aerodynamic particle sizer (APS) simultaneously to capture the smaller and larger ends of the particle size distribution, respectively. Reference to table 1 gives a clearer insight into the mathematics, whereby 10^9 particles of 10 nm diameter have the same mass as 1000 particles of 1 µm diameter or one particle of 10 µm diameter. Thus, the extremely numerous particles in the transient nuclei

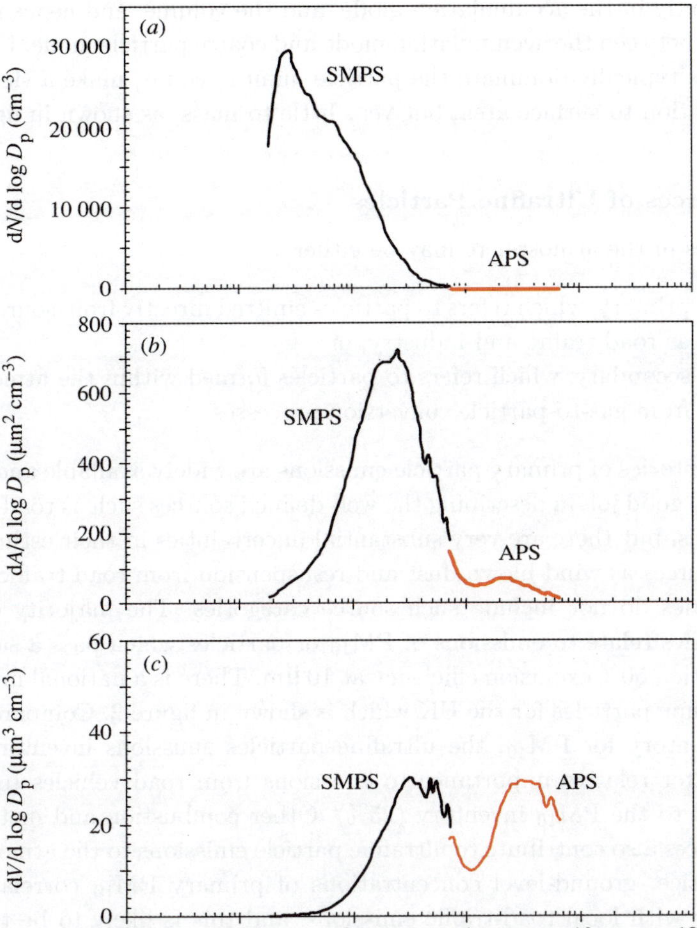

Fig. 1. A measured particle size distribution from suburban Birmingham, weighted by (*a*) number, (*b*) surface area, and (*c*) volume.

mode may represent only a very small proportion of total particle mass. Conversely, the very significant mass of particles in the coarse particle mode are very few in number. In simple terms it is likely that the bulk of the particle number is in the transient nuclei, the surface area is predominantly in the accumulation mode, and the volume, and hence mass, is divided between the accumulation mode and coarse particle mode. Ultrafine particles typically dominate the particle number count, make a significant contribution to surface area, but very little to mass, as shown in figure 1.

3. Sources of Ultrafine Particles

Particles in the atmosphere may be either

(a) primary, which refers to particles emitted directly from sources such as road traffic and industry; or
(b) secondary, which refers to particles formed within the atmosphere from gas-to-particle conversion processes.

Inventories of primary particle emissions are widely available and generally do a good job in describing the well-defined sources such as road-vehicle emissions, but there are very substantial uncertainties in their estimates of such sources as wind-blown dust and resuspension from road traffic. Many inventories do not include such source categories. The majority of such inventories relate to emissions of PM_{10} or particles which pass a sampling inlet with a 50% exclusion efficiency at 10 μm. There is a national inventory of ultrafine particles for the UK which is shown in figure 2. Compared with the inventory for PM_{10}, the ultrafine particles emissions inventory gives far greater relative importance to emissions from road vehicles (60%) as opposed to the PM_{10} inventory (25%). Other combustion and metallurgical sources also contribute to ultrafine particle emissions to the atmosphere. In practice, ground-level concentrations of primary PM_{10} correlate very strongly with local road-traffic emissions, and this is likely to be the case especially for ultrafine particles. Indeed, observations of particle number counts, which reflect primarily the abundance of ultrafine particles, have shown that such particles provide an excellent tracer of road-vehicle traffic emissions (Harrison *et al.* 1999*a*).

Homogeneous nucleation to form new particles is especially favourable in environments with a low pre-existing particle surface area, which acts as a competitive site for condensation. Therefore, homogeneous nucleation is expected to be important primarily in remote areas, and, indeed, quite spectacular bursts have been reported in the coastal zone, although the precise mechanisms are not fully understood (Allen *et al.* 1999; O'Dowd *et al.*

1999). There are also a small number of convincing observations of homogeneous nucleation to form new particles in the urban atmosphere. Current nucleation theories are limited in their ability to predict such phenomena, although the recent development of a theory for ternary nucleation may offer added insight (Korhonen *et al.* 1999).

4. Measurement Methods for Particulate Matter

A large variety of instruments is available, many of them through commercial suppliers, for the measurement and characterization of airborne particles. Within the scope of this article it would be possible to describe only a few of the most commonly used procedures. No endorsement of specific techniques or manufacturers is implied.

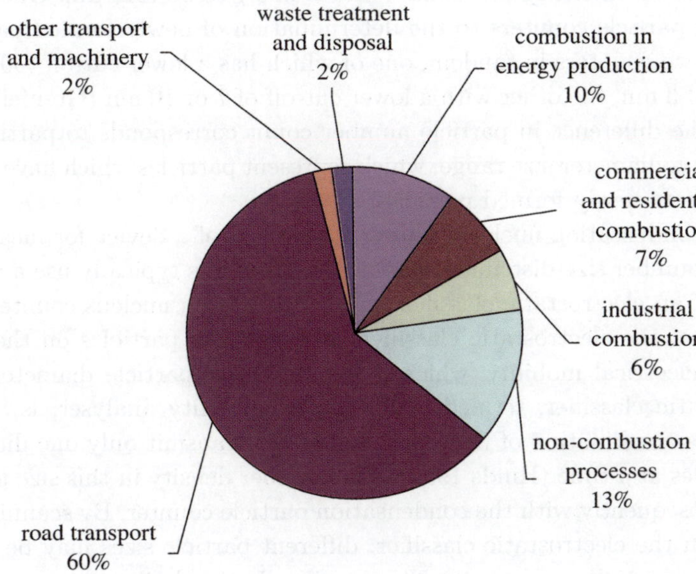

Fig. 2. Pie chart illustrating the sources of $PM_{0.1}$ emissions in the UK in 1996 (APEG 1999).

4.1. *Particle Number*

This is measured through use of condensation nucleus counters (CNCs), which have been available in manual form, such as the Nolan–Pollack counter, for many years, and more recently in the form of continuous devices. The continuous CNC instruments are typically based on drawing particles through a zone which is saturated with n-butanol vapour, which is subsequently cooled to cause condensation of the vapour on the particles (Stolzenburg & McMurry 1991). This causes the particles to grow to the order of 10 μm diameter, at which they are very effective light scatterers and they are monitored at low number density through counting the signals from particles as they pass through a light beam, or at higher number densities through a photometric mode which determines 90° scattered intensity of incident light. The lower size cut-off of such instruments is dependent on design and the degree of supersaturation achieved, but typically varies in commercial instruments from 3 nm in ultrafine particle counters to 10 or 20 nm in less-sophisticated devices. The upper size limit is determined by the aspiration efficiency of the inlet and is likely to be *ca.* 5 μm. One application of particle counters to the determination of newly formed particles is using two counters in tandem, one of which has a lower cut-off (50% efficiency) of 3 nm, the other with a lower cut-off of 7 or 10 nm (Grenfell *et al.* 1999). The difference in particle number count corresponds to particles in the 3–7 nm diameter size range, which represent particles which have grown slightly from newly formed particles.

The condensation nucleus counter forms part of a device for measuring particle number size distributions. Such instruments typically use a combination of an electrostatic classifier and condensation nucleus counter. The function of the electrostatic classifier is to separate particles on the basis of their electrical mobility, which is a function of particle diameter. The electrostatic classifier, termed a differential mobility analyser, is 'tuned' through a combination of flows and voltage to transmit only one diameter of particles at a time (Hinds 1999). The number density in this size is measured subsequently with the condensation particle counter. By scanning the voltage in the electrostatic classifier, different particle sizes may be transmitted sequentially and a complete size distribution built up over a period of a few minutes. Such instruments are now being deployed within a small UK measurement network.

4.2. *Particle Mass*

Classically, particle mass has been determined by collecting airborne particles by filtration and weighing the filter before and after particle collection.

In order to restrict the particles to a given size range, such as PM_{10}, $PM_{2.5}$ or $PM_{1.0}$, size-selective inlets are available which restrict the particles allowed access to the filter. Such samplers may be termed either high volume or low volume depending on the airflow rates. In UK networks, both PM_{10} and $PM_{2.5}$ are measured continuously using tapered element oscillating microbalance (TEOM) samplers. In these instruments, particles are collected on a small filter which is located on the tip of a tapered glass element which forms part of an oscillation microbalance. The oscillation frequency of the microbalance changes with the mass of particles collected on the filter. One facet of these instruments is that the inlet airstream is typically heated to 50 °C, which leads to an almost complete loss of semi-volatile particles, which in some situations can represent a significant proportion of particle mass. In the UK atmosphere, TEOM instruments typically measure concentrations of PM_{10} some 20–30% lower than the more conventional so-called gravimetric samplers (APEG 1999).

To date, no such instrument has been designed specifically for the determination of ultrafine particle mass. Estimates of the mass of particles less than 100 nm in diameter can be made through collecting particle samples in a size-fractionated manner using cascade impactors, which depend upon the inertial properties of particles to separate them into different size bands. Plotting of the full size distribution and making a cut at 100 nm would allow an estimate of ultrafine particle mass, although this is rarely measured. Loss of semi-volatile materials can be a major problem in the lower pressure impactors typically used for separating ultrafine particles.

4.3. *Surface Area*

This is rarely measured directly, although a device called an epiphaniometer has been described, which determines the Fuchs surface area of particles (Gaggeler *et al.* 1989). It does so by attaching a gaseous radionuclide to the particle surface and counting collected radioactivity. The physics of radionuclide attachment is not a simple single function of surface area across the entire particle size range, and particles significantly greater in diameter than the mean free path of the gas molecules offer a significant diffusion resistance to radionuclide attachment. Surface areas may also be estimated indirectly from measurements of particle size distribution, provided the particle geometry is known or assumed.

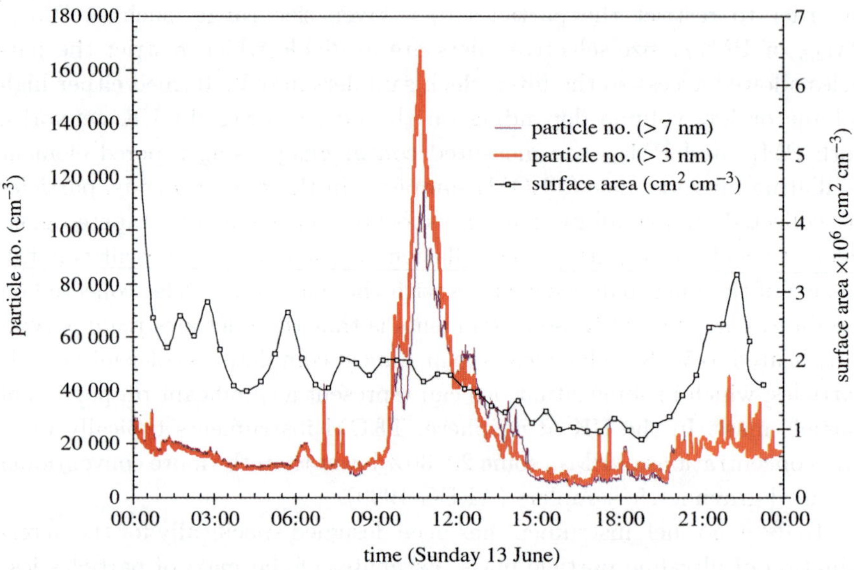

Fig. 3. Particle number count, greater than 7 nm and greater than 3 nm, and surface area on 13 June 1999 in suburban Birmingham.

5. Examples of Measurements of Atmospheric Ultrafine Particles

Four examples are given of measurements of particles in the ultrafine size range, illustrating a substantial range of environments.

5.1. *Mace Head, Ireland*

Mace Head is on the west coast of Ireland and frequently experiences relatively clean marine airmasses. Observations over a period of years have shown massive bursts in new particle formation at Mace Head, which, from their nature and from measurements offshore, are known to occur within the coastal zone (Allen *et al.* 1999). Particle number densities can well exceed 10^5 cm^{-3}, which is immense in comparison with background concentrations of particles in marine air at Mace Head, typically of the order of 100–500 cm^{-3}. Not only do particle number concentrations go to very high levels, a large proportion of the particles are in the 3–7 nm diameter size range and are, therefore, reflective of very newly formed particles. The mechanism of particle formation at Mace Head is not fully understood, but an intriguing observation is that the particle bursts appear to occur only in daytime and predominantly at low tide (Allen *et al.* 1999). It therefore seems likely that some substance, probably organic in nature, released from marine macroalgae at low tide, plays a major role in particle formation and/or growth. Most probably the nucleation process involves sulphuric acid, probably in combination with water vapour and ammonia, but rapid particle growth depends on the availability of low-volatility organic matter, probably originating from atmospheric oxidation of highly reactive organic compounds released from the macroalgae in the coastal zone. Similar observations have been made at other coastal sites, but appear to depend on a rocky coastline suitable for macroalgal growth (Mihalopoulos *et al.* 1992).

5.2. *Weybourne, North Norfolk*

Measurements in the summer of 1995 at the Weybourne site on the north Norfolk coast showed substantial bursts in particle number concentration which were not accompanied by an increase in surface area as measured by the epiphaniometer (Harrison *et al.* 2000). Closer examination of these data showed that, unlike Mace Head, there was no relationship to the tidal cycle. However, the nature of the coastlines is quite different, with no exposure of rocks and macroalgae at Weybourne at low tide. The largest bursts in new particle production occurred in polluted air travelling from the land towards the sea, and the onset of particle production corresponded to the increase in solar radiation capable of photolysing ozone to form excited state oxygen atoms, which in turn lead to the formation of hydroxyl radicals. These hydroxyl radicals are responsible for the oxidation of sulphur dioxide and also the rapid oxidation of many organic compounds, therefore leading to the production both of sulphuric acid vapour, capable of nucleation, and of oxidized organic compounds, which can contribute to particle growth.

5.3. *Suburban Birmingham*

Measurements with tandem particle counters and of size distributions using a scanning mobility particle counter, shown in figures 3 and 4, respectively, have shown clear evidence on a small number of occasions of new particle production within the suburban atmosphere of Birmingham. Thus, it is seen in figure 3 that between 09:30 and 12:00, there is a substantial burst in particle number concentration containing an appreciable proportion of particles in the 3–7 nm diameter range. The plot of size distribution versus time in figure 4 clearly shows that at around 10:15 the peak in the number distribution comes within range of the scanning mobility particle sizer at *ca.* 10 nm and steadily grows over a period of 2 h to *ca.* 27 nm. Such nucleation and growth processes are currently the subject of intensive research activity.

5.4. *Twin Site Measurements in London*

Measurements of PM_{10}, $PM_{2.5}$, traffic-related gases, particle number and size distribution are made on a continuous basis at two nearby sites in central London. London Marylebone Road is a site located adjacent to one of the busiest roads in central London, carrying *ca.* 70 000 vehicles per day. About 2 km distant, the site of London Bloomsbury is a central urban background site, where the monitoring station is located within the centre of an urban square. Data are archived on an hourly basis and differences calculated between the two sites, which are taken to be representative of the roadside increment due to traffic at the Marylebone Road location (APEG 1999). The data show a much elevated particle number concentration and a moderately elevated mass concentration at the Marylebone Road site, with the traffic increment in the particle size distribution having a mode in the number-weighted distribution at *ca.* 20 nm diameter, well below that recorded using conventional dilution tunnel methods in many of the studies of engine exhaust, which we have shown to give an unreliable estimate of size distribution (Shi & Harrison 1999). Particle number counts show substantial variations with time of day, reflective of road-traffic activity and prevailing meteorological conditions, as exemplified by figure 5, which shows a large morning rush-hour peak from 06:00 to 10:00.

6. Relationship of Particle Mass and Number within the Atmosphere

As noted above, particle number concentrations are dominated by particles in the transient nuclei mode, whereas the volume and, hence, mass lies predominantly in the accumulation and coarse particle ranges. There is therefore no necessity that the number and mass concentrations should be especially well correlated. If it were the case that the effects of particles on

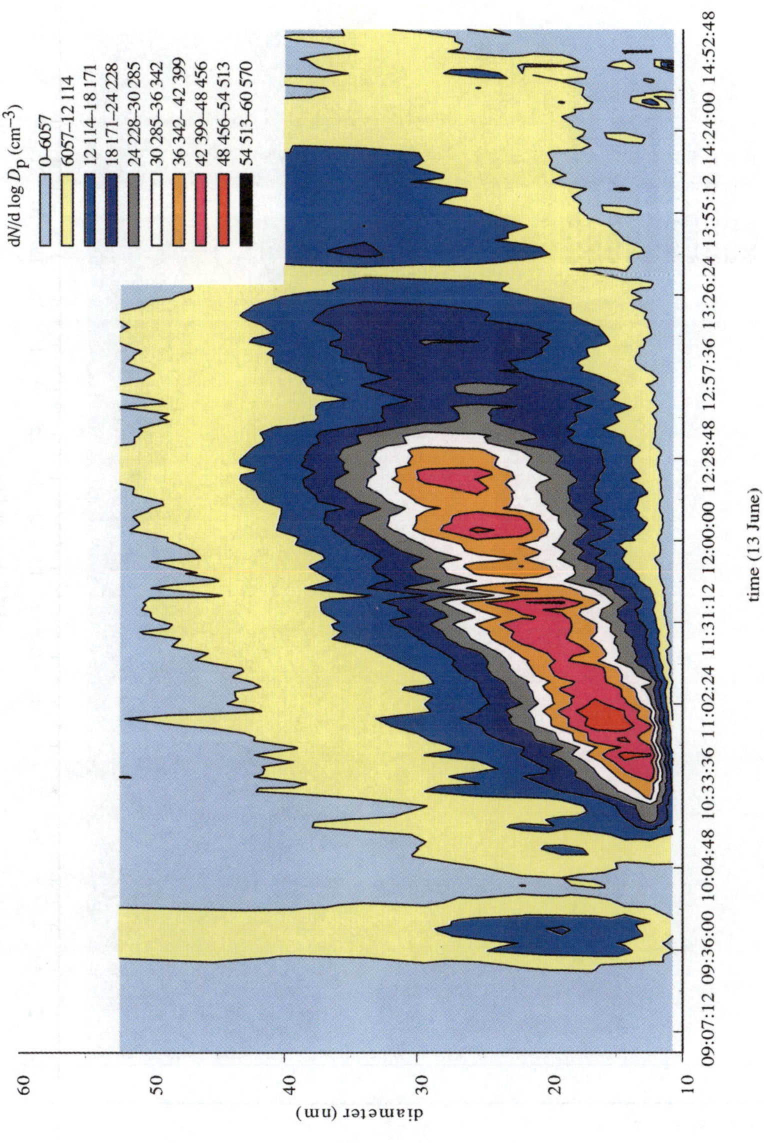

Fig. 4. The evolution of the particle size distribution during a nucleation event on 13 June 1999 (see figure 3).

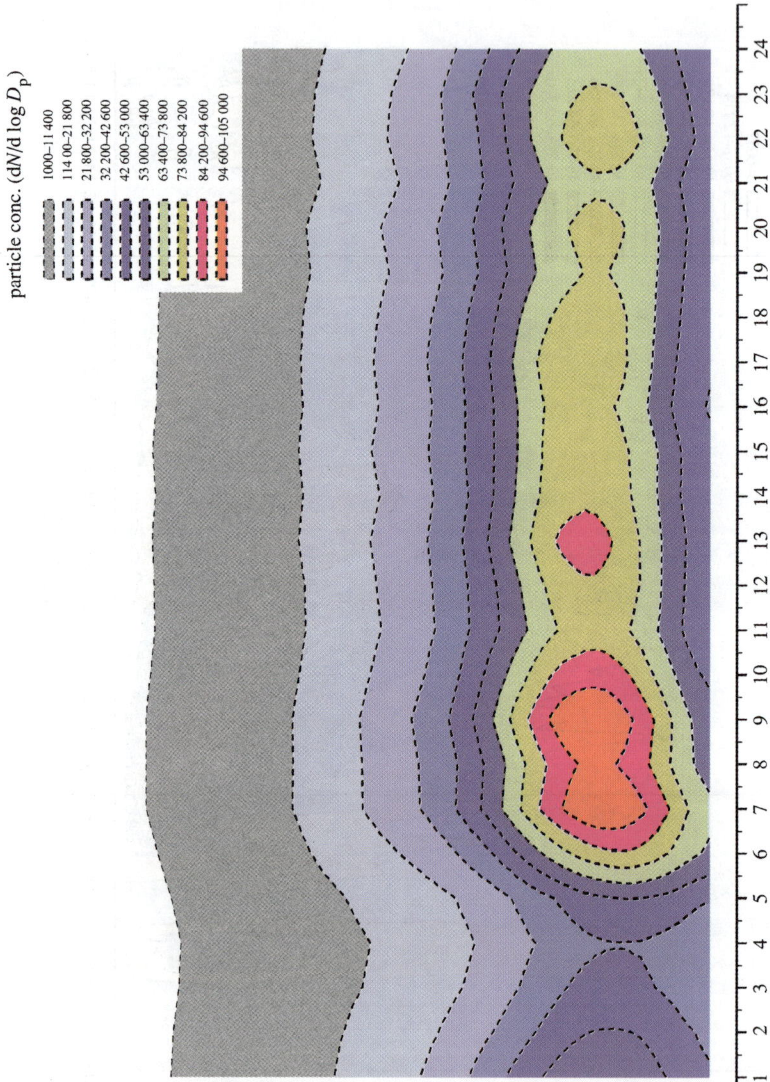

Fig. 5. Average diurnal plot of particle size distributions for April and May 1998 at London, Marylebone Road.

human health were driven by the ultrafine fraction, then it might be surprising if epidemiological studies using PM_{10} mass as the metric of particle concentration were to show a correlation with adverse health outcomes.

In practice, because fine particles (less than 2.5 μm) tend to form a relatively constant proportion of PM_{10} mass, and because the majority of such particulate matter arises from secondary sources and a small number of primary combustion-related sources, particle number and mass tend to be broadly correlated within the atmosphere. Thus, for example, at the London Marylebone Road site, particle number concentration and mass are quite strongly correlated in the road-traffic increment ($r = 0.72$), and in a study at a background site in central Birmingham a quite strong correlation between PM_{10} mass and number count ($r = 0.66$) was also observed (Harrison *et al.* 1999*b*). Such correlations are likely to be sufficiently strong as to obscure any ability to clearly differentiate through epidemiological studies between PM_{10} mass and particle number as the causal agent in driving adverse health outcomes, at least in the UK urban areas within which our measurements have been taken.

Acknowledgements

This research has been funded through grants and contracts with the Natural Environment Research Council, Department of Environment, Transport and the Regions, and the Perkins Engine Company Limited.

References

Allen, A. G., Grenfell, J. L., Harrison, R. M., James, J. & Evans, M. J. 1999 Nanoparticle formation in marine airmasses: contrasting behaviour of the open ocean and coastal environments. *Atmos. Res.* **51**, 1–14.

APEG (Airborne Particles Expert Group) 1999 Source apportionment of airborne particulate matter in the United Kingdom (ed. R. M. Harrison *et al.*). The First Report of the Airborne Particles Expert Group, Department of Environment, Transport and the Regions, London.

Charlson, R. J., Lovelock, J. E., Andreae, M. O. & Warren, S. G. 1987 Oceanic phytoplankton, atmospheric sulfur, cloud albedo and climate. *Nature* **326**, 655–661.

COMEAP (Committee on the Medical Effects of Air Pollutants) 1995 *Health effects of non-biological particles.* Department of Health, UK.

Donaldson, K. & MacNee, W. 1998 The mechanism of lung injury caused by PM_{10}. In *Issues in environmental science and technology* (ed. R. E. Hester & R. M. Harrison), no. 10, pp. 21–32. The Royal Society of Chemistry.

Gaggeler, H. W., Baltensperger, U., Emmenegger, M., Jost, D. T., Schmidt, O. H., Haller, P. & Hofmann, M. 1989 The epiphaniometer, a new device for continuous aerosol monitoring. *J. Aerosol Sci.* **20**, 557–564.

Grenfell, J. L. (and 16 others) 1999 An analysis of rapid increases in condensation nuclei concentrations at a remote coastal site in western Ireland. *J. Geophys. Res.* **104**, 13 771–13 780.

Harrison, R. M. & van Grieken, R. (eds) 1998 *Atmospheric particles.* Wiley.

Harrison, R. M., Shi, J. P. and Jones, M. R. 1999*a* Continuous measurements of aerosol physical properties in the urban atmosphere. *Atmos. Environ.* **33**, 1037–1047.

Harrison, R. M., Jones, M. & Collins, G. 1999*b* Measurements of the physical properties of particles in the urban atmosphere. *Atmos. Environ.* **33**, 309–321.

Harrison, R. M., Grenfell, J. L., Allen, A. G., Clemitshaw, K. C., Penkett, S. A. & Davison, B. 2000 Observations of new particle production in the atmosphere of a moderately polluted site in eastern England. *J. Geophys. Res.* **105**, 17 819–17 832.

Hinds, W. C. 1999 *Aerosol technology: properties, behavior and measurement of airborne particles.* Wiley.

Horvath, H. 1998 Influence of atmospheric aerosols on the radiation balance. In *Atmospheric particles* (ed. R. M. Harrison & R. van Grieken). Wiley.

Korhonen, P., Kulmala, M., Laaksonen, A., Viisanen, Y., McGraw, R. & Seinfeld, J. H. 1999 Ternary nucleation of H_2SO_4, NH_3 and H_2O in the atmosphere. *J. Geophys. Res.* **104**, 26 349–26 353.

Marti, J. J., Rodney, R. J. & McMurry, P. H. 1997 New particle formation at a remote continental site: assessing the contributions of SO_2 and organic precursors. *J. Geophys. Res.* **102**, 6331–6339.

Mihalopoulos, N., Nguyen, B. C., Boissard, C., Campin, J. N., Putaud, J. P., Belviso, S., Barnes, I. & Becker, K. H. 1992 Field study of dimethylsulfide oxidation in the boundary layer: variations of dimethylsulfide, methanesulfonic acid, sulfur dioxide, non sea-salt sulfate and Aitken nuclei at a coastal site. *J. Atmos. Chem.* **14**, 459–477.

O'Dowd, C. (and 11 others) 1999 On the photochemical production of new particles in the coastal boundary layer. *J. Geophys. Res. Lett.* **26**, 1707–1710.

QUARG (Quality of Urban Air Review Group) 1996 Airborne particulate matter in the United Kingdom (ed. R. M. Harrison *et al.*).

Seaton, A., MacNee, W., Donaldson, K. & Godden, D. 1995 Particulate air pollution and acute health effects. *Lancet* **345**, 176–178.

Shi, J. P. & Harrison, R. M. 1999 Investigation of ultrafine particle formation during diesel exhaust dilution. *Environ. Sci. Technol.* **33**, 3730–3736.

Stolzenburg, M. R. & McMurry, P. H. 1991 An ultrafine aerosol condensation nucleus counter. *Aerosol Sci. Technol.* **14**, 48–65.

Discussion

M. S. BINGLEY (*Cobham, UK*). Catalytic convertors take 10 min to warm up and start working. During this period, exhaust gas emissions will be rich ($\lambda < 1.0$). Will this add ultrafine particles to the atmosphere of urban areas during this period?

R. M. HARRISON. The operation of a petrol engine under fuel-rich conditions with a cold catalytic convertor is likely to lead to enhanced production of ultrafine particles, although I am not aware of any research into this issue. It should be borne in mind, however, that under ever-tightening emissions-control legislation, catalytic convertors are being designed to higher standards, such that they become effective far more rapidly than in the past.

M. WALLIS (*FoE Cymru, Cardiff, UK*). In the $PM_{0.1}$ inventory that you presented, the UK's industrial fraction of 10% surely has lower importance at street level. Particle lifetime and transport are relevant for estimating human exposure, so should we rate industrial sources as significant?

R. M. HARRISON. The industrial contribution to $PM_{0.1}$ emissions in the UK is greater than 10%. Industrial combustion and industrial processes together account for 19% of the inventory total, and a proportion of the 3% from waste treatment and disposal and 10% from combustion in energy production and transformation are also attributable to industry. Monitoring of the atmosphere clearly shows an elevation of ultrafine particle concentrations within plumes from higher temperature sources and I have no doubt that industrial sources are significant in many locations.

UNKNOWN SPEAKER. There appear to be anomalies in the PM_{10} data presented in the report by APEG (1999). In northeast Derbyshire in January 1997, figures approached 58 $\mu g\, m^{-3}$, which were found again in independent monitoring by another government quango in 1998. Yet, in the balance of 1997, figures went as low as zero to end up with an annual average of 26 $\mu g\, m^{-3}$, which just happens to be the figure presented to the EC for the UK. I have printouts of minus readings all over the UK, which appear as serious as a colleague's discovery of a $-17\ \mu g\, m^{-3}$ council monitor reading near a cement works when Environment Agency independent findings of 240 $\mu g\, m^{-3}$ had been published months earlier. Can you comment on the need for rigorous calibration of monitors and accuracy of computer analysis and modelling?

R. M. Harrison. It is not possible to comment on the specific 'anomalies' raised in the question without sight of the datasets being cited. Data in the APEG (1999) report were from reputable organizations and I have no reason to doubt their quality. Clearly, there is a need for rigorous calibration procedures and an awareness that instrumental analysers working on different principles measure different properties of the particles and, therefore, may give different measurements of 'mass'. The ultimate calibration of a truly gravimetric instrument can, however, be obtained through laboratory weighing of collected involatile particles. Numerical models are only as accurate as the input data and their physical descriptions and parametrizations of environmental dispersion processes. They are not able to give wholly accurate predictions, but nonetheless can be extremely useful in providing a guide to ground-level concentrations.

CHAPTER 2

THE CHEMICAL COMPOSITION OF ATMOSPHERIC ULTRAFINE PARTICLES

Glen R. Cass[1], Lara A. Hughes[2], Prakash Bhave[2], Michael J. Kleeman[3], Jonathan O. Allen[2] and Lynn G. Salmon[2]

[1]*School of Earth and Atmospheric Sciences, Georgia Institute of Technology, Atlanta, GA 30332, USA*
[2]*Environmental Engineering Science Department, California Institute of Technology, Pasadena, CA 91125, USA*
[3]*Department of Civil and Environmental Engineering, University of California, Davis, CA 95616, USA*

Atmospheric ultrafine particles (with diameter less than 0.1 μm) may be responsible for some of the adverse health effects observed due to air-pollutant exposure. To date, little is known about the chemical composition of ultrafine particles in the atmosphere of cities. Ultrafine particle samples collected by inertial separation on the lower stages of cascade impactors can be analysed to determine a material balance on the chemical composition of such samples. Measurements of ultrafine particle mass concentration made in seven Southern California cities show that ultrafine particle concentrations in the size range 0.056–0.1 μm aerodynamic diameter average 0.55–1.16 μg m^{-3}. The chemical composition of these ultrafine particle samples averages 50% organic compounds, 14% trace metal *oxides*, 8.7% elemental carbon, 8.2% sulphate, 6.8% nitrate, 3.7% ammonium ion (excluding one outlier), 0.6% sodium and 0.5% chloride. The most abundant catalytic metals measured in the ultrafine particles are Fe, Ti, Cr, Zn, with Ce also present. A source emissions inventory constructed for the South Coast Air Basin that surrounds Los Angeles shows a primary ultrafine particle emissions rate of 13 tonnes per day. Those ultrafine particle primary emissions arise principally from mobile and stationary fuel combustion sources and are estimated to consist of 65% organic compounds, 7% elemental carbon, 7% sulphate, 4% trace elements, with very small quantities of sodium, chloride and nitrate.

This information should assist the community of inhalation toxicologists in the design of realistic exposure studies involving ultrafine particles.

Keywords: ultrafine particles; atmospheric concentration; chemical composition; emissions rate

1. Introduction

Epidemiological studies suggest that mortality and morbidity rates increase on days with higher than usual airborne particle concentrations (Dockery *et al.* 1993; Pope *et al.* 1995). Statistical associations between increasing particle concentrations and adverse health effects are found for concentration increments as low as 10 μg m^{-3} at total particle concentrations below 100 μg m^{-3}. In contrast, toxicologists generally find that such low concentrations of chemically inert particles in particle sizes larger than about 0.5 μm in particle diameter do not produce serious adverse effects in laboratory inhalation studies (Schlesinger 1995). This suggests that there may be attributes of the airborne particle mixture that are capable of inducing damage to human health, while not contributing greatly to atmospheric particle mass concentrations.

Atmospheric ultrafine particles, defined here as those having diameters smaller than 100 nm (0.1 μm), are present at concentrations of about 10^4 particles per cm^3 of air in the atmosphere of cities. These ultrafine particles dominate atmospheric particle number concentrations, while at the same time making a negligible contribution to particle mass concentrations. One distinct possibility is that the number of inhaled particles is more important in producing the health effects associated with air pollution than is the inhaled particle mass concentration. Alternatively, perhaps a combination of the chemical composition and very small size of atmospheric ultrafine particles is critical in explaining the effect of airborne particles on human health. In order to assist the community of toxicologists in the design of realistic test atmospheres needed to further explore these hypotheses, it is useful to determine the chemical composition of atmospheric ultrafine particles.

At present, very little is known about the chemical composition of atmospheric ultrafine particles. To date, only a single study has been published on ultrafine particle chemical composition in the atmosphere of cities (Hughes *et al.* 1998). The purpose of the present paper is to expand the existing data on urban atmospheric ultrafine particle chemical composition, and to draw comparisons between the measured chemical composition of ultrafine particle emissions from sources and the chemical composition of atmospheric ultrafine particles. Atmospheric measurements made in seven cities in Southern California will be compared with a comprehensive inventory of the chemical composition of ultrafine particle emissions from sources based on a programme of source testing conducted in the South Coast Air Basin that surrounds metropolitan Los Angeles.

2. Atmospheric Ultrafine Particle Chemical Composition

Over the period 1995–1997, field experiments were conducted in Southern California to characterize the size distribution and chemical composition of the airborne particle mixture (Hughes *et al.* 1998, 1999, 2000; Allen *et al.* 2000). Particle number distributions were measured with differential mobility analyser/condensation nucleus counter combinations (TSI models 3071 and 3760), with electrical aerosol analysers (TSI model 3030), and with optical particle counters (Particle Measuring Systems model ASASP-X). Filter-based measurements of particle chemical composition were made in two particle size ranges: fine particles smaller than 2 μm in diameter, and total suspended particulate matter of all sizes. Of most importance to the present study, samples for particle chemical composition determination were collected in six narrow size ranges spanning the interval from 1.8 to 0.56 μm particle aerodynamic diameter using a pair of MOUDI cascade impactors (MSP Corp. model 100). These impactors were preceded by AIHL-design cyclone separators operated at a cut-point of 1.8 μm in order to remove coarse atmospheric particles that may contribute to particle bounce within the impactors (John & Reischl 1980). One impactor of each pair was loaded with Teflon collection substrates (Gelman Teflo, 47 mm in diameter), while the second impactor was loaded with aluminium foil impaction substrates (MSP Corp.). The impaction substrates were weighed before and after use in a temperature- and humidity-controlled environment (typically 22 °C and less than 48% relative humidity (RH) but with RH no more than ±3%) using a Mettler model M-55-A mechanical microgram balance. The Teflon impaction substrates were divided in half. One-half was analysed for sulphate, nitrate and chloride by ion chromatography (Mulik *et al.* 1976), using a Dionex model 2020i ion chromatograph, and for ammonium ions by an indophenol colorimetric technique (Bolleter *et al.* 1961), using an

Alpchem rapid flow analyser (model RFA-300). Trace-element concentrations were determined from the other half of the Teflon impaction substrates via instrumental neutron activation analysis (Olmez 1989). The concentration and size distribution of organic compounds and black light-absorbing elemental carbon (EC) was determined from samples collected on the aluminium foil impaction substrates by the thermal evolution and combustion procedure of Huntzicker *et al.* (1982), as modified by Birch & Cary (1996) and as adapted to impactor samples by Kleeman *et al.* (1999*a*). In the present work, measurements of organic carbon (OC) concentrations are converted into estimates of organic compound concentrations in the aerosol by scaling measured OC concentrations upward by a factor of 1.4 in order to account for the H, O, S and N in organic compounds typically found in the urban atmosphere.

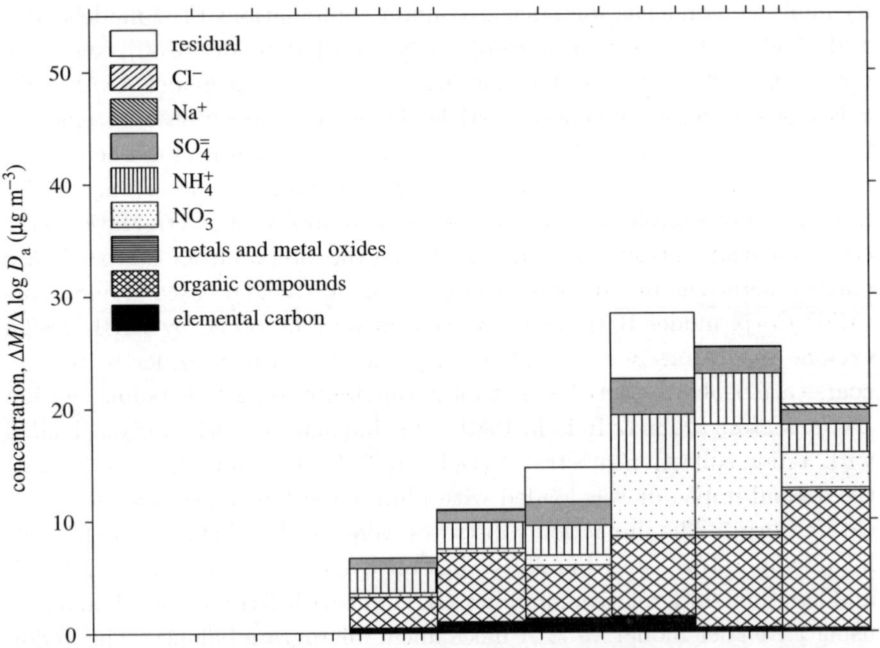

Fig. 1. The size distribution and chemical composition of atmospheric particles smaller than 1.8 μm aerodynamic diameter at Riverside, CA, 28 August 1997.

An illustration of the size and chemical composition distribution of chemical substances present in the airborne fine particle mixture in Southern California is shown in figure 1. These data show an accumulation mode aerosol in the particle diameter range less than 1.8 μm having a peak in the mass concentration at *ca.* 0.5 μm particle diameter. The fine particle mixture consists largely of elemental and organic carbon, ammonium sulphate and ammonium nitrate. The lower tail of the coarse particle soil dust and sea salt size distribution can often be seen in the larger fine particle size ranges as well.

Ultrafine particles, as presently defined by the community of health scientists, consist of particles smaller than 0.1 μm in diameter. Data on particle chemical composition in this size range are available from the lowest stage of the MOUDI impactor samples, as can be seen at the far left of figure 1. Samples collected for chemical determination over the interval 0.056–0.1 μm particle diameter, while falling in the ultrafine particle size range, are related to the lower tail of the accumulation mode particle size distribution as seen in figure 1. They probably represent the largest mass of the particles in the ultrafine particle size range but may be chemically different from the more numerous particles present at diameters smaller than 10 nm, for example.

Data on ultrafine particle mass concentrations and chemical composition for particles of the size 0.056–0.1 μm aerodynamic diameter at seven Southern California cities and for eight time intervals are shown in figure 2. Each chart represents an average over many samples taken during the time intervals shown.

Average ultrafine particle mass concentrations across all Southern California sites and all seasons studied lie in the fairly narrow range 0.55–1.16 $\mu g\, m^{-3}$. The average ultrafine particle mass concentration is *ca.* 0.8 $\mu g\, m^{-3}$ in the size range 0.056–0.1 μm aerodynamic diameter.

Organic compounds and elemental carbon contribute close to half of the ultrafine particle mass concentration at Pasadena, CA, in the winter months, with noticeable contributions from metal oxides and lesser amounts of sulphates and nitrates, as seen in figure 2*a* (Hughes *et al*. 1998). Samples taken at groups of air monitoring sites that are generally upwind/downwind of each other are illustrated in figure 2*b*, *c*, figure 2*d*, *e* and figure 2*f*–*h*. Fullerton is a city in Orange County, CA, located *ca.* 20 km inland from the Pacific Ocean along air parcel trajectories that often pass over the large

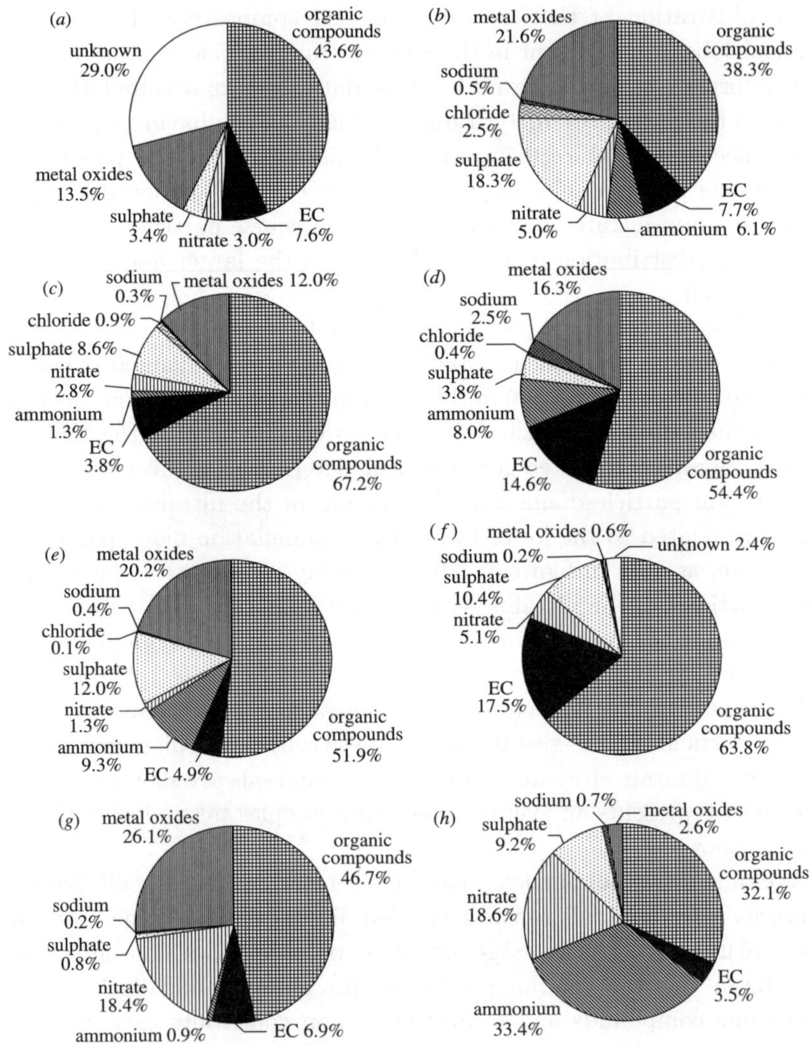

Fig. 2. The chemical composition of atmospheric ultrafine particles in the size range 0.056–0.1 μm particle aerodynamic diameter measured at cities in Southern California. (*a*) Pasadena, CA, January–February 1996 (0.82 μg m^{-3}); (*b*) Fullerton, CA, September–October 1996 (0.64 μg m^{-3}); (*c*) Riverside, CA, September–October 1996 (0.63 μg m^{-3}); (*d*) Los Angeles, CA, August 1997 (1.16 μg m^{-3}); (*e*) Azusa, CA, August 1997 (0.80 μg m^{-3}); (*f*) Diamond Bar, CA, September–November 1997 (0.55 μg m^{-3}); (*g*) Mira Loma, CA, September–November 1997 (0.58 μg m^{-3}); (*h*) Riverside, CA, August–November 1997 (0.91 μg m^{-3}).

industrial and harbour complex at Long Beach, CA, earlier in the day. Ultrafine particle composition at Fullerton is shown in figure 2*b*. Again as at Pasadena, carbonaceous aerosols are the largest contributors to the ultrafine particles at Fullerton, followed by metal oxides, sulphate, ammonium and nitrate ion, in that order. Riverside is located *ca.* 60 km inland from Fullerton and is directly downwind of Fullerton in the summer and early autumn months. Because of the slow wind speeds in Southern California, transport times from Fullerton to Riverside are often a day or longer. Riverside is largely a residential and commercial centre; the samples quantified in figure 2*c* were taken on the campus of the University of California at Riverside, where motor-vehicle traffic is the most obvious source of local particle emissions. Ultrafine particles at Riverside during September–October 1996 are likewise largely carbonaceous, with noticeable metal content and much less sulphate and nitrate aerosol than is associated with accumulation mode particles having diameters of several hundred nm at Riverside.

The ultrafine particle chemical composition in central Los Angeles, shown in figure 2*d*, is dominated by high-density motor-vehicle traffic. Sixty-nine per cent of the ultrafine particle mass concentration in central Los Angeles in August 1997 was carbonaceous, with a higher than usual proportion of black elemental carbon reflecting the high concentration of diesel engines in use at that location, not only in highway vehicles but also in railway locomotives and industrial diesel engines. Azusa, CA, is located *ca.* 30 km inland from central Los Angeles and is generally downwind of it in the summer months. As seen in figure 2*e*, 57% of the ultrafine particle mass at Azusa consists of organic compounds plus elemental carbon. Again, significant quantities of metal oxides are present, with noticeable amounts of ammonium and sulphate but little ammonium nitrate, even though aerosol nitrate is quite common in larger particles in the Los Angeles area.

Samples collected at the location of the South Coast Air Quality Management District Offices in Diamond Bar, CA, are shown in figure 2*f*. That site is located on a hill immediately adjacent to the intersection of the Pomona and no. 57 freeways. As in central Los Angeles, large quantities of organic compounds and elemental carbon are measured in the ultrafine particles at the Diamond Bar site. Mira Loma is located 10–20 km inland from Diamond Bar and it is immediately downwind from the Chino dairy area. The atmosphere at Mira Loma contains very high ammonia concentrations, well above the odour threshold for ammonia, as well as high

Table 1. Average ultrafine particle trace metal concentrations (central Los Angeles, CA, August–September, 1997).

trace metal	mean concentration (ng m^{-3})	range (ng m^{-3})
groups I and II		
Na	85	(bdl–249)
K	88	(bdl–93)
Cs	0.100	(bdl–0.34)
Ba	19	(bdl–19)
transition metals		
Sc	0.028	(bdl–0.054)
Ti	43	(bdl–43)
V	bbl	
Cr	6.7	(bdl–15)
Mn	bbl	(bbl–0.056)
Fe	186	(bdl–470)
Zn	3.8	(bbl–10)
Mo	0.48	(bdl–0.68)
Cd	0.19	(bbl–0.49)
Au	bdl	
Hg	0.09	(0.038–0.14)
lanthanides		
La	0.021	(bdl–0.021)
Ce	1.2	(bdl–2.3)
Sm	0.012	(bdl–0.019)
Eu	0.20	(bdl–0.37)
Yb	0.26	(bbl–0.50)
Lu	0.014	(0.0011–0.028)
actinides		
Th	bbl	
U	bdl	

bdl denotes 'below detection limits'. bbl denotes 'below blank levels'.

concentrations of inorganic nitrate due to nitric acid formation in photochemical smog on the downwind side of the Los Angeles area. As has been seen elsewhere, more than half of the ultrafine particle mass concentration at Mira Loma consists of organic compounds plus elemental carbon, with substantial metal oxides content. At Mira Loma we see our first example of relatively high nitrate aerosol content in ultrafine particles. This could be due to nitric acid reaction with the metal-containing particles seen in figure 2*g*. Alternatively, this could be due to ammonium nitrate formation accompanied by analytical problems with ammonium concentration measurement in the very small samples evaluated here. The ammonium measurement method is less accurate than sulphate and nitrate determination at these very small sample sizes. Clearly, there is both ammonium and nitrate present in the ultrafine particles collected at the Riverside site during approximately the same months, as shown in figure 2*g*. Riverside is 20–30 km downwind of Mira Loma. Overall, the chemical composition of ultrafine particles in Southern California is in the range 32–67% organic compounds, 3.5–17.5% elemental carbon, 1–18% sulphate ion, 0–19% nitrate ion, 0–9% ammonium ion (excluding one extreme outlier at Riverside), 1–26% metal oxides, 0–2% sodium and 0–2% chloride.

The concentrations of catalytic metals in atmospheric particles are of particular interest to the community of toxicologists, because catalytic metals deposited in the lung could catalyse oxidative damage to it. The mean trace element concentration (and range of concentrations) in the ultrafine particles in the size range 0.056–0.1 μm aerodynamic diameter in central Los Angeles is shown in table 1. The acronym 'bdl' in the table indicates that the lowest values were below the detection limits of the neutron activation analysis in some cases. The most abundant transition metals in the ultrafine particles in the central Los Angeles atmosphere were found to be Fe, Ti, Cr and Zn. The catalytic element Ce was also measured at concentrations of *ca.* 1 ng m^{-3}. Additional data on the trace metals content of ultrafine particles in the Pasadena, CA, atmosphere have been published previously by Hughes *et al.* (1998).

3. Emissions of Ultrafine Particles to the Southern California Atmosphere

The size distribution and chemical composition of fine particle emissions from the largest sources in Southern California have been measured by dilution source sampling (Schauer 1998). A pair of MOUDI cascade impactors was operated downstream of the dilution source sampler in order to measure the mass emissions rate, particle size and particle chemical composition by exactly the same methods as previously described for atmospheric particle samples. Sources tested in this way include catalyst-equipped petrol-powered cars and light trucks, non-catalyst petrol-powered cars and light trucks, medium-duty diesel trucks, fireplace combustion of hardwoods and soft wood, meat charbroiling, and cigarette smoke (Kleeman *et al.* 1999*a*, 2000). In previous years, the same source sampling system was used to measure the size distribution, mass emissions rate and bulk fine particle chemical composition of the emissions from natural gas combustion, distillate fuel oil combustion, tyre dust, brake lining wear dust, paved road dust, and plant fragments shed as leaves are rubbed together by the wind (Hildemann *et al.* 1991*a*, *b*). Data from these earlier source tests were used to supplement the more recent cascade impactor-based source tests under the approximation that the ultrafine particles counted in the earlier source tests had a chemical composition similar to the bulk fine particle (particle diameter less than 2 μm) samples collected from these sources. Finally, the source test data taken by our research group were combined with data on source activity (e.g. vehicle kilometres travelled, quantities of fuel burned) supplied by the California Air Resources Board as part of their inventory of particle emissions in Southern California. In essence, we took the State of California particulate matter emissions inventory for total suspended particulate matter and for particle mass smaller than 10 μm in diameter, replaced the emissions rate data by our own source measurements when available, and impressed the particle size and chemical composition distribution data from our source tests onto the resulting modified inventory. For the remaining minor sources in the inventory, the State of California emissions rate data, particle size and chemical composition data were retained. Our emissions inventory containing high-resolution size and chemical composition data was originally developed for use in air-quality models that predict the size distribution and chemical composition of the atmospheric particle complex in the presence of transport, chemical reaction in the atmosphere and dry deposition. In previous tests of those models, the model predictions compare quite favourably with measurements of atmospheric particle size and chemical composition (Eldering & Cass 1996; Kleeman *et al.* 1997, 1999*b*), so there is good reason to believe that the emissions inventory is reasonably accurate for fine particles and for PM_{10} (particles smaller than 10 μm aerodynamic diameter). Because that emissions inventory extends into the ultrafine particle size range, it is possible to extract and display the best presently available data on the mass emissions rate and chemical composition of ultrafine particle emissions from sources located within the South Coast Air Basin that surrounds metropolitan Los Angeles.

The primary particle mass emissions within the South Coast Air Basin of California in September 1996 in sizes smaller than 10 μm aerodynamic diameter (PM_{10}) total 380 tonnes per day. These PM_{10} emissions are dominated by close to 320 tonnes per day of mineral dust emissions from travel on paved and unpaved roads, from construction and agricultural activities, and dust raised due to erosion by the wind. For this reason, fugitive dust control has been selected as the usual response by government agencies when faced with meeting air-quality standards for PM_{10}.

New air-quality standards that limit the concentration of atmospheric particles smaller than 2.5 μm aerodynamic diameter ($PM_{2.5}$) have recently been proposed for the United States by the US Environmental Protection Agency. $PM_{2.5}$ emissions in the Los Angeles area total about 122 tonnes per day. Very importantly, from the point of view of the emissions control programme required to comply with a $PM_{2.5}$ air-quality standard, only about half of the primary fine particle emissions are due to fugitive sources, while the other half of the primary $PM_{2.5}$ emissions are directly emitted from stationary and mobile combustion sources. As seen in figure 1, much of the $PM_{2.5}$ in the Southern California atmosphere is also due to aerosol sulphates, nitrates and secondary organic aerosols that are formed by atmospheric chemical reactions.

The emissions inventory for ultrafine particles ($PM_{0.1}$) constructed for the Los Angeles area indicates a mass emissions rate of 13 tonnes per day in particle sizes smaller than 0.1 μm aerodynamic diameter. This emissions rate is approximately consistent with the measured $0.8\,\mu g\,m^{-3}$ ambient ultrafine particle concentration, indicating that most of the ultrafine particle mass in the Southern California atmosphere could well be due to primary particle emissions from sources. As shown in figure 3*a*, the largest sources are on-road motor vehicles (43%), stationary source fuel combustion (32%), non-highway mobile sources (10%, particularly diesel engines used in off-road vehicles and in mobile equipment such as refrigeration units), and other industrial processes (7%, of which the most significant is commercial food preparation, e.g. meat charbroiling). Of course, if one lives very close to a heavily travelled street, far more than 43% of the ultrafine particles would be from motor-vehicle traffic, as those emissions occur at ground level in the immediate vicinity of the person affected.

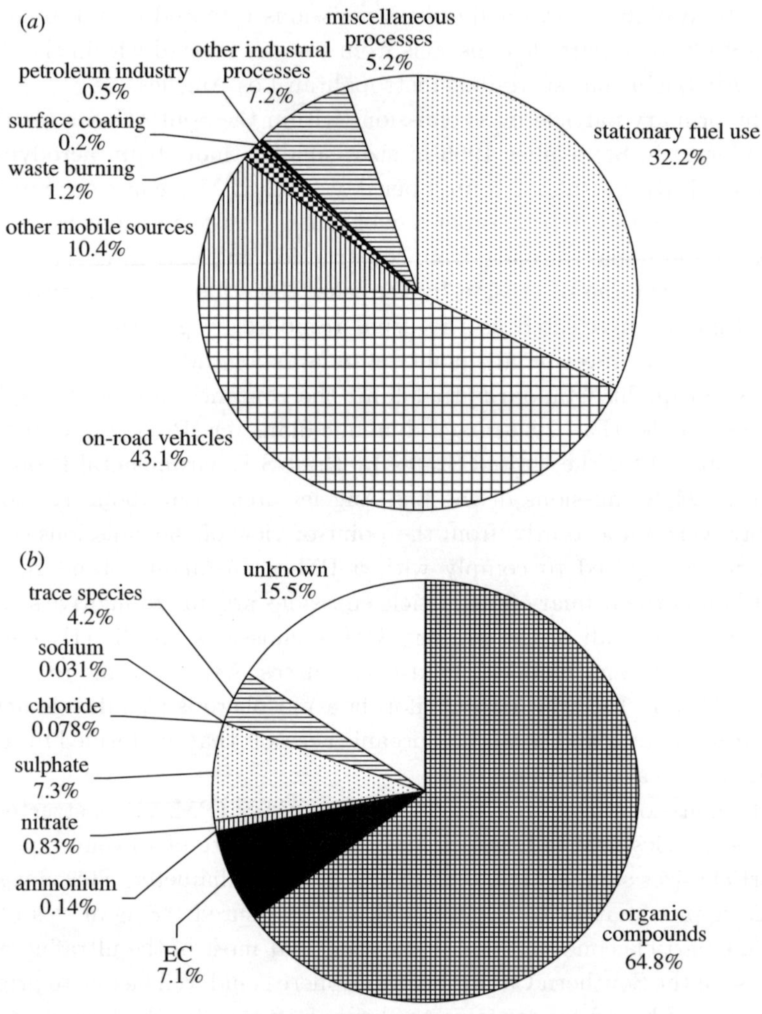

Fig. 3. Ultrafine particle emissions in California's South Coast Air Basin (1996) that surrounds Los Angeles: (*a*) source contributions to primary ultrafine particle emissions; (*b*) chemical composition of primary ultrafine particle emissions. Total $PM_{0.1}$ emissions are 13.25 tonnes per day.

The chemical composition of the primary ultrafine particle emissions in the Los Angeles area is indicated in figure 3*b*. The composition is 65% organic compounds, 7% elemental carbon, 7% sulphate, 4% trace elements, with very small quantities of sodium, chloride and nitrate. It may be important to note that the fuels used in the Los Angeles area are generally limited to very low sulphur content. If the trace elements were converted to the molecular mass of their common oxides, much of the unknown mass shown in figure 3*b* would be explained. These emissions data are quite similar to the chemical composition of the atmospheric samples described earlier. The average over the pie charts in figure 2 shows a mean ambient ultrafine particle chemical composition that is 50% organic compounds, 14% trace metal *oxides*, 8.7% elemental carbon, 8.2% sulphate, 0.6% sodium and 0.5% chloride. Ammonium and nitrate are significantly higher in the atmospheric samples than in the source emissions inventory, indicating that some ammonium nitrate formation is occurring on the atmospheric ultrafine particles studied here.

4. Conclusions

The average ultrafine particle mass concentration in the atmospheres of seven cities in Southern California in particles with aerodynamic diameter between 0.056 and 0.1 μm is in the range 0.55–1.16 $\mu g\ m^{-3}$. The chemical composition of these ultrafine particles is, typically, 32–67% organic compounds, 3.5–17.5% elemental carbon, 1–18% sulphate ion, 0–19% nitrate ion, 0–9% ammonium ion (excluding one extreme outlier at Riverside), 1–26% metal oxides, 0–2% sodium and 0–2% chloride. When averaged over all monitoring sites, average ambient ultrafine particle mass concentrations are *ca.* 0.8 $\mu g\ m^{-3}$, and the average chemical composition is 50% organic compounds, 14% trace metal *oxides*, 8.7% elemental carbon, 8.2% sulphate, 6.8% nitrate, 3.7% ammonium ion (excluding one outlier), 0.6% sodium and 0.5% chloride. The most abundant catalytic metals measured in the ultrafine particles were Fe, Ti, Cr, Zn, with Ce also present. A source emissions inventory constructed for particles smaller than 0.1 μm in diameter for California's South Coast Air Basin that surrounds Los Angeles identified primary ultrafine particle emissions equal to 13 tonnes per day, largely from motor-vehicle exhaust and from stationary-source fuel-combustion sources. The average chemical composition of the ultrafine particle emissions studied in Southern California consisted of 65% organic compounds, 7% elemental carbon, 7% sulphate, 4% trace elements, with very small quantities of sodium, chloride and nitrate. The mass emissions rate is sufficient to explain the 0.8 $\mu g\ m^{-3}$ ambient ultrafine particle concentration measured in the Los

Angeles area, and the chemical composition distribution in the emissions is generally similar to measured ambient ultrafine particle concentrations once the trace metals in the emissions are converted to the mass of their common oxides. The ambient ultrafine particles in the Southern California atmosphere in the size range 0.056–0.1 μm in diameter may be explained by primary particle emissions plus secondary aerosol nitrate formation at some places and times. This information should assist the community of laboratory toxicologists in the construction of realistic test atmospheres for particle-inhalation studies, and focuses attention on primary particles from combustion sources in the event that emissions controls for ultrafine particles are eventually needed.

Acknowledgements

This research was supported by the US Environmental Protection Agency under agreement R-827354-01-0.

References

Allen, J. O., Hughes, L. S., Salmon, L. G., Mayo, P. R., Johnson, R. J., Cass, G. R., Pastor, S. H. & Prather, K. A. 2000 Evolution of atmospheric aerosols affected by urban motor vehicle emissions in the Los Angeles air basin. *Environ. Sci. Technol.* (Submitted.)

Birch, M. E. & Cary, R. A. 1996 Elemental carbon-based method for monitoring occupational exposures to particulate diesel exhaust. *Aerosol Sci. Technol.* **25**, 221–241.

Bolleter, W. T., Bushman, C. T. & Tidwell, P. W. 1961 Spectrophotometric determination of ammonium as indophenol. *Analyt. Chem.* **33**, 592–594.

Dockery, D. W., Pope, C. A., Xu, X., Spengler, J. D., Ware, J. H., Fay, M. E., Ferris, B. G. & Speizer, F. E. 1993 An association between air pollution and mortality in 6 United States cities. *N. Engl. J. Med.* **329**, 1753–1759.

Eldering, A. & Cass, G. R. 1996 Source-oriented model for air pollutant effects on visibility. *J. Geophys. Res. Atmos.* **101**(D14), 19 343–19 369.

Hildemann, L. M., Markowski, G. R. & Cass, G. R. 1991*a* Chemical composition of emissions from urban sources of fine organic aerosol. *Environ. Sci. Technol.* **25**, 744–759.

Hildemann, L. M., Markowski, G. R., Jones, M. C. & Cass, G. R. 1991*b* Submicrometer aerosol mass distributions of emissions from boilers, fireplaces, automobiles, diesel trucks, and meat cooking operations. *Aerosol Sci. Technol.* **14**, 138–152.

Hughes, L. S., Cass, G. R., Gone, J., Ames, M., & Olmez, I. 1998 Physical and chemical characterization of atmospheric ultrafine particles in the Los Angeles area. *Environ. Sci. Technol.* **32**, 1153–1161.

Hughes, L. S. (and 13 others) 1999 Size and composition distribution of atmospheric particles in Southern California. *Environ. Sci. Technol.* **33**, 3506–3515.

Hughes, L. S., Allen, J. O., Salmon, L. G., Mayo, P. R., Johnson, R. J., Cass, G. R., Pastor, S. H. & Prather, K. A. 2000 Evolution of nitrogen-containing air pollutants along trajectories crossing the Los Angeles area. *Environ. Sci. Technol.* (Submitted.)

Huntzicker, J. J., Johnson, R. L., Shah, J. J. & Cary R. A. 1982 In *Particulate carbon, atmospheric life cycle* (ed. G. T. Wolff & R. L. Klimisch). New York: Plenum.

John, W. & Reischl, G. 1980 A cyclone for size-selective sampling of ambient air. *J. Air Pollut. Control Assoc.* **30**, 872–876.

Kleeman, M. J., Eldering, A. & Cass, G. R. 1997 Modeling the airborne particle complex as a source-oriented external mixture. *J. Geophys. Res. Atmos.* **102**, 21 355–21 372.

Kleeman, M. J., Schauer, J. J. & Cass, G. R. 1999*a* Size and composition distribution of fine particulate matter emitted from wood burning, meat charbroiling and cigarettes. *Environ. Sci. Technol.* **33**, 3516–3523.

Kleeman, M. J., Hughes, L. S., Allen, J. O. & Cass, G. R. 1999*b* Source contributions to the size and composition distribution of atmospheric particles: Southern California in September 1996. *Environ. Sci. Technol.* **33**, 4331–4341.

Kleeman, M. J., Schauer, J. J. & Cass, G. R. 2000 Size and composition distribution of fine particulate matter emitted from motor vehicles. *Environ. Sci. Technol.* **34**, 1132–1142.

Mulik, J., Puckett, R., Williams, D. & Sawicki, E. 1976 Ion chromatographic analysis of sulfate and nitrate in ambient aerosols. *Analyt. Lett.* **9**, 653–663.

Olmez, I. 1989 In *Methods of air sampling and analysis* (ed. J. P. Lodge), 3rd edn. Chelsea, MI: Lewis.

Pope, C. A., Dockery, D. W. & Schwartz, J. 1995 Review of epidemiological evidence of health effects of particulate air pollution. *Inhalation Toxicol.* **7**, 1–18.

Schauer, J. J. 1998 Source contributions to atmospheric organic compound concentrations: emissions measurements and model predictions. PhD thesis, California Institute of Technology, Pasadena, CA.

Schlesinger, R. B. 1995 Toxicological evidence for health effects from inhaled particulate pollution—does it support the human experience? *Inhalation Toxicol.* **7**, 99–109.

Discussion

C. F. Clement (*Wantage, Oxon, UK*). I would like to draw attention to the need to identify the primary particles corresponding to the chemical compositions shown. Some of the constituents, e.g. sulphates, could well arise from condensation on existing particles. In particular, the percentage of metal constituents seemed to vary considerably with location. It is not easy to form ultrafine metal aerosols; has the origin of the observed metals been identified?

G. R. Cass. A partial assessment of the trace metals sources has been conducted. The ultrafine particle emissions inventory used to construct figure 3 contains emissions estimates for over 30 trace elements emitted from 62 different types of emissions sources. Source chemical composition profiles showing the major chemical species are available for these sources but data on the minor species are absent in many cases. Petroleum refining, metallurgical industry fumes, engines and some sources processing mineral matter are indicated as sources of a significant fraction of the ultrafine particle iron, for example. Further source testing is needed to refine and extend the database.

R. Agius (*University of Edinburgh, UK*). You have shown that organic compounds comprise a substantial component of atmospheric ultrafine particles (up to 70%). Could you please shed some light on the chemical species in this important component?

G. R. Cass. We have not yet performed a detailed analysis of the individual organic compounds present in the ultrafine particles. We have, however, examined fine particle emissions ($d_p < 2\,\mu m$) from the most important sources of ultrafine organic aerosol, which include natural gas combustion, food cooking, woodsmoke, and motor vehicle exhaust. If the ultrafine particles resemble the fine particles from these sources chemically, then we would expect to find PAH, oxy-PAH, meat fat components, levoglucosan and related sugar derivatives, resin acids, substituted phenolic compounds, and heavy petroleum hydrocarbons (e.g. unburned motor oil).

M. S. Bingley (*Cobham, UK*). Could the high concentration of iron in your atmospheric particle analysis be due to bore wear in automobile engines?

G. R. CASS. Engine wear has not yet been established as the major source of the iron.

D. COSTA (*US EPA, NC, USA*). Do we have information on composition of indoor ultrafine particulate matter and to what extent do outdoor ultrafines penetrate indoors?

G. R. CASS. We have not yet studied ultrafine particles in indoor air.

CHAPTER 3

OVERVIEW OF METHODS FOR ANALYSING SINGLE ULTRAFINE PARTICLES

Andrew D. Maynard

US Department of Health and Human Services, Public Health Service,
Centers for Disease Control and Prevention,
National Institute for Occupational Safety and Health,
Division of Applied Research and Technology,
4676 Columbia Parkway, Cincinnati, OH 45226, USA

Increasing awareness that structures and attributes on a nanometre scale within aerosol particles may play a significant role in determining their behaviour has highlighted the need for suitable single ultrafine particle analysis methods. By adopting technologies developed within complementary disciplines, together with the development of aerosol-specific methods, a basis for characterizing single sub-100 nm (ultrafine) particles and features in terms of size, morphology, topology, composition, structure and physicochemical properties is established. Size, morphology and surface properties are readily characterized in the scanning transmission electron microscope (STEM), while high-resolution transmission electron microscopy (HRTEM) allows structural information on particles and atomic clusters to sub-0.2 nm resolution. Electron energy loss spectroscopy (EELS) and X-ray emission in the STEM allow the chemical analysis of particles and particle regions down to nanometre diameters. Scanning probe microscopy offers the possibility of analysing nanometre-diameter particles under ambient conditions, thus getting away from some of the constraints imposed by electron microscopy. Imaging methods such as atomic force microscopy and near field scanning optical microscopy (NSOM) offer novel and exciting possibilities for the characterization of specific aerosols. Developments in aerosol mass spectrometry are providing the means for chemically characterizing size-segregated ultrafine particles down to 10 nm in diameter on-line. By taking a multi-disciplinary approach, the compilation and development of complementary tools allowing both routine and in-depth analysis of individual ultrafine particles is possible.

Keywords: ultrafine; aerosol; single-particle analysis; particle collection; electron microscopy; scanning probe microscopy

1. Introduction

An aerosol is a complex material state that lies between a gas or vapour and a bulk material. At each extreme, analysis is simplified by compositional, and to a certain extent structural, homogeneity. However, an aerosol may consist of many orders of magnitude of discrete particles, each having the possibility of slightly different physical and chemical properties. The relevance of each particle's nature within the aerosol will depend on context, and in most systems some degree of simplification is possible. For instance, the motion in a gas of an aerosol consisting of chemically and structurally diverse particles with similar aerodynamic properties may be characterized by relatively few collective parameters. On the other hand, understanding interactions with the aerosol at a physical, biological or chemical level will require a more complex characterization. In aerosols where there is wide variation in particle size, shape, structure, composition, etc., and where the relevance of these parameters in determining aerosol behaviour is not well understood, the use of collective attributes such as gravimetric mass particle size distribution or overall chemical composition may not explain observed phenomena adequately. Within such systems, characterization on a particle-by-particle basis should be considered as the first step to understanding interaction mechanisms and simplifying monitoring requirements.

The investigation of single ultrafine particles is not a new or original venture. Commercial interest in the activity of nanometre-sized particles within heterogeneous catalysts, the role of ultrafine particles in determining microstructure within materials, development of quantum microdot technology, together with a more general fascination with the unique properties of nanometre-sized particles and atomic clusters, have collectively led to the application and development of a range of methods able to characterize individual particles in detail. However, few of these methods have found application in the analysis of environmental aerosols. This is perhaps understandable, given the complexity of most analysis methods, together with the hitherto relatively simple requirements of environmental aerosol analysis. However, data relating to the impact of fine (typically less than 5–10 μm) and ultrafine (typically smaller than 100 nm) aerosol particles on biological systems are becoming increasingly difficult to reconcile with simple mass-based analyses. Both epidemiology and toxicology studies indicate that biological response is mediated by factors other than mass and composition, although the nature of the underlying factors is by no means clear (Dockery *et al.* 1993; Oberdörster 1996; Donaldson *et al.* 1998). Published data in these fields alone justify a multi-disciplinary approach to environmental aerosol characterization, bringing methods and expertise from a variety of disciplines to bear on the problem of determining the role of specific particle attributes in initiating and mediating biological responses. However, given the unique nature of nanometre-sized particles, distinct from either the molecular or bulk state, it is likely that the application of ultrafine single-particle analysis methods to environmental aerosols will also shed light on aerosol interaction dynamics within other systems.

2. Single Ultrafine Particle Analysis Methods

Numerous methods have been applied to the analysis of single aerosol particles and have been well documented in a number of sources (Grasserbauer 1983; Fletcher & Small 1993; Ortner *et al.* 1998; De Bock & Van Grieken 1999). The vast majority of available methods are limited by spatial resolution and/or detection limits, and tend to be more applicable to the analysis of particles 0.5 μm to 1 μm in diameter and above. This includes many of the particle beam techniques such as particle-induced X-ray emissions (PIXE), electron probe micro analysis (EPMA) and secondary ion mass spectrometry (SIMS) (Maynard 1993). Electron microscopy has been used to characterize sub-100 nm diameter particles since the early days of its development (Drummond 1950), and for some time was considered the only method for investigating single particles in the nanometre region. Over the past decade, development of the resolution and analytical capabilities of the electron microscope has further increased its applicability to the study of ultrafine particles. The development of scanning force microscopes such as the scanning tunnelling microscope and atomic force microscope (AFM) have further added to the available instrumentation for nanometre particle analysis. Although still at a relatively early stage of development, methods involving mass spectrometry of vaporized and ionized particles are beginning to allow the size-related compositional analysis of single ultrafine particles *in situ*. These three technologies — electron microscopy, scanning probe microscopy, and particle vaporization-ionization/mass spectrometry — form the core of current single ultrafine particle analysis capabilities.

2.1. *Collection Methods*

Although *in situ* single-particle analysis methods allow direct sampling of an aerosol with little or no preparation, the more versatile off-line methods such as electron microscopy require the aerosol to be collected and presented in an appropriate manner. Suitable collection methods vary according to the size and nature of the particles under investigation. An applicable method must allow the particles to be presented as a homogeneous uniform deposit, while not altering the relevant particle characteristics significantly. Analysis of relatively large particles in the scanning electron microscope or environmental scanning electron microscope can be achieved with relatively little preparation, from particles collected onto a variety of substrates. At the opposite end of the spectrum, nanometre-diameter particles to be analysed in the transmission electron microscope or scanning transmission electron microscope must be presented without contaminants on a suitably thin electron-transparent support. Re-suspension (usually in liquid) and deposition of aerosols onto a suitable substrate has been a common approach used in the past for particle analysis, but the modification of aerosol particles from their native state is an inherent problem (Bérubé *et al.* 1999).

Inertial collection methods such as gravitational settling and centrifugal collection are suitable for relatively massive particles (e.g. greater than 1–10 μm in diameter), but are impractical to implement for ultrafine particles. Inertial deposition in impactors is achieved by increasing particle momentum in a high velocity air flow, and enabling inertial deposition onto a substrate by rapidly changing the flow direction. Use of low pressure stages in cascade impactors allows the collection of particles as small as 50 nm in devices such as the electrical low pressure impactor (Keskinen *et al.* 1992). Recent developments in nozzle design have led to hypersonic impactors capable of collecting particles down to 50 nm (Hering & Stolzenburg 1995), and focusing impactors capable in principle of operating below 10 nm (de Juan *et al.* 1998). However, deposition forces are necessarily high, leading to the possibility of particle damage. Aerosol samples collected by impaction are generally restricted to a small region of the substrate, thus increasing the probability of particle coincidence, and may be non-uniform with respect to particle size.

Electrostatic deposition allows relatively high deposition velocities, particularly at high particle charge-to-mass ratios. Where particles are unlikely to be damaged by the charging mechanism used or the electric fields encountered, relatively gentle and uniform deposition is possible. Assuming that particles are charged to their theoretical charge limit, electrostatic deposition velocities are relatively independent of particle size (Hinds 1999). However, this limit is difficult to achieve under practical sampling conditions. Under conditions where positive and negative ions may freely attach to aerosol particles, a charge equilibrium is reached that is highly size-dependent (characterized by a Boltzmann distribution). The fraction of nanometre-sized particles having a minimum of one charge drops off rapidly with decreasing size, leading to a dramatic fall in deposition velocity. Diffusional or photoelectric charging can be used to increase the average particle charge at small diameters, and as a general rule of thumb electrostatic precipitation can be used effectively for particles larger than 20 nm in diameter.

Below 10–20 nm, diffusion begins to dominate other deposition mechanisms. For particles smaller than 10 nm diffusion is ideally suited to obtaining uniform particle deposits on a range of sampler substrates, although samples will be highly biased towards smaller particles, and are unlikely to contain a significant fraction of particles larger than 20–30 nm.

Thermophoresis, the movement of aerosol particles in the presence of a temperature gradient, has the advantage that for a given particle composition, deposition velocity is constant below a size of *ca.* 100 nm (Talbot *et al.* 1980). Achievable deposition velocities are relatively low, but deposition is gentle and unlikely to influence the physical nature of the particles (although the thermal field may be detrimental to some temperature-sensitive particles). The technique has been used widely in the past; the Green and Watson thermophoretic precipitator formed a mainstay of occupational health aerosol sampling for many years in the mid-1900s (Watson 1937, 1958). Implementation of thermophoresis in a uniform temperature gradient between two horizontal surfaces has enabled uniform deposits of discrete particles from below 5 nm to nearly 1 μm directly on to transmission electron microscope support grids (Maynard 1995*b*).

2.2. *Electron Microscopy*

Electron microscopy is perhaps the most versatile tool for the analysis of single ultrafine aerosol particles. Scanning electron microscopes (SEMs) are

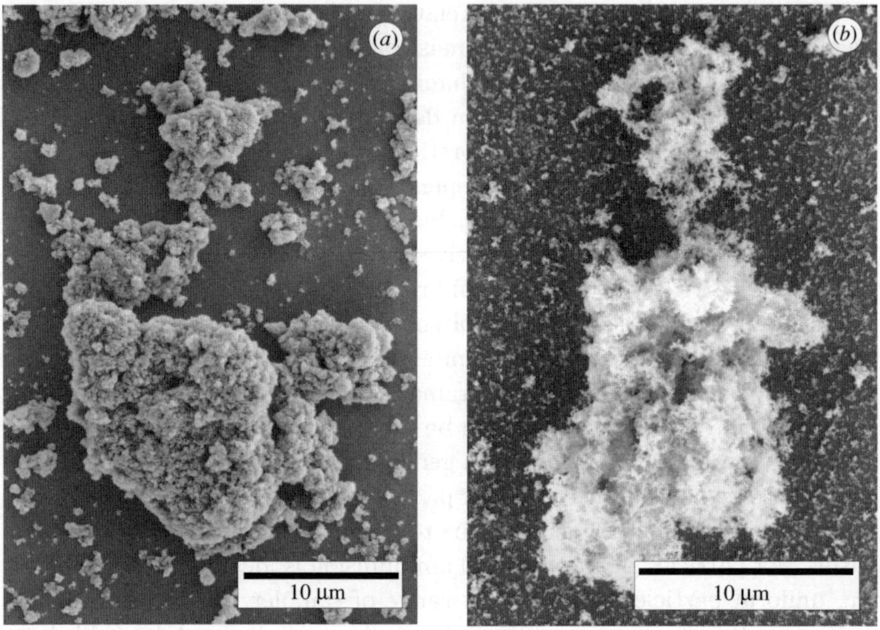

Fig. 1. Images of diesel exhaust particles taken in the FEG-SEM following precipitation (*a*) from liquid, following ultrasonic agitation, and (*b*) following direct deposition onto a SEM substrate. Reproduced with permission from Bérubé *et al.* (1999). © 1999 Elsevier Science Ltd.

routinely used for the analysis of micrometre-sized particles and above. Particles may be presented on a variety of substrates, provided they lie on the surface of the substrate, and are easily differentiable from it. Samples must be conducting to prevent localized charging, and this is achieved either by coating them with gold or carbon, or by using a conducting substrate. The latter leads to a deterioration in the imaging capabilities unless the particles themselves are sufficiently conducting. Samples are imaged by scanning a finely focused electron beam in a raster across their surface, and using the detection of resulting emissions such as backscattered or secondary emission electrons to modulate the intensity of a synchronized raster shown on a display device. In this manner, an image of the sample's surface is formed. Resolution is primarily a function of electron beam diameter and the area from which detected electrons are scattered or emitted, and approaches the diameter of the electron beam for secondary electron imaging. Low-energy secondary electron emissions are restricted to the sample's surface and allow detailed morphological imaging. Similarly, Auger electron emissions occur from the top few nanometres of the sample and may be used for surface layer elemental analysis. Current scanning Auger microscopy applications tend to have relatively poor lateral resolution, but may be adaptable to the surface analysis of ultrafine particles. Standard SEMs generally use a relatively low brightness tungsten electron source that provides insufficient beam current to obtain images with a resolution much below 50–100 nm. However, brighter sources such as LaB_6 filaments allow higher resolution imaging, and SEMs equipped with high brightness, high coherence field emission electron guns (FEG-SEMs) are able to image to a resolution of below 5–10 nm (Takasu *et al.* 1993; De Hosson *et al.* 1998; Van Cleempoel *et al.* 1998; Bérubé *et al.* 1999). The use of a bright electron source has the additional advantage of allowing imaging at lower accelerating voltages, thus reducing charging within poorly conducting samples. FEG-SEMs are able to provide size and surface-structure information on deposited nanometre particles, provided that there is sufficient contrast between the features of interest and the background. Bérubé *et al.* (1999) used the FEG-SEM to compare the morphology of diesel exhaust particles impacted directly onto a substrate with that of similar particles collected on a filter and deposited from an aqueous suspension onto a suitable substrate. The indirect collection method was found to alter the morphology and the size distribution of the particles significantly (figure 1). The effect of moisture on diesel exhaust particles has also been studied directly in the environmental SEM (ESEM) (Huang *et al.* 1994). A gas/vapour chamber above the sample in the ESEM allows sample analysis in a range of environments other than vacuum (see Donald & Thiel 1999). The possibilities of aerosol analysis in a 'natural' state before the removal of volatiles is clearly attractive, although the presence of the gas/vapour chamber within the ESEM currently restricts spatial resolution to *ca.* 100 nm at best (although this is dependent on the sample, and conditions within the microscope). Huang *et al.* were able to observe directly the alteration in morphology of diesel particles through a water condensation–evaporation cycle, as a function of particle sulphur content.

Higher spatial resolution is possible using transmission electron microscopy (TEM). Thin samples are mounted onto an electron-transparent substrate (usually a carbon film a few nanometres thick), which in most cases is held on a 3 mm diameter metal support grid. Spatial resolution is a function

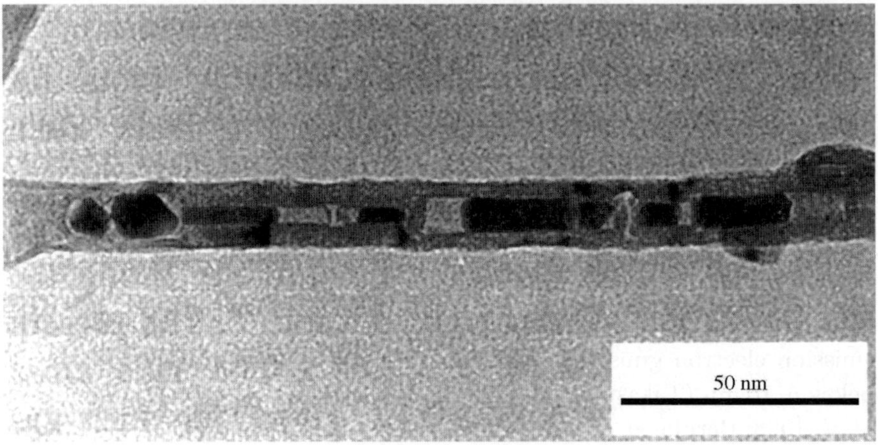

Fig. 2. HRTEM imaging of rare earth dicarbide crystals encapsulated in carbon nanocages. Reproduced with permission from Yosida (1997). © 1997 Elsevier Science BV.

of electron wavelength (determined by the accelerating voltage) and spherical and chromatic aberration within the microscope, although chromatic aberration can be minimized with the use of stable high-coherence electron sources. High-resolution TEM (HRTEM) is extensively used at resolutions below 0.2 nm to investigate the properties and nature of atomic clusters. For instance, Tanaka *et al.* (1993) have studied the behaviour of sub-nanometre tungsten clusters on a MgO film, using a 200 kV HRTEM. The application of HRTEM to studying internal particle structure is illustrated by Yosida's analysis of rare earth dicarbide crystals encapsulated in carbon nanocages of the order of 10–30 nm in diameter (Yosida 1997; figure 2). The relatively high proportion of surface atoms associated with nanometre particles and atom clusters has a profound effect on their reactivity and physicochemical behaviour in some cases (explaining their widespread use in heterogeneous catalysts). By providing insight into the atomic structure of such particles, HRTEM is able to contribute to the understanding of how particle behaviour in this size range differs from the bulk and free molecular regimes (Thölen 1990; Jefferson & Tilley 1999).

The scanning transmission electron microscope (STEM) offers an alternative configuration of transmission electron microscopy, and with it an extended range of analytical methods. In the STEM, as in the SEM, a finely focused electron beam is scanned across a raster on the specimen. Resultant signals used to image the specimen include the intensity of the transmitted beam, secondary electron emissions and elastically scattered electrons. TEMs are usually configurable as STEMs, although there is inevitably a degree of compromise with the electron optics, resulting in marginally reduced imaging and analysis capabilities. Spatial resolution in a dedicated STEM is typically better than 1 nm, and may approach *ca.* 0.3 nm in a high-resolution system. Resolution is limited by spherical aberration within the microscope, although current approaches to reducing spherical aberration (Krivanek *et al.* 1997) will allow significantly increased spatial resolution.

Imaging aerosol particles within the electron microscope, together with appropriate image analysis methods, provides a powerful tool for gaining information on particle size, morphology and structure. However, the analytical capabilities of the electron microscope extend far beyond imaging. Many analytical methods are highly specialized, and are only applicable to particle analysis in specific situations. However, a small number of methods are generally applicable to aerosol particles, and deserve inclusion here.

Selected area electron diffraction (SAED) within the TEM and STEM allows atomic order information within areas from tens of nanometres in diameter upwards. The method has been used to aid the identification of individual asbestos fibre types for some years, and has been used as an additional source of information for ambient aerosol identification in some instances (Sturges *et al.* 1989; Pósfai *et al.* 1994). Its application to ultrafine aerosol particle analysis is possibly more relevant to investigating the atomic arrangement within nanometre-sized particles and structural features, as this begins to have a significant effect on particle behaviour. Its applicability to ultrafine particles has been demonstrated in many investigations into metal and metal oxide ultrafine particle characteristics, usually within the context of heterogeneous catalysts. Structural information from a smaller specimen area is possible using convergent beam electron diffraction (CBED) in the STEM (Humphreys 1999). The area of analysis is defined by the electron beam width, allowing crystallographic information from particles, or regions of particle a few nanometres in diameter.

The use of X-ray emissions within the electron microscope is perhaps the most widely applied form of analytical electron microscopy within aerosol science (De Bock & Van Grieken 1999). Electrons interacting with the specimen excite inner shell atomic electrons, and the decay of these excited states leads to the emission of X-rays with energies characteristic of the element. Energy dispersive X-ray analysis (EDX) allows the quantification of elemental species of atomic number 6 (carbon) and above in the SEM, ESEM, TEM and STEM, although many detectors using a thin silicon protective window are limited to the detection of elements of atomic number 14 (silicon) and above. Analysis in the SEM is not ideal for ultrafine particles, as X-ray emissions from the holding substrate rapidly obscure those from particles under analysis. For the same reason, spatial resolution within the SEM is relatively low (of the order of 0.5–1 μm). Spatial resolution in the STEM and TEM approaches the electron beam width when using thin substrates or arranging for samples to be over a hole on the substrate. Sensitivity to high Z elements is sufficient for the identification of major elemental species in nanometre-diameter particles.

The sensitivity of EDX analysis in the TEM and STEM is limited by the relatively low detection efficiency for X-ray emissions. However, each core electron excitation within the specimen results in a corresponding energy loss within the electron beam. By extracting energy loss information from the beam using an energy-dispersive spectrometer, increased sensitivity to core electron excitations is achievable. Electron energy loss spectroscopy (EELS) within the STEM (and TEM in some configurations) is perhaps the most powerful analysis technique available for analysing single particles within the electron microscope. By recording and analysing the electron energy loss spectrum, details of specific inelastic interactions, and thus sample composition and structure, can be investigated. Energy losses below 50–100 eV are dominated by bulk electron excitations (plasmons) within the sample. At higher-energy losses, energy loss is characterized by atomic core electron excitations, appearing as 'edges' on a decreasing background. The position, amplitude and shape of each edge contain information on atomic core electron excitations, and the chemical environment surrounding the atom. The energy loss at which the edge occurs is related to the atomic electron transition, allowing identification of elemental components (Brown 1999).

Dedicated STEM/EELS systems are currently able to achieve an energy resolution of *ca.* 0.3 eV over a range of losses of up to 2 kV (Brown 1999). Serial detection systems scan the spectrum over a single detector, to build up a record of energy loss over a specific loss interval. Although such systems are effective, sample acquisition times can be long, restricting the speed of analysis, and increasing the risk of specimen damage within the electron beam. Parallel acquisition systems (parallel EELS or PEELS) allow the simultaneous collection of data over a range of energy losses, and are more suited to the analysis of single aerosol particles (Maynard 1995*a*). The analysis area is characterized by the electron beam width, and in principle a spatial resolution approaching that of the beam width is possible. Elemental analysis is possible in principle for most elements (Ahn & Krivanek 1983), although in practice quantification is most applicable to the lighter elements with atomic numbers greater than 3. Quantification using higher-energy edges is compromised by a complex edge shape in many cases. However, the edge structure contains valuable, if difficult to interpret, information on the chemical environment of an element. For instance, Sánchez López *et al.* (1998) have demonstrated the use of EELS near edge structure (ELNES) to distinguish the partitioning between Al and Al_2O_3 in passivated aluminium nanometre-sized particles (figure 3).

Although EELS spectra contain a wealth of information, analysis is not as straightforward as methods such as EDX. Limitations on the energy loss range that can be analysed at any one time and complexities in interpreting data, together with the difficulties of detecting edges against the background energy loss, result in EELS not being directly applicable to routine analysis using currently available systems. Most applications of EELS are to specimens where the constituent elements are known, and it is rare to see the method applied to a sample of unknown composition. However, the successful application of PEELS to the analysis of ambient aerosol particles has been demonstrated by detecting edges using a difference method to eliminate the background, and then quantifying elemental composition from each edge (Maynard 1995*a*). Semi-quantitative elemental analysis of particles down to 5 nm in diameter indicated a practicable relative mass detection limit of *ca.* 1–2% for elements as light as oxygen, with qualitative detection being possible at lower concentrations. Comparison of the results with EDX demonstrated the superior detection efficiency of PEELS for low Z elements, although there were clear advantages in using both methods for identifying and analysing higher Z components (figure 4).

Although the electron microscope is a versatile tool for the analysis of single ultrafine aerosol particles, it has a number of limitations. The high vacuum environment (up to 10^{-11} Torr) and high current density electron beam used in the majority of microscopes has implications for the prepa-

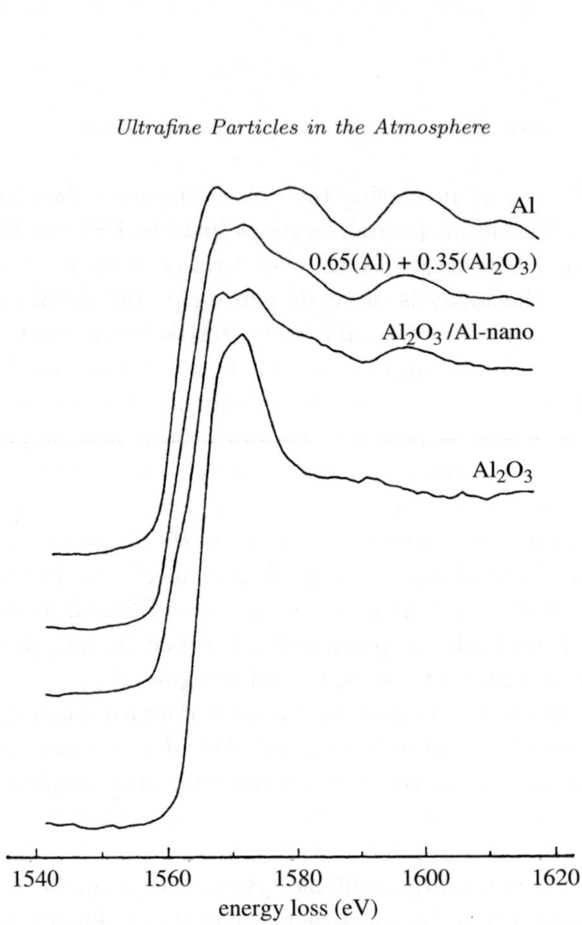

Fig. 3. EELS spectra of the Al K edge taken from Al, Al_2O_3, and passivated ultrafine Al particles, demonstrating the use of near-edge structure to investigate chemical environment. Reproduced with permission from Sánchez López *et al.* (1998). © 1998 Elsevier Science Ltd.

ration of samples, and their stability under analysis. To maintain the high vacuum in a TEM or STEM, samples must be free from volatile species that will degrade the vacuum. Removal of such 'contaminants' is commonly carried out by heating the sample in a vacuum, and thus particles that contain volatile components, or change structure or chemistry at elevated temperatures, are likely to be damaged prior to imaging and analysis. Once in the microscope, susceptible materials may be easily damaged within the electron beam, particularly if the beam is held in the same place for EELS or EDX analysis for an appreciable length of time. Analysis in the TEM and STEM is also time consuming, unlike many emerging SEM systems where automation has led to increasingly rapid analysis of simple speci-

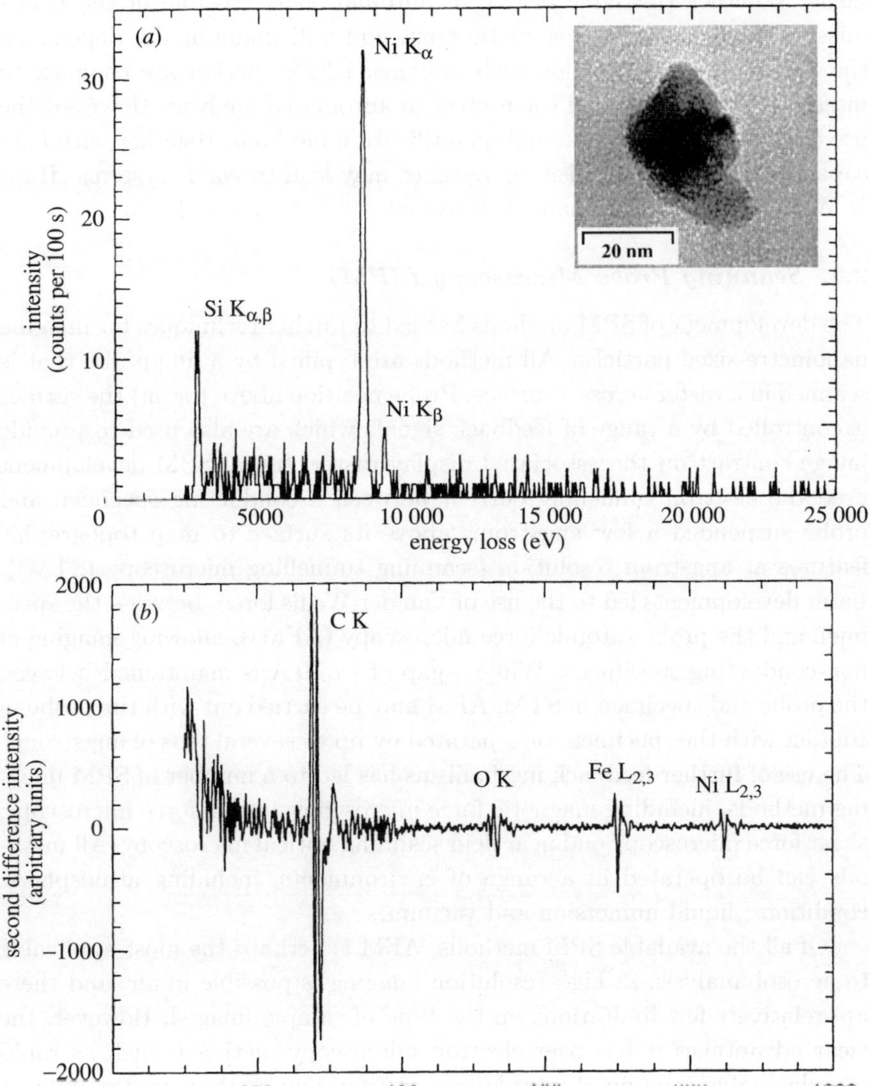

Fig. 4. Comparison of double difference PEELS spectra and EDX spectra for identifying elemental species in an ultrafine ambient particle. Si and Ni peaks in the EDX spectrum are attributable to the Ni support grid, and the EDX detector Si window. © 1995 Elsevier Science Ltd.

mens. Whether the same degree of automation is possible in the transmission microscope has yet to be seen, and will undoubtedly depend on the commercial demand for such systems. EELS spectra are complex to interpret, and do not lend themselves to automated analysis. However, the use of novel edge detection and quantification methods, together with high capacity, rapid data acquisition systems, may lead to viable systems (Hunt & Williams 1991; Kundmann & Krivanek 1991; Maynard 1995*a*).

2.3. *Scanning Probe Microscopy (SPM)*

The development of SPM methods has led to further techniques for imaging nanometre-sized particles. All methods are typified by a fine probe that is scanned in a raster across a surface. Probe position above (or on) the surface is controlled by a range of feedback signals which are also used to provide image contrast on the associated display raster. Initial SPM development used the electron tunnelling current between a conducting specimen and probe suspended a few angstroms above its surface to map topographic features at angstrom resolution (scanning tunnelling microscopy (STM)). Later developments led to the use of Van der Waals forces between the specimen and the probe (atomic force microscopy (AFM)), allowing imaging of non-conducting specimens. While a gap of *ca.* 10 Å is maintained between the probe and specimen in STM, AFM may be carried out with the probe in contact with the specimen, or separated by up to several tens of angstroms. The use of further feedback mechanisms has led to a number of SPM imaging methods, including magnetic force microscopy, lateral force microscopy, shear force microscopy and near field scanning optical microscopy. All methods can be operated in a range of environments, including atmospheric conditions, liquid immersion and vacuum.

Of all the available SPM methods, AFM is perhaps the most applicable to aerosol analysis, as high-resolution imaging is possible in air, and there are relatively few limitations on the type of sample imaged. However, the clear advantages it has over electron microscopy methods, such as rapid sample analysis, minimal sample preparation, and analysis under ambient conditions, are somewhat balanced by a lack of clarity concerning image interpretation and applicability. Friedbacher *et al.* (1995) have successfully applied AFM to the analysis of ultrafine environmental particles collected on a polyester foil using a low pressure cascade impactor. The substrate was found to have a suitably flat surface (root mean square roughness of

1 nm over 4 μm^2) to allow the identification and sizing of sub-30 nm particles. The AFM-derived size distribution agreed well with that expected from the impactor stage cut-off. However, the presence of large-diameter particles with very little height in samples indicated that there was some degree of particle modification subsequent to sampling, bringing into question the direct interpretation of aerosol size distribution from the AFM data. The assumption was made that these particles were the result of droplet deposition followed by evaporation, leaving a residue. Interestingly, the ability to differentiate by height gave the analysis method an advantage over TEM imaging, where differentiation between droplet residues and solid particles isn't always straightforward. Although it is likely that these particles resulted from a loss of volatile components, Köllinsperger *et al.* (1999) were able to demonstrate that the AFM may be used to image environmental particles prior to the loss of volatiles. They were also able to demonstrate the use of automated image analysis in the AFM with environmental particles, allowing rapid characterization of the aerosol size distribution. However, the samples analysed were from the lower stages of a cascade impactor, and thus did not contain large particles that may have caused complications. Cohen *et al.* (2000) have used the AFM to detect and size ultrafine acid particles deposited onto an iron film a few nanometres thick. The reaction between the acid component of the particles and the ion substrate was found to lead to distinctive raised features around the deposition site, with an overall reaction site diameter several times that of the original particle (in many ways the technique is similar to the use of Liesegang rings described by Podzimek & Podzimek (1999)). By detecting and sizing these features using an AFM, Cohen *et al.* were able to rapidly analyse the number and size distribution of 100 nm diameter sulphuric acid-coated carbon particles.

Although SPM can resolve horizontal and vertical details to fractions of a nanometre, it is unable to deal with large changes in vertical profile occurring over a few nanometres. Köllinsperger *et al.* (1997) estimated errors arising from convolutions between the scanning tip and the relatively sharp vertical gradients at the edges of nanometre-sized particles to be of the order of 10%. There is also some concern over the degree to which scanning probe analysis alters the distribution of particles on a substrate. Friedbacher found no alteration of the distribution of environmental particles on a polyester substrate after repeated scans. However, Schleicher *et al.* (1993) reported the removal of *ca.* 8 nm diameter silver particles from highly oriented pyrolytic graphite during STM analysis. Cohen *et al.* also reported the removal of particles while using AFM in contact mode.

Near-field scanning optical microscopy (NSOM or SNOM) is an SPM technique that has some potential benefits for the analysis of ultrafine particles. Conventional optical microscopy is limited to a theoretical spatial resolution of $\lambda/2$. However, if a specimen is illuminated through a sub-wavelength sized aperture held to within a few angstroms of its surface (the near-field), spatial resolution approaching the diameter of the aperture is possible (Synge 1928). By using SPM methods to scan a fine aperture over a sample, optical imaging with a resolution below 100 nm can be achieved. The aperture is usually formed at the tip of a drawn glass fibre coated with aluminium to form a light pipe, and is held a few angstroms from the specimen using non-contact AFM or shear force microscopy feedback methods (Pohl *et al.* 1984; Betzig *et al.* 1991, 1992). Although resolution does not extend far into the ultrafine region, the possibilities for applying optical analysis and detection methods to isolated nanometre diameter particles are of interest.

2.4. *Laser Desorption/Ionization of Ultrafine Particles*

Mass spectrometry (MS) of vaporized then ionized single particles has gained increasing recognition over the past few years as a viable method for analysing the size-resolved compositional make-up of aerosols in near real-time. The aerosol is first formed into a particle beam and transported to a high-vacuum region (*ca.* 10^{-4} Torr), using a series of differentially pumped orifices (see, for example, Liu *et al.* 1995*a*, *b*). Particle acceleration can be related to aerodynamic diameter in the expanding flow fields, and time-of-flight measurements may be used to size particles larger than *ca.* 0.3 μm. Formation of a particle beam in vacuum is followed by particle vaporization and ionization, and detection of ions in a mass spectrometer (Prather *et al.* 1994). Flash vaporization on a resistively heated surface may be used for ion formation, but has limitations at small particle sizes. Laser desorption/ionization (LDI) of individual particles is an alternative vaporization method that is finding increasing use in single particle mass spectrometry (Johnson & Wexler 1995). In a typical system, particles entering the final analysis zone within the instrument are detected using scattered light pulses from a continuous wave laser. These are used to trigger the firing of a second high-energy laser, which vaporizes them in flight. A commercial aerosol time-of-flight mass spectrometer (ATOFMS) is now available, based in the work of Prather *et al.*, that allows single particle size and compositional measurements down to 0.3 μm diameter (TSI Inc. Model 3800 ATOFMS).

Although LDI and MS are in principle applicable to particles of nanometre diameters, the use of optical scattering to trigger vaporization becomes impractical for particles smaller than 0.3 μm. Reents *et al.* (1995) have developed a system capable of analysing particles as small as 20 nm in diameter by using a laser pulsed at between 10 and 30 Hz, independently of the presence of particles. However, the reduction in particle size is at the expense of detection frequency. Reents *et al.* were interested in monitoring contaminant particles in the semiconductor industry. Carson *et al.* (1997) extended the technique down to 12 nm diameter particles for the analysis of size-selected aerosol particles. Size differentiation was on the basis of electrical mobility, using a differential mobility analyser (DMA). Analysis of sodium chloride, ammonium nitrate, potassium chloride and anthracene particles demonstrated that chemical speciation is feasible for nanometre-sized single particles, and that positive ion and free electron production tends to dominate for ultrafine particles with the UV excimer laser used. Ion peak area relative to particle mass increased for smaller particles, with the implication that higher ion yields were being observed at smaller particle sizes. Zhaozhu *et al.* (1998) developed the instrument used by Carson *et al.* and carried out a feasibility study into the analysis of single multi-component ultrafine aerosol particles. Analysis of NaCl/KCl particles of *ca.* 50 nm diameter indicated that detection of a relative mass of KCl of *ca.* 0.06% was possible in a single particle (corresponding to *ca.* 10^{-20} g KCl in the particle). Analysis of 60 nm particles containing traces of several metal salts indicated that detection at mole fractions *ca.* 1% is possible for metal species (corresponding to an absolute mass of the order of 10^{-17} g for each metal), and that for some species the detection limit may be significantly lower (figure 5).

3. Summary

Single-particle analysis has rarely been a valid surrogate for collective particles analysis; perhaps even more so in the case of ultrafine particles, where characterized particles may represent a small fraction of a per cent of a given aerosol. However, in many cases the role of individual particle properties

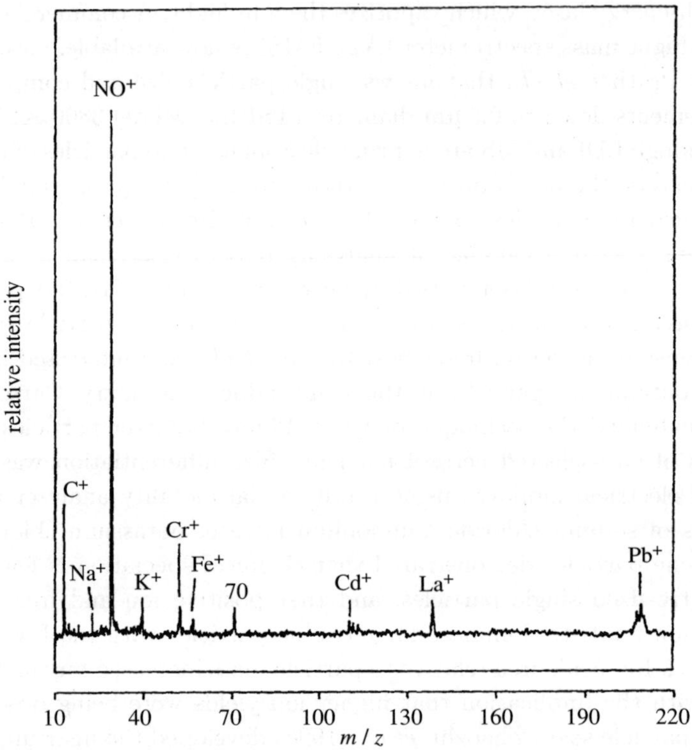

Fig. 5. Mass spectrometry of an individual 60 nm particle doped with *ca.* 1% concentrations of a number of metal species, and vaporized using LDI. Reproduced with permission from Zhaozhu *et al.* (1998). © 1998 American Chemical Society.

must be understood prior to the selection of appropriate collective analysis methods, and this is where the ability to characterize an aerosol at the single particle level is invaluable. Electron microscopy is perhaps the most generally applicable method. Size and morphology are readily characterized in the FEGSEM, TEM and STEM. HRTEM allows structural information on particles and atomic clusters to sub-0.2 nm resolution, while EELS and EDX analysis in the STEM allow the chemical analysis of particles down to nanometre diameters. By combining analysis methods, investigation of particle size, shape, structure, composition and surface properties is in principle possible. However, the analysis environment is harsh, and only suited to robust particles with low volatility. Analysis in the ESEM overcomes some of the analysis environment restrictions and allows in principle the characterization of particles with a significant volatile component, although its application is currently restricted to particles larger than *ca.* 100 nm. SPM offers the possibility of analysing nanometre-diameter particles under ambient conditions, thus getting away from some of the constraints imposed by electron microscopy. Imaging methods such as AFM and NSOM offer novel and exciting possibilities for the characterization of specific aerosols. For instance, the use of NSOM to identify, size and count fluorescently tagged ultrafine particles would seem applicable to identifying particle transport and deposition characteristics within biological systems. While SPM is currently limited in the information that can be obtained from ultrafine aerosol samples, the uniqueness of the information available should allow it to be developed as a complementary tool to electron microscopy. While electron microscopy and SPM are confined to the analysis of collected samples, and are constrained by the limitations of the collection and preparation systems used, developments in aerosol mass spectrometry are providing the means for chemically characterizing size-segregated ultrafine particles on-line. Current technology allows the speciation of individual particles *ca.* 10 nm in diameter, and as this is reduced still further, the resulting methods should provide invaluable complementary data to off-line methods.

By adopting technologies developed within complementary disciplines, together with the development of aerosol-specific methods, it is possible to develop a basis for characterizing single sub-100 nm particles and features in terms of size, morphology, topology, composition, structure and physicochemical properties. The methods available provide complementary means to characterize single ambient particles in depth. Currently, with few exceptions, they are complex, time-consuming to use, and in many cases still at a developmental stage. As such they are not ideally suited to the routine analysis of aerosols. However, by adopting a multi-disciplinary approach, the potential is there to develop complementary tools that will provide routine and detailed information on the particles that influence the environment we live and work in.

4. Disclaimer

Mention of company names and/or products does not constitute endorsement by the Centers for Disease Control and Prevention (CDC).

Acknowledgements

This review was supported in part by the Health and Safety Executive, UK. My thanks to Professor L. M. Brown of the University of Cambridge for providing advice on electron microscopy methods.

References

Ahn, C. C. & Krivanek, O. L. 1983 *EELS atlas*. Gatan Inc., USA.

Bérubé, K. A., Jones, T. P., Williamson, B. J., Winters, C. & Morgan, A. J. 1999 Physicochemical characterisation of diesel exhaust particles: factors for assessing biological activity. *Atmos. Environ.* **33**, 1599–1614.

Betzig, E., Trautman, J. K., Harris, T. D., Weiner, J. S. & Kostelak, R. L. 1991 Breaking the diffraction barrier: optical microscopy on a nanometric scale. *Science* **251**, 1468–1470.

Betzig, E., Finn, P. L. & Weiner, J. S. 1992 Combined shear force and near field scanning optical microscopy. *Appl. Phys. Lett.* **60**, 2484–2486.

Brown, L. M. 1999 Electron energy loss spectrometry in the electron microscope. Part I. Introduction. In *Impact of the electron and scanning probe microscopy on materials research* (ed. D. G. Rickerby, G. Valdrè & U. Valdrè), pp. 209–230. Dordrecht: Kluwer.

Carson, P. G., Johnston, M. V. & Wexler, A. S. 1997 Laser desorption/ionization of ultrafine aerosol particles. *Rapid Commun. Mass Spectrometry* **11**, 993–996.

Cohen, B. S., Li, W., Xiong, J. W. & Lippmann, M. 2000 Detecting H^+ in ultrafine ambient aerosol using iron nano-film detectors and scanning probe microscopy. *Appl. Occup. Environ. Hyg.* **15**(1), 80–89.

De Bock, L. A. & Van Grieken, R. E. 1999 Single particle analysis techniques. In *Analytical chemistry of aerosols* (ed. K. R. Spurney), pp. 243–275. Boca Raton, FL: Lewis.

De Hosson, J. T. M., De Haas, M. & Teeuw, D. H. J. 1998 High resolution scanning electron microscopy oberservations of nano-ceramics. In *Impact of electron and scanning probe microscopy on materials research* (ed. D. G. Rickerby, G. Valdrè & U. Valdrè), vol. 364, pp. 109–134. Dordrecht: Kluwer.

de Juan, L., Fernández, J. & de la Mora, J. F. 1998 Sizing nanoparticles with a focussing impactor: effect of collector size. *J. Aerosol Sci.* **29**, 589–599.

Dockery, D. W., Pope, C. A., Xu, X., Spengler, J. D., Ware, J. H., Fay, M. E., Ferris, B. G. & Speizer, F. E. 1993 An association between air pollution and mortality in six US cities. *N. Engl. J. Med.* **329**, 1753–1759.

Donald, A. M. & Thiel, B. L. 1999 ESEM image contrast and applications to wet organic materials. In *Impact of the electron and scanning probe microscopy on materials research* (ed. D. G. Rickerby, G. Valdrè & U. Valdrè), pp. 209–230. Dordrecht: Kluwer.

Donaldson, K., Li, X. Y. & MacNee, W. 1998 Ultrafine (nanometer) particle mediated lung injury. *J. Aerosol Sci.* **29**, 553–560.

Drummond, D. G. 1950 The practice of electron microscopy. *J. R. Microsc. Soc.* **70**, 1–141.

Fletcher, R. A. & Small, J. A. 1993 Analysis of individual collected particles. In *Aerosol measurement. Principles, techniques and applications* (ed. K. Willeke & P. A. Baron), pp. 260–295. New York: Van Nostrand Reinhold.

Friedbacher, G., Grasserbauer, M., Meslmani, Y., Klaus, N. & Higatsberger, M. J. 1995 Investigation of environmental aerosol by atomic-force microscopy. *Analytical Chem.* **67**, 1749–1754.

Grasserbauer, M. 1983 Micro and surface analysis for environmental studies. *Mikrochim. Acta* **III**, 415–448.

Hering, S. V. & Stolzenburg, M. R. 1995 On-line determination of particle size and density in the nanometer size range. *Aerosol Sci. Technol.* **23**, 155–173.

Hinds, W. C. 1999 *Aerosol technology: properties, behavior, and measurement of airborne particles.* Wiley.

Huang, P.-F., Turpin, B. J., Pipho, M. J., Kittelson, D. B. & McMurry, P. H. 1994 Effects of water condensation and evaporation on diesel chain-agglomerate morphology. *J. Aerosol Sci.* **25**, 447–460.

Humphreys, C. J. 1999 Convergent beam electron diffraction. In *Impact of the electron and scanning probe microscopy on materials research* (ed. D. G. Rickerby, G. Valdrè & U. Valdrè), pp. 325–337. Dordrecht: Kluwer.

Hunt, J. A. & Williams, D. B. 1991 Electron energy loss spectrum imaging. *Ultramicroscopy* **38**, 47–73.

Jefferson, D. A. & Tilley, E. E. M. 1999 The structural and physical chemistry of nanoparticles. In *Particulate matter: properties and health effects* (ed. R. L. Maynard & C. V. Howard), pp. 63–84. Oxford: BIOS.

Johnson, M. V. & Wexler, A. S. 1995 Mass spectrometry of individual aerosol particles. *Analytical Chem.* **67**, 721A–726A.

Keskinen, K., Pietarinen, M. & Lehtimaki, M. 1992 Electrical low pressure impactor. *J. Aerosol Sci.* **23**, 353–360.

Köllinsperger, G., Friedbacher, G., Grasserbauer, M. & Dorffner, L. 1997 Investigation of aerosol particles by atomic force microscopy. *Fresenius J. Analytical Chem.* **358**, 268–273.

Köllinsperger, G., Friedbacher, G., Krammer, A. & Grasserbauer, M. 1999 Application of atomic force microscopy to particle sizing. *Fresenius J. Analytical Chem.* **363**, 323–332.

Krivanek, O. L., Dellby, N., Spence, A. J., Camps, R. A. & Brown, L. M. 1997 *Aberration correction in the STEM.* EMAG97, Cambridge, UK. London: IOP Publishing.

Kundmann, M. R. & Krivanek, O. L. 1991 Automated processing of automated parallel-detection EELS data. *Microsc. Microanal. Microstruct.* **2**, 245–256.

Liu, P., Ziemann, P. J., Kittelson, D. B. & McMurry, P. H. 1995*a* Generating particle beams of controlled dimensions and divergence. I. Theory of particle motion in aerodynamic lenses and nozzle expansions. *Aerosol Sci. Technol.* **22**, 293–313.

Liu, P., Ziemann, P. J., Kittelson, D. B. & McMurry, P. H. 1995*b* Generating particle beams of controlled dimensions and divergence. II. Experimental evaluation of particle motion in aerodynamic lenses and nozzle expansions. *Aerosol Sci. Technol.* **22**, 314–324.

Maynard, A. D. 1993 The collection and analytical electron microscopy of ultrafine aerosol particles. PhD thesis, University of Cambridge, UK.

Maynard, A. D. 1995*a* The application of electron energy-loss spectroscopy to the analysis of ultrafine aerosol particles. *J. Aerosol Sci.* **26**, 757–777.

Maynard, A. D. 1995*b* The development of a new thermophoretic precipitator for scanning-transmission electron-microscope analysis of ultrafine aerosol-particles. *Aerosol Sci. Technol.* **23**, 521–533.

Oberdörster, G. 1996 Significance of particle parameters in the evaluation of exposure dose-response relationships of inhaled particles. *Particulate Sci. Technol.* **14**, 135–151.

Ortner, H. M., Hoffman, P., Stadermann, F. J., Weinbruch, S. & Wentzel, M. 1998 Chemical characterization of environmental and industrial particulate samples. *Analyst* **123**, 833–842.

Podzimek, J. & Podzimek, M. 1999 Liesegang ring technique applied to the chemical identification of atmospheric aerosol particles. In *Analytical chemistry of aerosols* (ed. K. R. Spurny), pp. 231–242. Boca Raton, FL: Lewis.

Pohl, D. W., Denk, W. & Lanz, M. 1984 Optical stethoscopy: image recording with resolution $\lambda/20$. *Appl. Phys. Lett.* **44**, 651–653.

Pósfai, M., Anderson, J. R. & Buseck, P. R. 1994 *Atmos. Environ.* **28**, 1747–1756.

Prather, K. A., Nordmeyer, T. & Salt, K. 1994 Real-time characterization of individual aerosol particles using aerosol-time-of-flight mass spectrometry. *Analytical Chem.* **66**, 1403–1407.

Reents, W. D., Downey, S. W., Emerson, A. B., Mujsce, A. M., Muller, A. J., Siconolfi, D. J., Sinclair, J. D. & Swanson, A. G. 1995 Single particle characterization by time-of-flight mass spectrometry. *Aerosol Sci. Technol.* **23**, 263–270.

Sánchez López, J. C., Caballero, A. & Fernandez, A. 1998 Characterisation of passivated aluminium nanopowders: an XPS and TEM/EELS study. *J. Eur. Ceramic Soc.* **18**, 1195–1200.

Schleicher, B., Jung, T. & Burtscher, H. 1993 Characterization of ultrafine aerosol-particles adsorbed on highly oriented pyrolytic-graphite by scanning tunneling and atomic-force microscopy. *J. Colloid Interface Sci.* **161**, 271–277.

Sturges, W. T., Harrison, R. M. & Barrie, L. A. 1989 Semi-quantiative X-ray diffraction analysis of size fractionated atmospheric particles. *Atmos. Env.* **23**, 1083–1098.

Synge, E. H. 1928 A suggested method for extending microscopic resolution into the ultra-microscopic region. *Phil. Mag.* **6**, 356–362.

Takasu, Y., Kaneko, F., Tsutsui, T., Nakagawa, M., Yamada, M. & Yahikozawa, K. 1993 Comparison of high-resolution scanning electron-microscopy with transmission electron-microscopy for the characterization of ultrafine palladium particles embedded on active-carbon. *Bull. Chem. Soc. Japan* **66**, 2419–2421.

Talbot, L., Cheng, R. K., Schefer, R. W. & Willis, D. R. 1980 Thermophoresis of particles in a heated boundary layer. *J. Fluid Mech.* **101**, 737–758.

Tanaka, N., Kitagawa, T. & Kuzuka, T. 1993 High-resolution electron-microscopy of tungsten and C-60 clusters supported on single-crystal MgO films. *Mater. Sci. Engng* B **19**, 53–60.

Thölen, A. R. 1990 Electron-microscope investigation of small particles. *Phase Transitions* **24-6**, 375–406.

Van Cleempoel, A., Joutsensaari, J., Kauppinen, E., Gijbels, R. & Claeys, M. 1998 Aerosol synthesis and characterization of ultrafine fullerene particles. *Fullerene Sci. Technol.* **6**, 599–627.

Watson, H. H. 1937 A system for obtaining from mine air, dust samples for physical, chemical and petrological examination. *Trans. Inst. Min. Metall.* **46**, 155–240.

Watson, H. H. 1958 The sampling efficiency of the thermal precipitator. *Br. J. Appl. Phys.* **9**, 78–79.

Yosida, Y. 1997 A new type of ultrafine particles: rare earth dicarbide crystals encapsulated in carbon nanocages. *Physica* B **229**, 301–305.

Zhaozhu, G. E., Wexler, A. S. & Johnson, M. V. 1998 Laser desorption/ionization of single ultrafine multicomponent aerosols. *Environ. Sci. Technol.* **32**, 3218–3223.

Discussion

T. Benham (*Volvo Technical Development, Sweden*). With respect to the picture of particles prepared by two different techniques (figure 1), how is it possible to identify which one is the correct representation of the particles?

A. D. Maynard. Intuitively, the particles with the least preparation—in this case those simply collected via impaction on a substrate, with no further processing other than being given a conductive coating—will be most representative of the airborne particles. In this case, we also know a great deal about what we expect the particles to look like (agglomerates of very small primary particles) from a large body of published data, and so we can be reasonably confident that the impacted particles are the closest representation of the airborne particles.

M. S. BINGLEY (*Cobham, UK*). I would like to remind people that greater resolution can be obtained with immersion objectives with numerical apertures of 1.524 instead of the usual 'schoolboy' objective, 1.25 NA $\times$ 100 that is supplied to scientists. I have use one of these, so they really do exist!

The use of mono-brom-naphthalein immersion fluid enables numerical apertures of 1.6 to be realized. The late Horace Dall, master lens and instrument maker, made the front element of an immersion objective out of diamond and reached an aperture of 1.9 NA. These techniques might provide more information on ultrafine particles. The microscope makers should be pressed to make some decent lenses again!

A. D. MAYNARD. The resolving power of an optical objective lens is $X = 0.61\lambda/\mathrm{NA}$, where X is the distance between two just-resolved points, λ is the illumination light wavelength and NA is the lens numerical aperture. Thus with a numerical aperture of 1.6 and using illuminating light with a wavelength of 400 nm, it is in principle possible to achieve a resolution of 150 nm (a 1.25 NA objective would raise the resolution limit to 200 nm). These limits are theoretical limits, and in practice, will be dependent on a number of other factors, including illuminating conditions and specimen contrast. Although this resolution limit is sufficient to observe fine details on larger sub-micrometre particles, it is insufficient for the detailed analysis of particle smaller than 100 nm in diameter.

C. V. HOWARD (*Fetaltoxico-Pathology, University of Liverpool, UK*). Have you considered the use of partial vacuum electron microscopy? For example, Mike Gorringe in Oxford, among others, has been showing video images of catalyst particles in motion on a substrate by using an environmental cell. Could this approach be used in your study?

A. D. MAYNARD. The use of partial vacuum electron microscopy (environmental SEM) is particularly attractive to the study of ultrafine particles that may have an appreciable mass of volatile material, and have a physical structure that changes with the loss of volatile material. However, there is a trade-off within the ESEM between resolution and gas pressure, that renders this type of analysis somewhat difficult. Current ESEMs are able to image at a resolution of *ca.* 100 nm at pressures of a few pascals.

CHAPTER 4

PARTICLES FROM INTERNAL COMBUSTION ENGINES — WHAT WE NEED TO KNOW

N. Collings and B. R. Graskow
Department of Engineering, University of Cambridge, Trumpington Street, Cambridge CB2 1PZ, UK

Internal combustion (IC) engines are a major contributor to the total particulate emissions inventory, especially in urban areas. Recent epidemiological studies suggesting links between fine particles and negative health effects have sparked an increased interest in this subject. While particulate emissions from IC engines have been the focus of research for many years, a great deal of information crucial to our understanding of this subject still remains unknown. In this paper the authors address some of these unknowns, focusing primarily on the process and consequences of aerosol dilution strategy. The thermodynamics of dilution are considered, and the inadequacy of conventional constant-volume sampling dilution tunnels for ultrafine particle characterization are demonstrated using experimental data. Finally, time-resolved data demonstrating the variation in concentration of pollutants in a vehicle moving in traffic are used as an example of the difficulties in setting legislation aimed at controlling exposure to ultrafine particles.

Keywords: nanoparticle; aerosol; dilution; thermodynamics; nucleation; condensation

1. Introduction

Without doubt the most important fact to establish with regard to particulate emissions from internal combustion (IC) engines is the exact relationship between emission, exposure and subsequent health effects. This relationship is not very well understood currently, and it seems unlikely to be resolved in the near future, especially with regard to the long-term health effects of exposure to ultrafine particulate matter. In the case of

IC engine emissions, debate continues with regard to what the important parameters are which should be examined (e.g. particle size, number, surface area, mass, composition, etc.) in terms of measurement, control and legislation. No general agreement yet exists as to what the most appropriate techniques and equipment (if they even exist) for the dilution and subsequent measurement of IC engine exhaust particulate matter might be. It is unclear whether much of the data collected to this point are representative or even applicable for the characterization of particulate exposure with respect to health effects. Consequently, there is a low level of confidence concerning our ability to determine appropriate limits for the legislative control of vehicle particulate emissions.

What do we know at present?

(a) There is no doubt that IC engines are responsible for a significant fraction of total particulate matter present in the atmosphere, especially in urban areas (UK QUARG 1996). The fraction attributable to mobile sources depends on how the particles are counted, a recurring dilemma.
(b) It is now accepted that both the manner in which a vehicle's exhaust is diluted (if at all) and the technique and instrumentation used for subsequent measurement of the exhaust aerosol can (and usually do) have profound effects on the measured character (e.g. size, number, composition) of the aerosol.
(c) It can be exceedingly difficult to establish adequate repeatability between repeat tests, let alone different laboratories, particle levels sometimes exhibiting a large, seemingly random, element.
(d) The widespread use of particle traps will reduce diesel vehicle particulate emissions very significantly.
(e) Low-sulphur fuel will reduce particle emissions.

At present, vehicle particle emissions legislation is based solely on the mass emission (i.e. PM_{10} and $PM_{2.5}$) of particles collected on a filter directly from a standard constant-volume dilution tunnel. This is a standard of measure which, in effect, exempts ultrafine particles from legislated control. This 'exemption' of ultrafines is due to the negligible contribution of ultrafines to total particle mass, and is compounded by the poor representation of atmospheric dilution obtained with conventional constant-volume dilution systems used in making such measurements. Ultrafine particles in the atmosphere that result from vehicle emissions can form both on short time-scales (e.g. through condensation and nucleation as exhaust gas exits the vehicle tailpipe and mixes with the atmosphere) and on much longer time-scales, e.g. due to photochemical processes.

In this context it is tempting to simply require emission levels to be as low as technically feasible, on the basis that there is no threshold level at which harm is zero; and this, in essence, is the route being followed as far as gaseous vehicle emissions. There are significant and fundamental reasons for not following this route with respect to particle emissions. Perhaps the most compelling as far as IC engines are concerned is the persistent difficulty associated with the definition of appropriate test procedures for dilution and sampling.

In the case of gaseous emissions, the composition of legislated species (e.g. HC, CO, NO_x) remains virtually unchanged during mixing and dilution in the atmosphere, while secondary processes (e.g. formation of photochemical smog) occur on relatively long time-scales. On the other hand, it is well known that upon leaving the tailpipe, the particle size spectrum and composition changes dramatically on time-scales ranging from milliseconds to days, depending on a number of factors including rate of dilution, final dilution ratio, atmospheric conditions (temperature, humidity, background particle levels, etc.; see Abdul-Khalek *et al.* (1999, 2000) and Graskow *et al.* (2000)). In the case of exposure to nanoparticles ($d_p < 50$ nm), the dose rate that the individual is subject to decreases strongly as the separation (both in terms of distance and time) between the individual and the emission source is increased (due to agglomeration and convective and diffusive dispersion and dilution). Consequently, exposure to vehicle-borne nanoparticles may vary wildly from person to person, depending on an individual's 'lifestyle'.

Progress in assessing the importance that should be attached to such considerations can only be made if the relative health effects are better understood. Recent studies (Donaldson *et al.* 1996, 1998; Ferin *et al.* 1992) have highlighted the possibility that current legislation may actually increase health risks if it was found that nanoparticles were significantly more harmful than large particles, since it is essentially only the latter that are controlled, and minimizing their mass may result in an increase in total number due to the reduction in surface area of carbon adsorbate. Leaving the important questions of whether we are monitoring the appropriate parameters in the environment and how legislation should be progressed, let us focus our attention on the issue of dilution and standard test procedures.

2. Current Test Procedures

The current standard test procedures (i.e. those required by law to be used in assessment of particulate emissions) are designed specifically for the measurement of particle mass emissions. In the current standard method, the entire engine exhaust flow is directed into a constant-volume dilution tunnel, where the exhaust is mixed with particle-free dilution air. In such systems, the total flow of the engine exhaust and dilution air is held constant. Since the amount of engine exhaust flow changes according to different engine operating conditions (exhaust flow increases roughly in proportion to engine speed for diesel engines, and in proportion to power in gasoline engines), the dilution ratio also changes; typical dilution ratios for constant-volume dilution systems range from 3 to 15, depending on the engine and operating condition. A sample of this diluted aerosol is then collected on a filter, which can subsequently be analysed to determine the mass and soluble organic fraction of the emitted particles. If particle losses in the dilution and sampling systems are ignored, then the mass collected is largely independent of the details of the dilution process. This is because the total mass of particles collected in such a way is overwhelmingly dominated by relatively large ($d_p > 100$ nm) carbonaceous particles that are formed within the combustion chamber, and which therefore remain virtually unchanged by the dilution process. Consequently, the current methods are adequate for determining compliance with the current mass-based legislation (so long as sufficient care is taken with regard to particle losses and gains during the measurement).

However, if it proves to be correct that much of the negative health impact from particles is due to ultrafines (Donaldson *et al.* 1996, 1998; Ferin *et al.* 1992), it seems that a change in legislative emphasis away from mass and toward number or surface-area weighting is likely. The current method for measuring mass-based particle emissions is totally inappropriate for assessing emission of ultrafine particles, since these particles contribute negligible mass, even if present in extremely high concentrations. If such a change in standards were to occur, then it is difficult to overemphasize the difficulties in defining consistent, representative test procedures. The lack of reproducibility in measurements pertaining to the smallest particle size ranges between laboratories is legion.

The reason for this variability in results is not difficult to explain. Nanoparticles are largely generated during the dilution process as the hot exhaust gas mixes with cool ambient air. The primary mechanism for nanoparticle formation during dilution is homogeneous nucleation of sulphuric acid, onto which either volatile organic compounds or their oxidation products condense. All of these components normally remain in the gas phase at exhaust gas temperatures. Nucleation is, of course, a notoriously nonlinear process, which can be exquisitely sensitive to a number of variables in the dilution process. A number of these variables have been shown to have a significant effect on nanoparticle production, including overall dilution ratio, rate of dilution, turbulence intensity, mixing length-scales, dilution air temperature, humidity, and background particle concentration (Abdul-Khalek *et al.* 1999, 2000). In addition to the variability introduced by dilution, particle formation is also significantly affected by the exact nature of the exhaust gas (temperature, amount and composition of particles and gas-phase particle precursors), which itself is strongly dependent on the fuel and lubricating oil used, as well as the engine operating conditions, mechanical condition of engine components, etc.

Given this, and the practical difficulties of producing an appropriate test procedure that determines the representative nanoparticle production of a given engine/vehicle combination, it may be worthwhile to examine the possibility of developing a standard predictive model, the input for which is the undiluted exhaust gas composition itself. If the exhaust gas composition relevant to its particle-forming potential can be measured, then one might use those data to apply predictive models based on a wide range of ambient and dilution conditions (assuming that these can be modelled effectively). This would yield emissions information for a much wider range of conditions than is practical to test experimentally, either in the laboratory or on the road.

In light of the extreme dependence of particle formation on the dilution process, one must take great care in creating a dilution system which provides dilution conditions that are representative of those which occur in the real world. While the standard constant-volume dilution method described above is adequate for making mass-based particle measurements (due to the insensitivity of large particles to dilution conditions), such a system is completely inappropriate for use with number-based (e.g. scanning mobility particle sizer (SMPS) and condensation particle counter (CPC)) measurements. As stated earlier, typical dilution ratios for constant-volume dilution systems range from 3 to 15. Under such circumstances the dilution ratio never reaches a point where the processes of nucleation, condensation or agglomeration are effectively arrested, as they are in the atmosphere. In fact, whereas the nuclei mode is normally prominent in measurements of emissions from modern diesel engines measured both on the road and using non-constant-volume sampling (non-CVS) dilution systems, the nuclei mode is often conspicuously absent in measurements made using the CVS dilution system (see experimental section below). This is because under many conditions, the dilution process in CVS tunnels is insufficient to trigger nucleation, which would normally occur during the process of dilution in the atmosphere. While this makes little difference in terms of measured particle mass emissions, it can have an overwhelming influence on particle number, potentially resulting in several orders of magnitude error in the estimated particle number emissions.

In addition to the low dilution ratios, it seems unlikely that the dilution process in the current standard dilution tunnels are comparable with those encountered in atmospheric dilution; the resulting change in the mixing process may have profound implications for the thermodynamics of particle formation (see discussion below). Finally, since the dilution ratio changes between different operating conditions, the resulting particle size distribution may change from condition to condition, even if exhaust-gas composition remains identical. Consequently, such a system is unsatisfactory for number-based measurement of ultrafine particles.

Abdul-Khalek *et al.* (1999, 2000) highlighted these issues by the design of a two-stage dilution system, in which the effect of dilution ratio, temperature, relative humidity and residence time between dilution stages can be studied. The results showed that total measured particle number emissions from a diesel engine at a set operating condition could be changed by two orders of magnitude as a result of modest changes in dilution conditions.

In summary, it is argued that the production of nanoparticles is primarily driven by nucleation of one or more precursor species (of which, sulphate is believed to be most important) that exist in the gas phase in hot engine exhaust. Particles are formed as these species are forced into a state of supersaturation during dilution as exhaust is cooled and mixes with ambient air. After the dilution ratio has exceeded some critical value, dynamic particle formation and growth processes (nucleation, condensation, agglomeration) effectively cease to operate due to a lack of driving potential at high dilution ratios. The total number of particles produced during dilution is extremely nonlinear and is highly sensitive to detailed conditions of the dilution process itself. Based on a simple examination of particle formation and the dilution process, it can be clearly seen that standard constant-volume dilution systems are unsuitable for measurement of ultrafinc particulate matter. Such systems may drastically under-represent the total number of ultrafine particles that is likely to be emitted from IC engines under real-world dilution conditions.

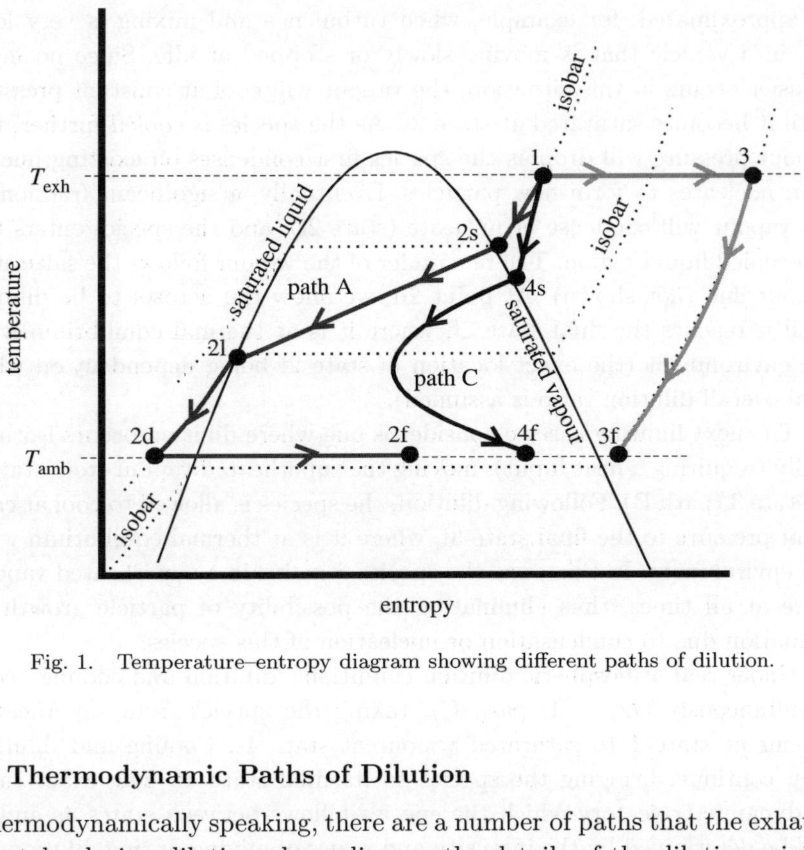

Fig. 1. Temperature–entropy diagram showing different paths of dilution.

3. Thermodynamic Paths of Dilution

Thermodynamically speaking, there are a number of paths that the exhaust can take during dilution, depending on the details of the dilution mixing process. The details of the thermodynamic path taken can have profound consequences in terms of particle formation in the diluting exhaust. Figure 1 is a temperature–entropy (T–s) diagram which considers the thermodynamic state of a particle precursor species (e.g. sulphate) during dilution. State 1 on this diagram represents the species at its starting point, as a superheated gas in the exhaust exiting the tailpipe. We will begin by examining the fate of this species under two limiting cases of dilution based on different mixing strategies.

In the first case (path A), convective mixing is poor, and consequently the aerosol is cooled but not diluted by the ambient air. This is the case when the Lewis number (Le, defined as the ratio of thermal diffusivity to mass diffusivity) is much greater than unity ($Le \gg 1$). This situation may be approximated, for example, when turbulence and mixing is very low, e.g. in a vehicle that is moving slowly or stopped at idle. Since no mass transfer occurs in this situation, the vapour will cool at constant pressure until it becomes saturated at state 2s. As the species is cooled further, the vapour pressure will drop as the species first condenses on existing nuclei, then nucleates to form new particles. Eventually, a significant fraction of the vapour will condense or nucleate (state 2l), and the species enters the sub-cooled liquid region. The remainder of the vapour follows the saturated vapour line (not shown). At point 2d, we allow the aerosol to be diluted until it reaches the final state 2f, where it is at thermal equilibrium with the environment (the exact location of state 2f being dependent on what final overall dilution ratio is assumed).

The next limiting case to consider is one where dilution occurs isothermally (requiring a heat input), moving the superheated vapour from state 1 to state 3 (path B). Following dilution, the species is allowed to cool at constant pressure to the final state 3f, where it is at thermal equilibrium with the environment. In this case, the species remains in a superheated vapour state at all times, thus eliminating the possibility of particle growth or formation due to condensation or nucleation of this species.

Under real atmospheric dilution conditions, dilution and cooling occur simultaneously ($Le \sim 1$; path C), taking the species from superheated vapour at state 1 to saturated vapour at state 4s. Cooling and dilution then continue, bringing the species to its final state 4f. The exact thermodynamic trajectory which the species follows between states 4s and 4f will be determined by the intensity and scales of mixing during dilution; as mixing intensity increases, the trajectory will shift to the right, resulting in lower peak saturation ratios. This reflects the reduction in diffusive heat loss between pockets of diluting exhaust precursors and the cooler dilution air (Davenne *et al.* 2000). Under conditions of intense mixing, dilution may become adiabatic, possibly avoiding a saturated vapour state altogether.

Although the illustration of the three dilution scenarios described above is highly qualitative, it does serve to demonstrate how the thermodynamic state of a particle precursor species may be affected by the dilution process. Obviously, if one could dilute in such a way as to avoid reaching the saturated vapour stage (e.g. with isothermal dilution or very intense mixing), then nucleation could be prevented altogether, resulting in a dramatic reduction in nanoparticle emissions. Even if it is impossible to completely avoid a saturated vapour state, one may still manipulate the mixing process in such a way as to minimize saturation ratio and, thus, particle formation. One important point to note regarding figure 1 is that between the initial and final states (i.e. state 1 to 2f/3f/4f) the species is not likely to be in a state of thermodynamic equilibrium. Consequently, condensation and nucleation processes may become time-limited. If mixing intensity is high, then the amount of time which a species spends at high saturation ratios may be short enough to avoid nucleation (or condensation) completely.

Curiously enough, entropy maximization is key here: the number of particles formed during dilution is inversely proportional to the final entropy of the diluted aerosol. Because the species has a lower state of entropy in particle (liquid) form than it does in gas form (from figure 1, we see that saturation ratios increase in the vapour dome as entropy decreases), nucleation of particles will result in a net decrease in entropy. Consequently, if we assume that the final dilution ratio is identical for all dilution scenarios, then the dilution path with the highest final entropy should produce the fewest particles. The price which is paid for higher entropy is the energy required to supply heat (isothermal dilution, path B) or to produce rapid turbulent mixing (path C).

The actual mixing process is always, in practice, turbulent (in experimental as well as real-world dilution), and inhomogeneous on a micro scale (Davenne *et al.* 2000). A picture of a turbulent plume (figure 2) illustrates this point; some of the vapour at the extremities of an eddy is micro mixing with short time-scales, while other vapour at the centre of an eddy is cooling before mixing. This is an essential difficulty, and experiments that

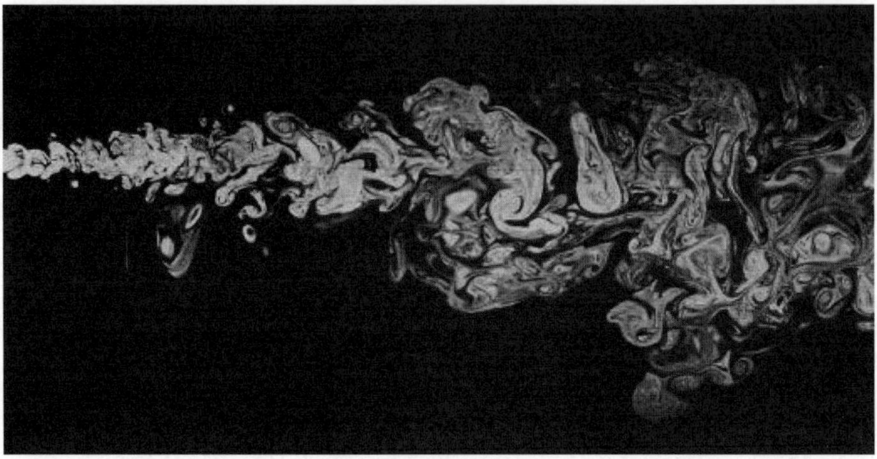

Fig. 2. Mixing plume of a turbulent jet (Van Dyke 1982).

attempt to mimic real-world dilution processes have to address the issue of mixing scaling. In addition, the question of exactly what level of mixing is representative of atmospheric mixing arises; given that vehicles operate under a wide range of speeds and under widely varying atmospheric conditions (temperature, humidity, wind, background particle concentration, etc.), this is likely to be a question for which there is no simple answer.

It might be argued that as current significant particle emitters (diesel engines, perhaps some types of gasoline engines, maybe other engine types depending on how the standards progress) will in the future operate with low sulphur fuel and be fitted with particle traps, IC engines will cease to be a significant contributor to the particle emissions inventory. There is no doubt that particle mass will be very significantly reduced by these and other means (higher injection pressures in the case of diesel engines for example), though the effect of these measures on number emissions may not be so dramatic. Indeed, nanoparticle formation is itself very significantly reduced by any reduction in fuel sulphur, though where a finite level of engine oil consumption occurs (as in the diesel engine), sulphur-related fine particle generation is likely to remain significant. Such considerations focus attention back onto identifying test procedures that will appropriately assess the fine particle problem.

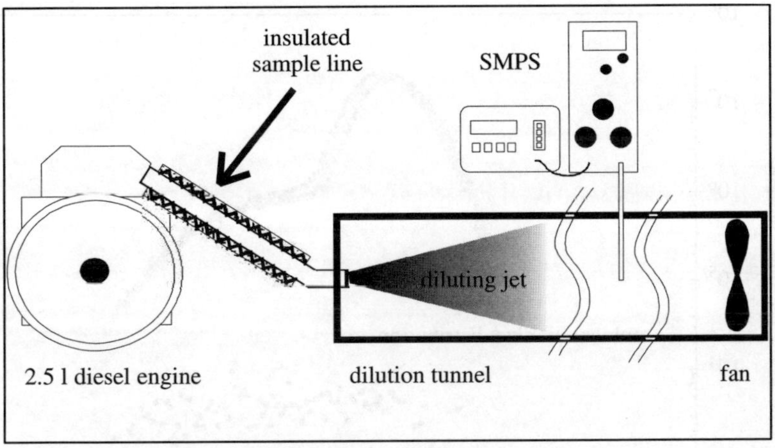

Fig. 3. Dilution tunnel experiment schematic.

4. Experimental

We present here some results obtained during an investigation to examine real-world dilution processes. Particle size distributions emitted from a diesel engine have been measured using a dilution system which has a dilution ratio sufficient to approach atmospheric dilution conditions. Figure 3 shows the apparatus used for this. The engine used was a 2.5 l displacement direct injection four-cylinder diesel, using standard 300 ppm sulphur diesel fuel; engine loading was accomplished through the use of a water brake dynamometer. The exhaust from one of the cylinders was taken to the dilution tunnel via a short (1 m) heated sample line, which was maintained at 200 °C. The exhaust from a single cylinder was used, as this ensured a minimum overall final dilution ratio of approximately 100:1 in the tunnel that was used. When mixing is complete at this dilution ratio, particle evolution (nucleation, condensation, agglomeration) can be assumed to effectively be halted. The dilution tunnel itself had a square cross-section, of side 0.4 m, with a dilution air flow rate of 0.3 $m^3\ s^{-1}$ (unfiltered ambient air was used), resulting in a tunnel Reynolds number of 10^5. The diluted sample was extracted at a location 4.0 m downstream from the tunnel entrance, allowing the aerosol to become fully mixed with the dilution air.

The objective of this exercise was to measure the resultant particle size distributions under dilution conditions that mimicked those occurring in

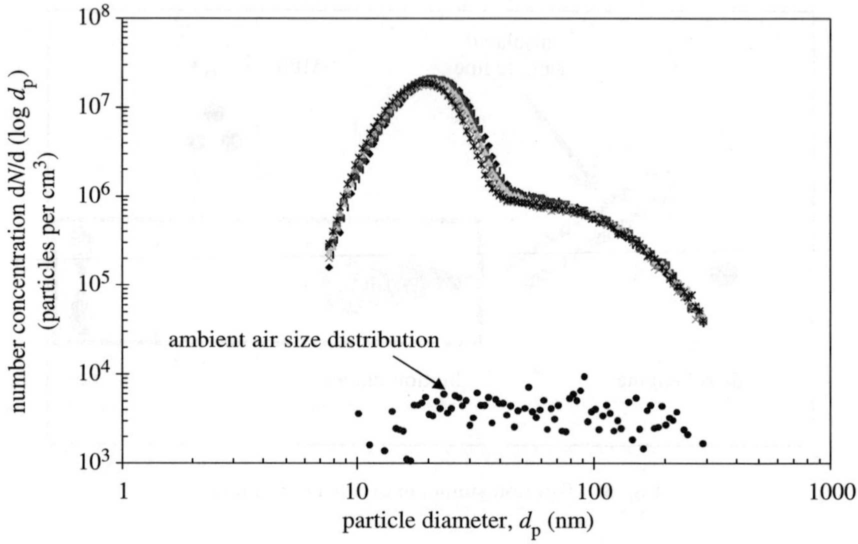

Fig. 4. Tunnel-diluted diesel exhaust particle size distribution.

the environment. Typical results are consistent with those which might be expected on the basis of Abdul-Khalek *et al.*'s experiments (1999, 2000): a significant nuclei mode, not typically observed in a conventional CVS dilution tunnel for such an engine (Rickeard *et al.* 1996; Greenwood *et al.* 1996; McAughey 1997). Figure 4 shows several repeat number-weighted size distributions (measured using an SMPS) measured at a typical engine operating condition (1500 rpm, 7.5 kW). This nuclei mode was consistently displayed over a wide range of operating conditions.

Although this type of experiment is reasonably well controlled, the 'dilution tunnel' was quite massive, even though only 25% of the gas from this (smallish) engine was used. If similar dilution methods were to be used in some sort of standard test method, it seems practical that only a partial sample of the total exhaust flow can be used, and then, as mentioned in the previous section, the question of experiment scaling arises.

In a companion study, an SMPS was mounted in a chase vehicle, and a series of vehicles were followed on a motorway (*ca.* 100 km h^{-1}) at an approximate distance of 100 m. The aerosol was measured at the exit of the chase vehicle's cabin ventilation system, thus the aerosol measured is the

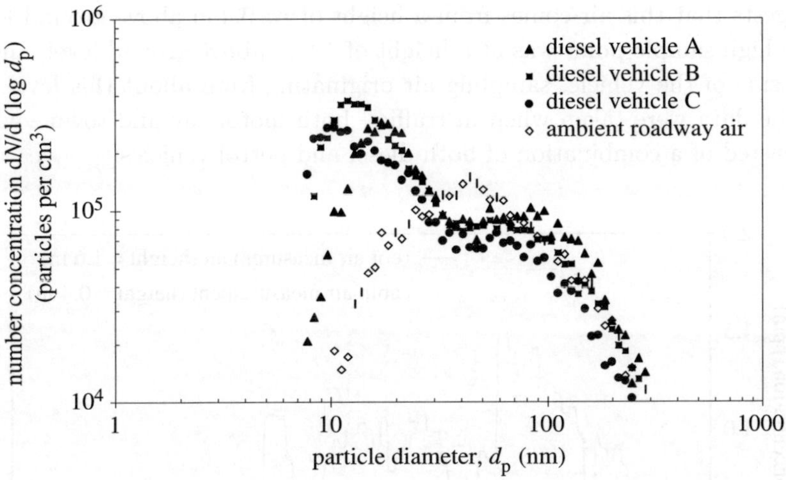

Fig. 5. Motorway measurement of diesel vehicle aerosol size distributions (100 km h^{-1}).

same aerosol to which vehicle occupants would be exposed. Of course, such experiments must necessarily sacrifice controlled, experimentally reproducible conditions for real-world data, but the results are revealing (see figure 5). The size distributions measured behind diesel-powered vehicles are similar to those measured in the laboratory using the dilution tunnel (figure 4); as the data in figures 4 and 5 are not corrected for dilution ratio, no inferences should be made on the basis of differences in the absolute value of measured number concentration. Significantly, aerosol emitted from the diesel vehicles contains a prominent nuclei mode; again, not what would typically be measured if the vehicle was tested using a conventional CVS dilution system. Size distributions measured for gasoline-powered vehicles were not discernibly different from the ambient size distribution.

Though not directly concerned with particle measurements, a similar study looking at on-road NO_x emissions showed that there is a significant variation of this pollutant with height above the road surface. In this study, two chemiluminescent NO_x detectors were fitted to the chase vehicle; one detector was used to measure the NO_x concentrations in the air entering the cabin compartment, and the other sampled from a point at the side, and top, of the vehicle. The cabin air was taken from a grill between the bonnet and the front windscreen, and reference to air flow studies over similar vehicles suggests that this air comes from a height of *ca.* 0.4 m above ground level. The high sample point was at a height of 1.6 m above ground level, and to the side of the vehicle, sampling air originating from about this level. All of the data were taken when in traffic—both motorway and town—which consisted of a combination of both diesel and petrol vehicles.

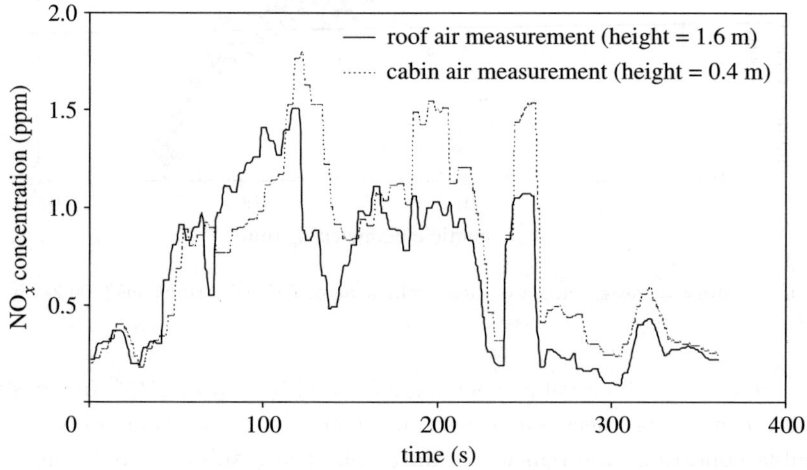

Fig. 6. Variation in NO_x concentration above the roadway surface in traffic.

A typical set of results for these experiments is given in figure 6. As this figure shows, while there is a good degree of correlation between the trends in NO_x, the absolute concentration levels exhibit significant variation. Quite significant is the fact that the concentration of NO_x was generally much higher in the cabin ventilation air than it was at a point at the top of the chase vehicle, reflecting the increase in dilution ratio as height above the roadway surface increases. Presumably, similar variations would be observed in terms of particle concentration at different heights above the roadway. Clearly, this type of anecdotal data begs many questions, but it would seem to have ramifications for vehicle occupants in traffic.

Given that significant fluctuations in exhaust gas composition are observed on very short time-scales both on the road and in the laboratory (Peckham *et al.* 1998; Sutela *et al.* 1999), the question of measurement time response arises. In addition, there is a good deal of evidence suggesting that particle emissions from engines may also exhibit significant fluctuations under steady-state operating conditions (Graskow *et al.* 1998). The problem in measuring short time-scale variations in engine-out emissions is exacerbated by the short time-scales on which particles form and evolve. Particle sizing instruments currently available have time responses which vary from a few seconds upwards (usually much longer); typical aerosol sampling and dilution systems significantly degrade time response even further. If we are to examine particle emission, formation and evolution properly both for steady-state and transient engine operating conditions (e.g. for currently legislated standard drive cycles), we will require instruments capable of measuring particle size on very fast (sub-second) time-scales; such instruments are currently under development.

5. Conclusions

Again, the single most important piece of information that we need to understand is how exposure to particles affects the health of individuals, and the public-at-large in general. This needs to be determined for both acute (short-term) and chronic (long-term) exposures, with the effects of particle size, morphology (surface area) and composition considered. Related to this is the need to understand typical exposure conditions with regard to particle size, number, composition, etc. Exposure will further need to be assessed subject to the dramatic variations that different individuals may be exposed to based on their lifestyle and activities.

In terms of measurement and control of particulate emissions, we need to better understand the process of dilution and how it can affect particle formation. One of the first steps toward understanding this is the study of real atmospheric dilution, outside of the laboratory. Given the wide variety of vehicle designs, vehicle operating conditions, and atmospheric conditions, it will be important to establish the variability in dilution which will need to be simulated in the laboratory. Of course, closely controlled experimental laboratory studies are also needed in order to establish methods whereby these representative conditions of true atmospheric dilution can be simulated practically.

An intimate knowledge of the relationship between dilution and particle formation may also yield insight that will allow particle formation and emissions to be reduced by active on-vehicle manipulation of the atmospheric dilution process itself. Development of instrumentation for the fast measurement of particles will be very important for the study of such formation and dilution phenomena.

References

Abdul-Khalek, I. S., Kittelson, D. B. & Brear, F. 1999 The influence of dilution conditions on diesel exhaust particle size distribution measurements. International Society of Automotive Engineers technical paper 1999-01-1142.

Abdul-Khalek, I. S., Kittelson, D. B. & Brear, F. 2000 Nanoparticle growth during dilution and cooling of diesel exhaust: experimental investigation and theoretical assessment. International Society of Automotive Engineers technical paper 2000-01-1515.

Davenne, T. R., Graskow, B. R., Collings, N. & Britter, R. E. 2000 A study of the parameters affecting dilution induced particle formation. *J. Aerosol Sci.* (Submitted.)

Donaldson, K., Beswick, P. H. & Gilmour, P. S. 1996 Free radical activity associated with the surface of particles: a unifying factor in determining biological activity? *Toxicol. Lett.* **88**, 293–298.

Donaldson, K., Li, X. Y. & MacNee, W. 1998 Ultrafine (nanometer) particle mediated lung injury. *J. Aerosol Sci.* **29**, 533–560.

Ferin, J., Oberdörster, G. & Penney, D. P. 1992 Pulmonary retention of ultrafine and fine particles in rats. *Am. J. Resp. Cell Mol. Biol.* **6**, 535–542.

Graskow, B. R., Kittelson, D. B., Abdul-Khalek, I. S., Ahmadi, M. R. & Morris, J. E. 1998 Characterization of exhaust particulate emissions from a spark ignition engine. International Society of Automotive Engineers technical paper 980528.

Graskow, B. R., Kittelson, D. B., Ahmadi, M. R. & Morris, J. E. 2000 Size and concentration of particles emitted from a spark ignition engine: fuel and dilution effects. International Society of Automotive Engineers technical paper 2000-01-1516.

Greenwood, S. J., Coxon, J. E., Biddulph, T. & Bennett, J. 1996 An investigation to determine the exhaust particulate size distribution for diesel, petrol and compressed natural gas fuelled vehicles. International Society of Automotive Engineers technical paper 961085.

McAughey, J. J. 1997 Regional lung deposition and dose of ambient particulate in humans by particle mass and number. Research report, AEA Technology, Aerosol Science Centre, Oxfordshire, UK.

Peckham, M. S., Collings, N., Schurov, S. M., Burrell, J. D. & Hands, T. 1998 Real-time in-cylinder and exhaust NO measurements in a production SI engine. International Society of Automotive Engineers technical paper 980400.

Rickeard, D. J., Bateman, J. R. & Yeong, K. K. 1996 Exhaust particulate size distribution: vehicle and fuel additives in light duty vehicles. International Society of Automotive Engineers technical paper 961980.

Sutela, C. J., Collings, N. & Hands, T. 1999 Fast response CO_2 sensor for automotive exhaust gas analysis. International Society of Automotive Engineers technical paper 1999-01-3477.

UK QUARG (United Kingdom Quality of Urban Air Review Group) 1996 Airborne particulate matter in the United Kingdom, 3rd status report.

Van Dyke, M. 1982 *An album of fluid motion*, p. 97. Stanford, CA: Parabolic.

Discussion

M. S. BINGLEY (*Cobham, Surrey, UK*). Driving in an open-topped sports car demonstrated that exhaust gas went forward over the car. This demonstrated that it was very difficult to determine the manner of exhaust gas dilution in a car. Nevertheless, were you going to conduct experiments, sampling exhaust gas, on a car, on the road?

N. COLLINGS. The car involved in the tests was not open top. The measurements of NO_x concentrations were those that occupants would be exposed to.

C. F. CLEMENT (*Oxon, UK*). The time-scale and sequence of the physical processes of cooling and dilution will control the nature of the aerosol emerging from internal combustion engines. If the cooling occurs before the expansion, vapour will condense on pipe walls and aerosol formed may have time to coagulate out of the ultrafine size range. With the process of coagulation, it is the condensed mass concentration and, therefore, the amount and rate of dilution which determines the aerosol size. The faster the dilution rate, the more likely the aerosol is to be 'frozen' into ultrafine sizes.

N. COLLINGS. I agree with this comment, and it is an important aspect of the sensitivity of the final spectrum to the dilution trajectory. However, very fast dilution might lead to a smaller ultrafine component if the time for nucleation was short enough. Whether in actual exhausts such very fast dilution is practical is an open issue; a very large quantity of gas (air) would be required to get very fast 100:1 dilution, the order of magnitude required to freeze the processes. As the comment suggests, cooling before dilution might be a better route to ultrafine particle suppression, though the rate of condensation on walls might be less important than coagulation/agglommeration processes.

M. Williams (*DETR, London, UK*). If ultrafine particles are found to be important for regulation, given the sensitivities of their production in exhaust emissions, do you feel that rather than regulating number concentration in emissions, it may be more profitable to regulate their precursors?

N. Collings. I agree with the suggestion that measurement of the ultrafines will be very problematic in practice, and that regulation of the precursors may be the only viable option. A concept of 'ultrafine-particle-forming potential' would seem to be appropriate, much as ozone-forming potential is presently used, where measurements of the exhaust gas composition, especially the different hydrocarbon species, leads, via a model, to the ground-level ozone-forming potential.

L. M. Brown (*Cavendish Laboratory, Madingley Road, Cambridge, UK*). If I understand it right, the dilution effects that you are describing will not affect thermodynamically stable particles. For example, are metal wear particles from the engine unaffected by dilution?

N. Collings. All particles, whether vapour or solid, are subject to growth (due to condensation, agglomeration, etc.) during the dilution process, so the 'stable' particle spectrum will still be a strong function of dilution conditions, whatever form they originate in.

CHAPTER 5

SIZE DISTRIBUTIONS OF 3–10 NM ATMOSPHERIC PARTICLES: IMPLICATIONS FOR NUCLEATION MECHANISMS

Peter H. McMurry[1], Keung Shan Woo[1], Rodney Weber[2], Da-Ren Chen[1] and David Y. H. Pui[1]

[1] *125 Mechanical Engineering, University of Minnesota, 111 Church Street S.E., Minneapolis, MN 55455, USA*

[2] *School of Earth and Atmospheric Sciences, Georgia Institute of Technology, 221 Bobby Dodd Way, Atlanta, GA 30332, USA*

The formation of new atmospheric particles by gas-to-particle conversion leads to enhanced concentrations of nanoparticles. We have studied the formation and growth of new particles in urban Atlanta and in the remote atmosphere in locations ranging from the North Pole to Mauna Loa, Tasmania and the South Pole. Key to this work was our development of new measurement techniques for freshly formed nucleation mode particles between 3 and 10 nm. In this paper we show that measured aerosol size distributions in the 3–10 nm diameter range often increase with decreasing size down to our minimum detectable size of 3 nm, presumably because nucleation was occurring during the measurement. Furthermore, we show that the Atlanta nucleation mode size distributions are consistent with a collision-controlled nucleation process in which accommodation coefficients for all collisions between condensing molecules and molecular clusters and between molecular clusters are assumed to be equal to one, and in which evaporation from molecular clusters is neglected, as would be expected for a highly supersaturated vapour.

Keywords: atmospheric aerosol; ultrafine aerosol; homogeneous nucleation; sulphuric acid; nanoparticles; nucleation mode; gas-to-particle conversion

1. Introduction

The recent development of instrumentation for measuring particles as small as 3 nm (see, for example, Stolzenburg & McMurry 1991; Winklmayr *et al.* 1991; Saros *et al.* 1996; Reischl *et al.* 1997; Chen *et al.* 1998) has led to the discovery of a new mode of atmospheric particles in the nanometre size range (Covert *et al.* 1996*a*). We refer to these 3–10 nm particles as 'nucleation mode' aerosols, as they are almost certainly produced by recent nucleation from the gas phase. Other modes that have been previously documented include the nuclei or Aitken mode (typically *ca.* 20–50 nm mean size), the accumulation mode (between 0.1 and 1.0 μm) and the coarse particle mode (greater than 1 μm) (Whitby 1978). This paper briefly describes the instruments that we have developed and used to measure nucleation mode aerosols and discusses some results of those measurements in urban Atlanta and in the remote troposphere.

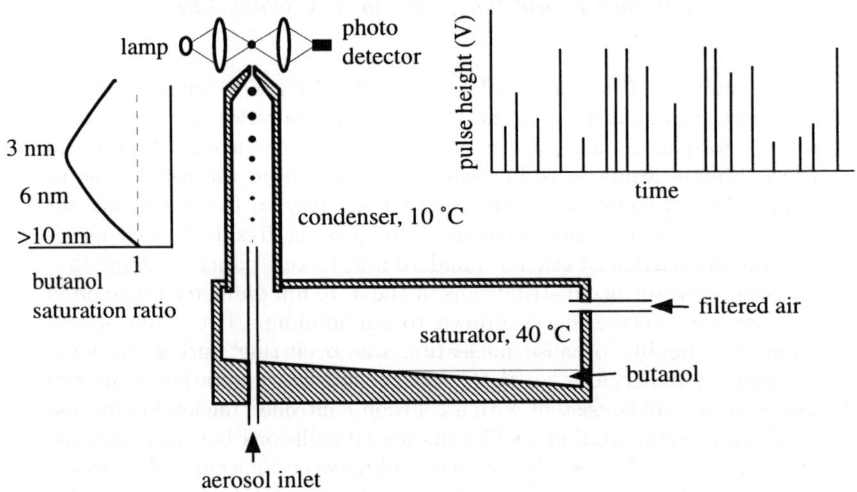

Fig. 1. Schematic of the UCPC-PHA instrument. The alcohol saturation ratio required to initiate condensational growth increases sharply with decreasing size below 10 nm. Therefore, small particles must travel further into the condenser before they begin to grow, and they grow to a smaller ultimate size. The initial particle size is inferred from measurements of the final droplet size measured with the white light optical detector.

Nucleation mode aerosols have been observed in several characteristic situations in a wide variety of locations. The appearance of 3–10 nm particles sometimes follows regular diurnal patterns, with peak particle production rates occurring near midday, when solar radiation is most intense (Bradbury & Meuron 1938; Went 1964; Hogan 1968; Koutsenogii & Jaenicke 1994). At other times, nucleation occurs in response to atmospheric perturbations, such as the removal of pre-existing aerosol by cloud processing (Hegg *et al.* 1990; Radke & Hobbs 1991; Perry & Hobbs 1994; Clarke *et al.* 1998) or the addition of gas phase reactants from a surface source. For example, measurements downwind of the coast at Mace Head, Ireland, have shown the rapid production of exceedingly high concentrations of very small particles during on-shore flow (McGovern *et al.* 1996; McGovern 1999). These events occur only in sunlight and during low tide. Nucleation was also detected in the remote marine atmosphere downwind of penguin colonies on Macquarie Island (Weber *et al.* 1998*a*). Weber and co-workers hypothesized that new particles were produced when ammonia, or perhaps some other gas emitted by these colonies, reacted with sulphuric acid that was present in the air flowing over the island to produce new particles. Nucleation has been observed on mountains (Shaw 1989; Marti 1990; Weber *et al.* 1995, 1997; Raes *et al.* 1997; Wiedensohler *et al.* 1997), in the boreal forests of Finland (Mäkelä *et al.* 1997; Kulmala *et al.* 1998) and in northern Finland (Pirjola *et al.* 1998) and in moderately polluted continental air in Germany (Birmili & Wiedensohler 1998). Measurements of aerosol composition suggest that the freshly nucleated particles in the Finnish boreal forest are enriched with dimethyl amine (Mäkelä *et al.* 1999). Evidence of nucleation in the marine boundary layer (MBL) has been reported (Covert *et al.* 1992; Hoppel *et al.* 1994; Clarke *et al.* 1998). Several groups (Covert *et al.* 1996*a*, *b*; Wiedensohler *et al.* 1996) have argued that nucleation mode particles detected in the MBL are probably produced aloft in free tropospheric cloud outflows and transported to the surface. Evidence suggests that nucleation in the upper tropical troposphere is a significant global source of atmospheric particles (Clarke 1993; Brock *et al.* 1995; Clarke *et al.* 1998).

While nucleation in a wide variety of circumstances is now well documented, we do not yet have validated models for predicting nucleation rates. Evidence suggests that sulphuric acid vapour may often participate in nucleation (Eisele & McMurry 1997; Clarke *et al.* 1998). Observed nucleation rates are occasionally consistent with predictions of the binary theory for sulphuric acid and water (Weber *et al.* 1999), but rates of particle formation are often orders of magnitude higher than can be explained by the binary theory. Furthermore, growth rates of freshly nucleated particles are typically two to ten times higher than can be explained by the condensation of sulphuric acid and its associated water and ammonia (Weber *et al.* 1996, 1997, 1998*a*, 1999).

The present paper reports on measurements of freshly nucleated 3–10 nm aerosol size distributions. We show that in systems where nucleation is occurring or has recently occurred, size distributions exhibit an increasing trend with decreasing particle size. Our measurements were made possible by our development of new instrumentation for measuring size distributions in the 3–10 nm diameter range. We also show that 3–10 nm size distributions measured in Atlanta are consistent with the theoretical predictions for collision-controlled nucleation.

2. Advances in Instrumentation

The measurements described in this paper use two different instruments for measuring nucleation mode size distributions: the ultrafine condensation particle counter pulse height analysis method (UCPC-PHA) (Saros *et al.* 1996) and the nanometre scanning mobility sizer (nano-SMPS). The nano-SMPS uses the new nano-DMA (Chen *et al.* 1998) to classify particles according to electrical mobility. In this section we briefly summarize the relative strengths of these techniques.

The UCPC-PHA technique uses the instrument described by Stolzenburg & McMurry (1991), the prototype of the TSI 3025 UCPC. A schematic of this instrument is shown in figure 1. The sampled aerosol enters the condenser, where it is surrounded by an annular filtered sheath flow that has been saturated with butanol at 40 °C. Because the flow in the condenser is laminar, the particles remain on axis as they flow through it. The butanol rapidly diffuses into the aerosol, where it becomes supersaturated as a result of heat transfer from the 10 °C walls. Saturation ratios along the axis increase from the inlet value of 1.0 until, due to condensation on the cool walls, they decrease after reaching a peak value about two-thirds of the way through the condenser. The saturation ratio that is required to initiate condensational growth increases with decreasing size due to the effect of curvature on equilibrium vapour pressure (Thomson 1871). The highest saturation ratio that is achieved in the condenser is sufficient to initiate condensational growth on *ca.* 3 nm particles, but these particles have a relatively short time to grow. Growth of larger particles is activated at lower saturation ratios. The final droplet size at the exit from the condenser decreases with decreasing growth time. Therefore, very small particles grow to a smaller final droplet size and scatter less light than do larger particles. The PHA technique involves measuring the voltage pulses produced by individual droplets as they flow through the optical detector. The measured distribution of pulse heights can be inverted to obtain information about size distributions in the 3–10 nm diameter range (Weber *et al.* 1998*b*).

The advantage of the UCPC-PHA technique is that all 3–10 nm particles in the sampled flow are detected. This enables fast measurements at very low concentrations. The lower limit on concentration is determined by counting statistical uncertainties. The aerosol flow rate is 0.5 $cm^3\ s^{-1}$. The number of nanoparticles counted, therefore, is

$$C_{3\text{–}10} = 0.5 N_{3\text{–}10} t, \tag{1}$$

where $N_{3\text{–}10}$ is the number concentration of 3–10 nm diameter particles and t is the counting time in seconds. The Poisson counting statistical uncertainty varies as the square root of the number of counts. If we assume that particle counts are equally distributed among 10 size (pulse height) bins, then the relative uncertainty in any one bin is, approximately,

$$\left(\frac{\Delta C}{C}\right)_{\text{PHA Bin}} \approx \left(\frac{20}{N_{3\text{–}10} t}\right)^{0.5}. \tag{2}$$

Assuming that an acceptable relative uncertainty for measurements is *ca.* 10%, it follows that

$$N_{3\text{–}10} t \geqslant 2000\ \text{cm}^{-3}\ \text{s}^{1}. \tag{3}$$

Therefore, a typical counting time of 20 s permits measurements of 3–10 nm size distributions when $N_{3\text{–}10}$ exceeds *ca.* 100 cm^{-3}. Coincidence errors occur when more than one particle is simultaneously present in the optical detector, and this happens when total concentrations of all particles exceed *ca.* 4000 cm^{-3}. It is necessary to dilute the aerosol prior to measurement when sampling aerosols with concentrations higher than this.

The nano-SMPS was designed for optimal measurements of electrical mobility distributions of particles in the 3–50 nm diameter range. This instrument system consists of a TSI 3080N nano-DMA with a TSI 3025 UCPC detector. For the measurements discussed in this paper, aerosols are exposed to a bipolar ion cloud, where they are brought to a known charge distribution (Wiedensohler 1988). At the exit from this bipolar charger,

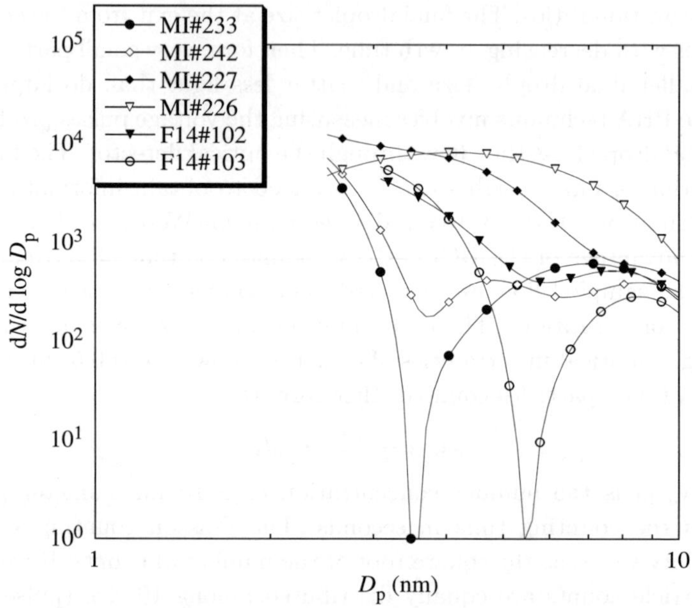

Fig. 2. Examples of nucleation mode size distributions measured with the UCPC-PHA in the remote marine troposphere. The symbols on the curves are shown to identify the measurements, and do not correspond to 'size bins' for the PHA-UCPC.

1.27% of 3 nm particles and 5.03% of 10 nm diameter particles contain −1 elementary charges. The aerosol then is classified according to electrical mobility by the nano-DMA and counted with the UCPC, which samples at 0.5 $cm^3\ s^{-1}$ downstream of the nano-DMA. The relationship between the concentration detected by the UCPC and the size distribution for particles much smaller than the mean free path of air is (Knutson 1976)

$$N_{\mathrm{UCPC}} \approx 0.5 f_{\mathrm{aerosol}} \varphi \frac{\mathrm{d}N}{\mathrm{d}\ln D_{\mathrm{p}}} \approx 0.5 f_{\mathrm{aerosol}} \varphi \frac{N_{3\text{–}10}}{\ln(10/3)}, \tag{4}$$

where f_{aerosol} is the ratio of the aerosol to sheath air flow rates into and out of the nano-DMA, φ is the fraction of aerosol at a given size that carries an elementary charge, and $\mathrm{d}N/\mathrm{d}\ln D_{\mathrm{p}}$ is the value of the measured aerosol size distribution function. Using arguments similar to those discussed previously for the UCPC-PHA technique, we conclude that measurements providing

10% accuracy at a given nano-DMA classifying voltage will be obtained when

$$
\begin{aligned}
N_{3\text{–}10} t_{\text{nano-DMA voltage setting}} &\geqslant \frac{5000}{\varphi}\ \mathrm{cm^{-3}\,s}, \\
&\sim 3.7 \times 10^5\ \mathrm{cm^{-3}\,s} \quad \text{(for 3 nm particles)}, \\
&\sim 1.0 \times 10^5\ \mathrm{cm^{-3}\,s} \quad \text{(for 10 nm particles)}.
\end{aligned} \tag{5}
$$

The total time required to measure the size distribution between 3 and 10 nm varies in proportion to the number of DMA classifying voltages employed. Thus, for a given counting time the UCPC-PHA can measure size distributions that are about a factor of 500–1000 times lower than can be measured with the nano-DMA. However, the time response of the nano-DMA could be improved by about a factor of five to ten by using a higher flow-rate detector, and an additional factor of ten by using a unipolar charger in place of the bipolar charger (Chen & Pui 1999).

The sizing resolution of the nano-DMA is superior to that of the UCPC-PHA. For example, laboratory calibrations with monodisperse calibration aerosols show that particles that vary by ±50% in diameter can produce the same pulse height with the UCPC-PHA. In contrast, particles that vary by about ±4% in diameter will exit the nano-DMA at a given classifying voltage under operating conditions used in our studies. Therefore, the nano-DMA is the instrument of choice if concentrations are sufficiently high to permit measurements in a reasonable period of time. Because the UCPC-PHA can rapidly measure size distributions at low concentrations, it has advantages for measurements in the clean troposphere.

3. Tropospheric Measurements

We have used the PHA-UCPC to study nucleation in the remote marine and continental troposphere. The PHA-UCPC measurements were done as part of short-term (four to six weeks) intensive field programmes at locations including the Arctic Ocean (Covert *et al.* 1996*a*), Mauna Loa, HI (Weber *et al.* 1995), the Rocky Mountains of Colorado (Weber *et al.* 1997), the South Pole and over the Southern Ocean during ACE-1 (Weber *et al.* 1998*a*, 1999). Examples of nucleation mode size distributions measured over the Southern Ocean are shown in figure 2. The data labelled 'MI' were measured downwind of Macquarie Island, the site of a large penguin colony. Distributions 233 and 242 were measured *ca.* 1 km downwind of the

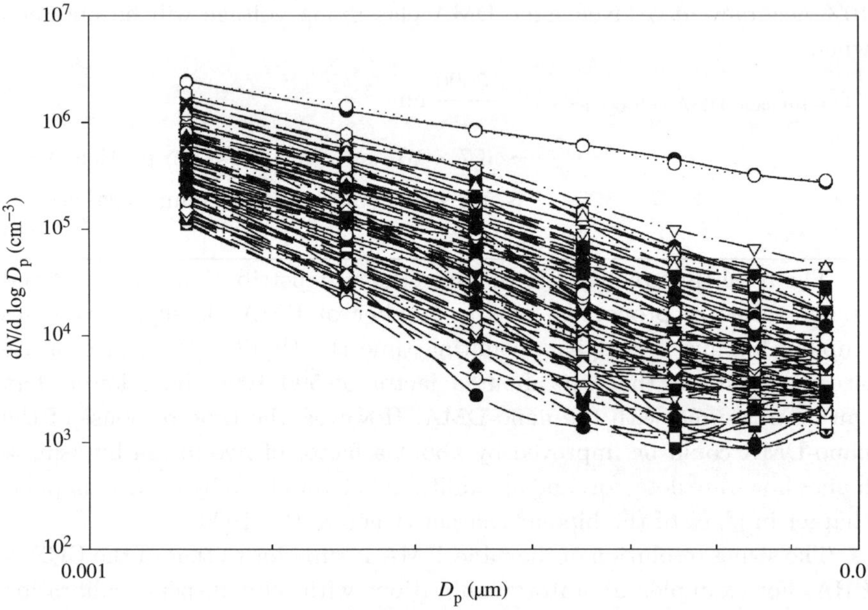

Fig. 3. Nucleation mode size distributions measured with the nano-DMA in urban Atlanta, GA. These measurements represent all data during a one-year period where significant concentrations of 3–10 nm particles were present.

island, and distributions 227 and 226 were measured 21 and 32 km downwind respectively. We believe nucleation occurred during these measurements when marine air entrained emissions (possibly ammonia or amines) from the island. Distributions 102 and 103 were measured in the outflow regions of convective clouds.

Several interesting features can be observed in the size distribution functions shown in figure 2. Most significantly, observed distributions often increase with decreasing diameter at the bottom end of our measurement range, especially shortly after nucleation is first observed. For example, distributions 233 and 242, which were measured immediately downwind of Macquarie Island, show a pronounced increasing trend with decreasing size. We believe this is because nucleation was occurring during our measurements, and there was a continual flux of particles into the measured size range at these times. This rising trend often disappeared after nucleation had proceeded for some time (see, for example, distribution 226). The sharp minima in several of the distributions are also intriguing. Due to inherent limits in size resolution of the UCPC-PHA technique, we are not absolutely certain these minima are real. However, our analyses suggest that it should be possible to recover such minima when inverting UCPC-PHA data (Weber *et al.* 1998*b*). Furthermore, it will be shown in the following section that theory shows such minima can occur during nucleation.

The nano-DMA was used to measure size distributions of 3–50 nm particles in Atlanta (Woo *et al.* 2000). In this study we also used a conventional SMPS for particles between 20 and 250 nm, and a PMS LASAIR optical particle counter for particles between 0.1 and 2 μm. These measurements are being carried out as part of the ARIES aerosol health-effects experiment. Measurements began in August 1998, and size distributions have been measured every 12 min all year round. During the first year of this study we measured 85 hourly-average size distributions that increased with decreasing size in the 3–10 nm diameter range. These size distributions are shown in figure 3.

The data in figure 3 exhibit several interesting features. Size distributions again increase with decreasing size down to our minimum detectable size, suggesting that nucleation was occurring. Also, while the magnitudes of the distributions vary by about a factor of 10 at any given size, the slopes are quite linear with a mean value of −3.5 and fall within the range,

$$-1.19 > \frac{\mathrm{d}}{\mathrm{d}\log D_\mathrm{p}} \log\left(\frac{\mathrm{d}N}{\mathrm{d}\log D_\mathrm{p}}\right) > -5.64. \tag{6}$$

The magnitudes of the nucleation mode distribution functions measured in Atlanta exceed those measured in the remote troposphere by typically one to two orders of magnitude. As with the distributions measured in the remote troposphere, minima are occasionally observed, although the minima for Atlanta tend to occur at somewhat larger sizes and are not as pronounced.

4. Discussion

In previous work we calculated numerically time-dependent size distributions of nucleation mode aerosols in systems where a condensable species (the 'monomer') is produced at a constant rate, R. The calculations take account of monomer production by gas phase chemical reactions, condensation of monomer on newly formed molecular clusters and on pre-existing aerosol, coagulation between molecular clusters, evaporation of monomer from molecular clusters, and coagulation between molecular clusters and pre-existing aerosol. Following the approach typically used with nucleation theory, we assumed that evaporation rates can be calculated using the capillarity approximation. We assumed that the size distribution of the pre-existing aerosol is not significantly altered by condensation during the nucleation event, and that the mass accommodation coefficient of monomer on pre-existing aerosol or molecular clusters equals one, which is justified by both theory and measurement (Clement *et al.* 1996; Jefferson *et al.* 1997). These analyses showed that the time-dependent nucleation mode size distributions depend on three dimensionless variables, E, A and L, defined as

$$E = N_{\mathrm{s}}\left(\frac{\beta_{11}}{R}\right)^{1/2}, \tag{7}$$

$$A = 4\left(\frac{\pi}{6}\right)^{1/3}\frac{v_1^{2/3}\sigma}{k_{\mathrm{B}}T}, \tag{8}$$

$$\sqrt{L} = \left(\frac{k_{\mathrm{B}}T}{2\pi m_1}\right)^{1/2}\frac{I}{\sqrt{\beta_{11}R}}. \tag{9}$$

Variables are defined in the nomenclature. The parameters E and A determine the significance of monomer evaporation (Rao & McMurry 1989); calculated distribution functions are highly sensitive to these parameters. The surface tension parameter A depends only on temperature and properties of the condensing species. For systems in which the vapour is produced at a very high rate (i.e. R is large) or for which the monomer saturation vapour pressure is small (N_{s} is very small), E approaches zero and the monomer evaporation terms become negligible. We refer to this as the 'collision-controlled regime'. McMurry (1980) found that total number concentrations and size distributions of aerosols larger than 10 nm produced by the photo-oxidation of SO_2 in smog chambers at reaction rates exceeding *ca.* 10^6 molecules $\mathrm{cm^{-3}\,s^{-1}}$ are in good agreement with predictions of the collision-controlled theory. L determines the significance of monomer condensation and the coagulation of freshly formed particles onto pre-existing aerosol (McMurry 1983). Note that $\sqrt{L}$ varies in proportion to the Fuchs integral, I, which equals the aerosol surface area for particles that are much smaller than the gas mean free path. For typical transition regime atmospheric aerosols, the Fuchs integral is a bit smaller than the surface area.

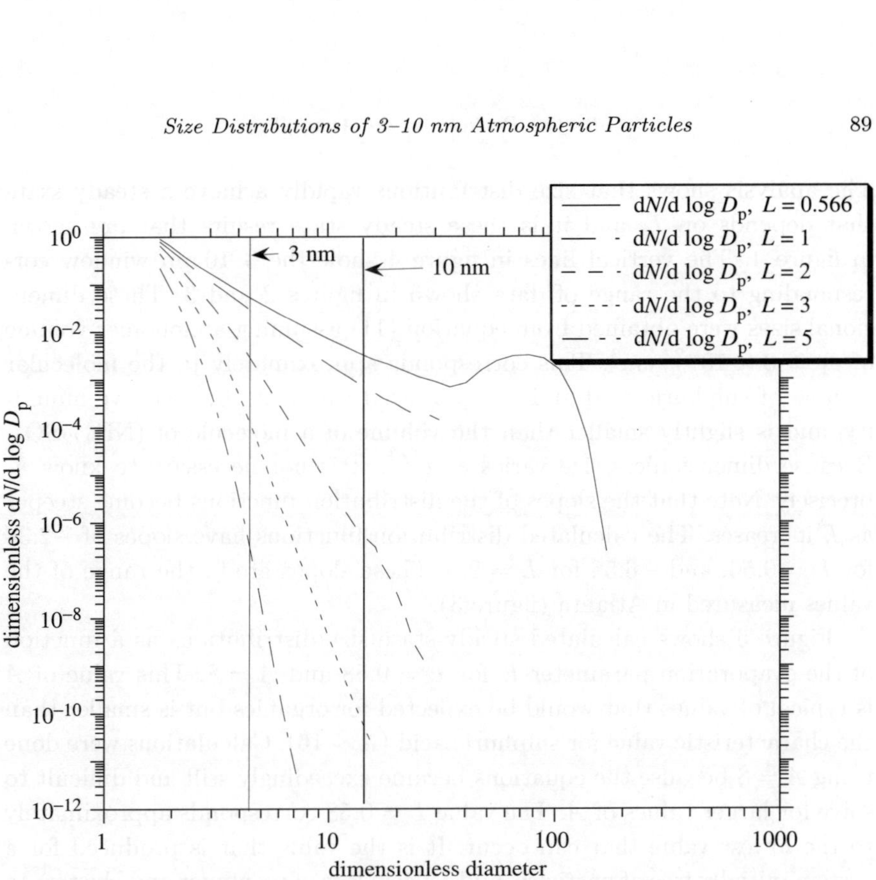

Fig. 4. Calculated steady-state aerosol size distributions when condensable vapour is produced at a constant rate in a collision-controlled system. The parameter L increases with increasing aerosol surface area of particles larger than *ca.* 10 nm.

Although atmospheric nucleation is undoubtedly heteromolecular, this theory treats the process as a quasi-single-component process, with the growth or evaporation of the molecular clusters rate limited by a single, low-vapour-pressure species.

Figure 4 shows calculated size distributions as a function of the heterogeneous loss parameter, L, for collision-controlled nucleation ($E = 0$). The results shown are dimensionless. The relationship between the dimensional and dimensionless size variables is

$$\left.\frac{\mathrm{d}N}{\mathrm{d}\log D_\mathrm{p}}\right|_\mathrm{dimensional} = \left(\frac{R}{\beta_{11}}\right)^{1/2} \left.\frac{\mathrm{d}\tilde{N}}{\mathrm{d}\log \tilde{D}_\mathrm{p}}\right|_\mathrm{dimensionless}, \tag{10}$$

$$D_\mathrm{p}|_\mathrm{dimensional} = v_1^{1/3} \tilde{D}_\mathrm{p}|_\mathrm{dimensionless}. \tag{11}$$

The analysis shows that size distributions rapidly achieve a steady state that depends on L, and it is these steady state results that are shown in figure 4. The vertical lines in figure 4 show the 3–10 nm window corresponding to the range of data shown in figures 2 and 3. These dimensional sizes were obtained from equation (11) assuming a monomer volume of $v_1 \cong 3 \times 10^{-22}$ cm^3. This corresponds approximately to the molecular volume of sulphuric acid and its associated water at 50% relative humidity, and is slightly smaller than the volume of a molecule of $(NH_4)_2SO_4$. Because dimensionless size varies as $v_1^{1/3}$, it is not necessary to know v_1 precisely. Note that the slopes of the distribution functions become steeper as L increases. The calculated distribution functions have slopes of -2.38 for $L = 0.56$, and -6.56 for $L = 2.0$. These slopes are in the range of the values measured in Atlanta (figure 3).

Figure 5 shows calculated steady-state size distributions as a function of the evaporation parameter E for $L = 0.58$ and $A = 8$. This value of A is typical of values that would be expected for organics but is smaller than the characteristic value for sulphuric acid ($A \sim 16$). Calculations were done using $A = 8$ because the equations became exceedingly stiff and difficult to solve for larger values of A. The value $L = 0.58$ corresponds approximately to the lowest value that can occur. It is the value that is produced for a system initially free of particles, and reflects loss of monomer and clusters to particles larger than *ca.* 10 nm that were produced by nucleation. Note that size distributions are highly sensitive to E for $E > 0.02$. Also, for large values of E, minima are predicted for particles in the 3–10 nm diameter range. These trends would be even more pronounced for larger values of A. The size distributions for values of E up to 0.02 are qualitatively consistent with observed size distributions in Atlanta. The measured size distributions are quite different from size distributions calculated for larger values of E, however. The size distributions observed for larger values of E might be consistent with the PHA-UCPC measurements in the remote troposphere (figure 2).

If we assume that nucleation is collision-controlled, two approaches can be used to find the monomer production rate, R, for the Atlanta data (figure 3). The slope of the measured distribution function provides a value for the dimensionless scavenging rate parameter, L (see figure 4). R is then evaluated from equation (9), where the Fuchs integral, I, is calculated from measured size distributions. We refer to this value of R as R_L.

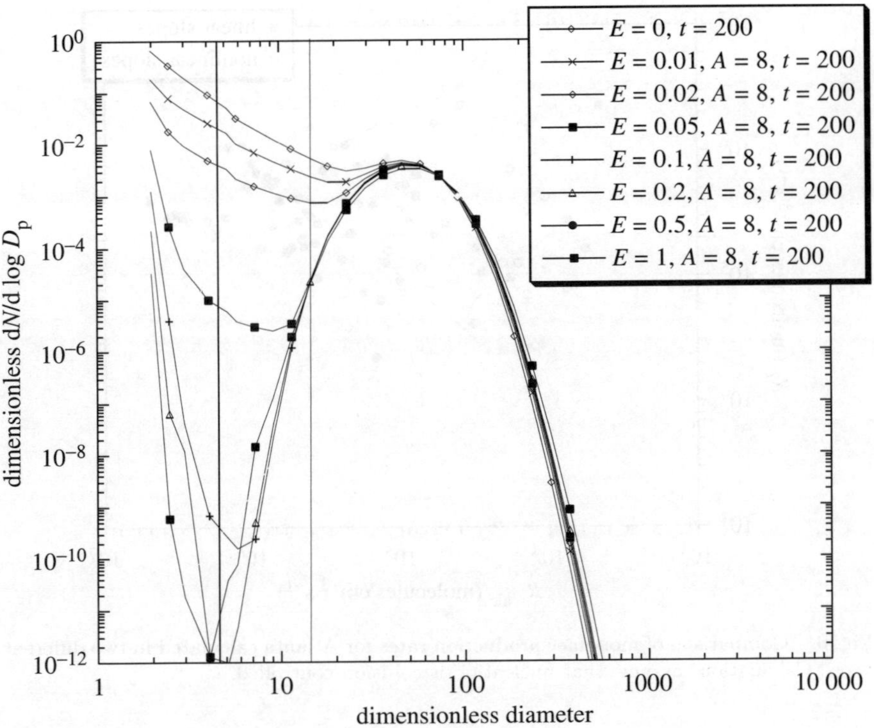

Fig. 5. Calculated steady-state aerosol size distributions for several values of the evaporation parameter E. Calculations were done using $A = 8$ and $L = 0.58$.

Alternatively, the slope of the measured size distribution is used to find the value of the dimensionless size distribution at the minimum detectable size (see figure 4), and the value of R that scales the dimensionless to the dimensional size distribution is evaluated from equation (10). We refer to this value of R as R_{scale}. R_L and R_{scale} are compared in figure 6. As was shown in figure 3, most of the size distributions measured in Atlanta had linear slopes, but some did not. The open circles in figure 6 apply to data with nonlinear slopes. Note that for *ca.* 90% of our measurements, the values of R calculated in these ways agreed to within a factor of 10. Several of the outliers apply to measurements with nonlinear slopes. Several of these measurements were made early in the morning or late in the evening, when the assumption that nucleation mode aerosol size distributions are

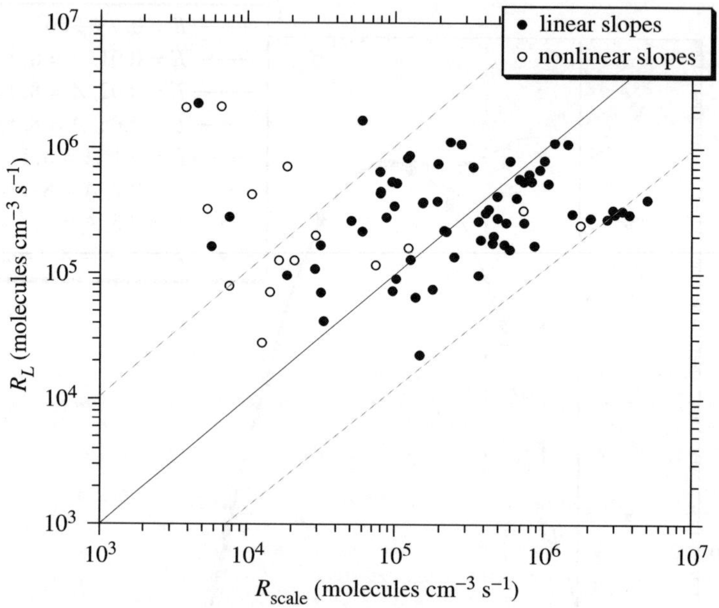

Fig. 6. Comparison of monomer production rates for Atlanta calculated in two different ways. Calculations assume that nucleation is collision-controlled.

at steady state would be invalid. The values of R determined by these two approaches are not systematically different, although there is significant scatter. Measured and theoretical collision-controlled (i.e. $E = 0$) size distributions are shown for one typical measurement in figure 7. Theoretical size distributions corresponding to the values of R_L and R_{scale} obtained for this measurement (2.5×10^5 molecules cm^{-3} s^{-1} and 5.7×10^5 molecules cm^{-3} s^{-1}, respectively) are shown.

The results shown in figures 6 and 7 are based on the assumption that nucleation in Atlanta was collision-controlled. A more rigorous testing of this hypothesis would require solutions of the cluster balance equations for $E > 0$ over a wider range of L and for values of A applicable to the nucleating aerosols; the results shown in figure 5 were done for L=0.58, and we have not carried out calculations for other values of L. However, if the results in figure 5 are characteristic of those for other values of L, it would appear unlikely R_L and R_{scale} would have been comparable in magnitude

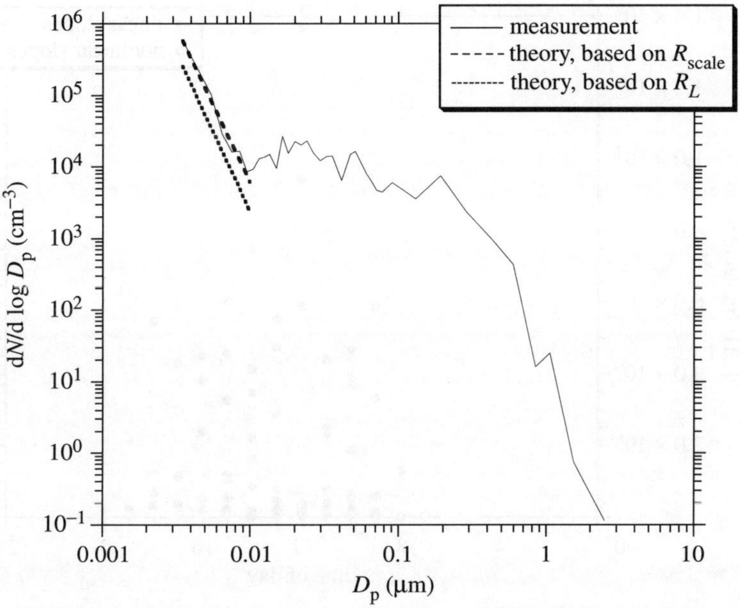

Fig. 7. Comparison of measured and theoretical size distributions for 3–10 nm aerosols for one typical Atlanta measurement. The theory assumes collision-controlled nucleation ($E = 0$). Theoretical results are shown for the monomer production rate, R, calculated in two different ways.

(figure 6) if evaporation from clusters had played a significant role. The results of figure 5 show that slopes of the distribution are comparable for $E = 0$, $E = 0.01$ and $E = 0.02$, but the 3.5 nm intercepts vary by more than a factor of ten. If the true value of E had been 0.02 (rather than 0 as was assumed above), then R_{scale} would have been more than a factor of 100 higher than was found for collision-controlled nucleation ($E = 0$). It is likely that this discrepancy would be even larger if calculations had been done for larger values of A and L, as might be appropriate for these atmospheric aerosols. Because the slopes of these curves are similar, however, R_L would be changed by only a small amount. Therefore, values of E as small as 0.01 or 0.02 would have led to $R_{\text{scale}} \gg R_L$. The results shown in figure 6 show that this is not the case. Furthermore, for values of $E > 0.02$, theory shows that the slope of the distribution function would not have been linear as was experimentally observed in Atlanta, further supporting our argument that

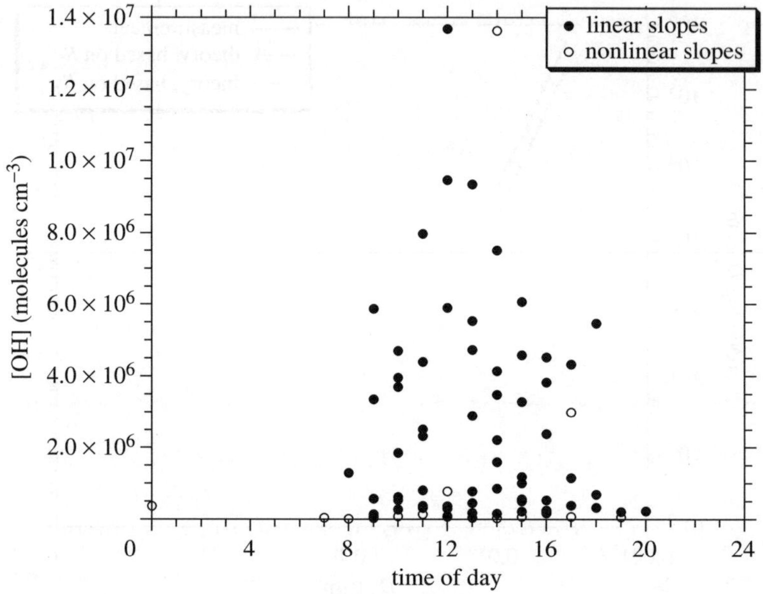

Fig. 8. Hydroxyl radical concentrations required to produce calculated monomer production rates for measured concentrations of sulphur dioxide.

nucleation was collision-controlled. The remote tropospheric distribution functions (figure 3), however, do not have linear slopes. This could reflect the importance of cluster evaporation during these measurements. In order to explain such size distributions with theory, it will be necessary to fit the measured distribution function to a theoretical function that is similar in shape. We have not yet attempted to do this.

It is instructive to speculate on species that might be responsible for the observed nucleation. For collision-controlled nucleation, the dimensionless monomer concentration is insensitive to L, ranging from 0.58 for $L = 0.6$ to 0.49 for $L = 2$ (McMurry 1983). As an approximation we assume a typical value of 0.5. The monomer concentration is therefore (McMurry 1983)

$$N_1 = 0.5\left(\frac{R}{\beta_{11}}\right)^{0.5} \approx 2.24 \times 10^4 \sqrt{R} \text{ molecules cm}^{-3}. \qquad (12)$$

Based on the values of R shown in figure 6 (similar results are obtained with either R_L or R_{scale}), we find that N_1 falls below 1.2×10^7 molecules

cm^{-3} for 50% of our measurements and below 2.2×10^7 molecules cm^{-3} for 90% of our measurements. Based on our previous studies (see, for example, Eisele & McMurry 1997), we believe that sulphuric acid may participate in nucleation. Sulphuric acid vapour was not measured during the Atlanta study. In our previous studies in the remote troposphere, however, sulphuric acid vapour concentrations measured during nucleation events occasionally reached levels as high as 2×10^7 molecules cm^{-3} (Weber *et al.* 1996), but covered the range $1 \times 10^4 < [H_2SO_4] < 2 \times 10^7$ molecules cm^{-3} with an average value of *ca.* 1×10^6 molecules cm^{-3}. Thus, the calculated monomer concentrations for collision-controlled nucleation in Atlanta are somewhat higher (up to a factor of 10) than the sulphuric acid concentrations that have been measured in the remote troposphere when nucleation is occurring. If similar species were involved with nucleation in both locations, then the evaporation terms in the cluster balance equations would certainly be less significant in Atlanta since the supersaturation of the nucleating species was approximately a factor of ten higher. It is likely that other species that participate in nucleation (ammonia, amines, etc.) are more abundant in Atlanta than in the urban troposphere. This could also lead to reduced sulphuric acid vapour pressures and lead to nucleation that is more nearly collision-controlled.

An upper limit for the saturation vapour concentration of the condensing species, N_s, can be estimated from equation (7). Based on the above arguments, we assume that during nucleation in Atlanta, the E was less than 0.01. Because 90% of the calculated monomer production rates, R_{scale}, were below 1.3×10^6 molecules cm^{-3} s^{-1}, we conservatively conclude that the saturation vapour concentration was below 5×10^5 molecules cm^{-3}. It would be equally justifiable to use a low value of R_{scale} to estimate the upper limit for N_s, since measured distribution functions were also found to be linear for small values of R_{scale}. We found that R_{scale} was below *ca.* 1.4×10^4 molecules cm^{-3} s^{-1} for *ca.* 10% of our observations. The corresponding upper limit for N_s is 5×10^4 molecules cm^{-3}. Saturation vapour concentrations of sulphuric acid vapour above solid ammonium sulphate aerosol particles of *ca.* 2.5×10^4 molecules cm^{-3} were reported by Marti *et al.* (1997). It follows that our calculated N_s values are in a reasonable range.

Another argument in support of the hypothesis that sulphuric acid participated in nucleation in Atlanta is our observation that sulphur dioxide concentrations were typically elevated during the nucleation events (Woo *et al.* 2000). To test the plausibility of the monomer production rates shown in figure 6, we have calculated the hydroxyl radical concentrations that would have been required to produce the calculated monomer production rates. The calculated hydroxyl radical concentrations were obtained from the following equation:

$$[\mathrm{OH}] = \frac{R_{\mathrm{scale}}}{8.5 \times 10^{-13}[\mathrm{SO_2}]} \text{ molecules cm}^{-3}, \tag{13}$$

where $[SO_2]$ is the measured concentration of sulphur dioxide in molecules per cm^3 and the second-order rate constant for the SO_2–OH reaction is 8.5×10^{-13} cm^3 $molecule^{-1}$ s^{-1} (DeMore *et al.* 1992). Values of [OH] calculated in this way are plotted versus time of day in figure 8. The calculated hydroxyl radical concentration follows a reasonable diurnal variation, with peak values occurring near noon. Half of the calculated hydroxyl concentrations are below *ca.* 8×10^5 molecules cm^{-3} and 90% are below *ca.* 8×10^6 molecules cm^{-3}. These values are in a reasonable range for an urban area (W. Chamiedes, personal communication), but hydroxyl radical concentrations have not been measured in the Atlanta atmosphere, and we have not attempted to compare our results with models applicable to our measurement periods.

5. Conclusions

Two instruments were used to measure size distributions of 3–10 nm diameter aerosols when nucleation was occurring. One of these systems (the UCPC-PHA), which measures the amount of light scattered by individual particles downstream of the condenser of an ultrafine condensation particle counter, is best suited for measurements where concentrations are low and measurements must be made quickly. For example, this instrument is well suited for aircraft measurements in the remote troposphere. The other system (the nano-SMPS) determines size with a new electrostatic classifier that was specially designed for particles as small as 3 nm and concentration with an ultrafine condensation particle counter. The nano-SMPS provides better sizing resolution than the UCPC-PHA but requires more time to complete a measurement. We used the nano-SMPS for measurements in Atlanta where concentrations were high and accurate measurements could be carried out in a few minutes.

Both instrument systems showed that aerosol size distribution functions increase with decreasing size at the minimum detectable particle size particle size (*ca.* 3 nm) when nucleation was occurring. We are not aware that this trend has been observed previously. Theory predicts that this should occur.

About 70 of the 85 observed hourly-averaged 3–10 nm diameter size distributions measured during nucleation in Atlanta over a period of one year can be expressed as

$$\frac{\mathrm{d}N}{\mathrm{d}\log D_{\mathrm{p}}} = A(D_{\mathrm{p}})^{B}; \qquad -5.64 < B < -1.19; \qquad B_{\mathrm{average}} = -3.5. \quad (14)$$

The magnitude of the distribution function at $D_p = 3.5$ nm (the midpoint of the smallest size range) ranged from *ca.* 10^5 to 2×10^6 cm^{-3}, which was one to two orders of magnitude higher than distribution functions measured in the remote troposphere. Also, the remote tropospheric distribution functions did not obey this simple functional relationship.

The Atlanta data are consistent with theoretical predictions for collision-controlled nucleation. The key assumptions of collision-controlled nucleation theory are that all condensing molecules stick together when they collide, and that evaporation from molecular clusters does not occur. We find that the monomer (i.e. condensing molecule) production rates that are required to produce the observed size distributions are in reasonable expectations with values that would be expected for the gas phase oxidation of sulphur dioxide by the hydroxyl radical. The collision-controlled analysis suggests that the vapour pressure of the condensing species is less than 50 000 molecules cm^{-3}.

Clearly, more work is required to verify the above hypotheses. It will be necessary to definitively identify the condensing species and to show experimentally that its concentration is equal to the value predicted theoretically. Furthermore, because the calculated monomer concentrations and equilibrium vapour concentrations are far below values that would be expected for sulphuric acid according to the classical binary theory, the process must involve species in addition to sulphuric acid and water. It is important that these species be identified.

Nomenclature

A	surface tension parameter (see equation (8))
D_{p}	particle diameter
E	evaporation rate parameter (see equation (7))
I	$\dfrac{4\pi}{3}\displaystyle\int_{\substack{\text{pre-existing}\\ \text{aerosol}}} D_{\mathrm{p}}^2\left(\frac{Kn+Kn^2}{1+1.71Kn+1.33Kn^2}\right)\frac{\mathrm{d}N}{\mathrm{d}\log D_{\mathrm{p}}}\,\mathrm{d}\log D_{\mathrm{p}}$
k_{B}	Boltzmann's constant
L	dimensionless scavenging rate parameter (see equation (9))
m_1	monomer mass
N	aerosol number concentration
N_{s}	saturation concentration of nucleating vapour
R	monomer production rate (molecules volume^{-1} time^{-1})
T	temperature
v_1	monomer volume
Kn	$2\lambda/D_{\mathrm{p}}$
λ	mean free path
β_{11}	monomer collision frequency function $= 4\sqrt{2}\left(\dfrac{3}{4\pi}\right)^{1/6}\left(\dfrac{6k_{\mathrm{B}}T}{\rho_{\mathrm{p}}}\right)^{1/2} v_1^{1/6}$
σ	surface tension

Acknowledgements

This research was supported by EPRI Agreement WO9181-01 'Fine and Ultrafine Aerosol Size Distributions in Atlanta' and by DOE grant no. DE-FG02-98ER62556, 'Composition of Freshly Nucleated Aerosols'. We gratefully acknowledge this support.

References

Birmili, W. & Wiedensohler, A. 1998 The influence of meteorological parameters on ultrafine particle production at a continental site. *J. Aerosol Sci.* **29**, S1015–S1016.

Bradbury, N. E. & Meuron, H. J. 1938 The diurnal variation of atmospheric condensation nuclei. *Terr. Magn.* **43**, 231–240.

Brock, C. A., Hamill, P., Wilson, J. C., Honsson, H. H. & Chan, K. R. 1995

Particle formation in the upper tropical troposphere: a source of nuclei for the stratospheric aerosol. *Sci.* **270**, 1650–1653.

Chen, D.-R. & Pui, D. Y. H. 1999 A high efficiency, high throughput unipolar aerosol charger for nanoparticles. *J. Nanoparticle Res.* **1**, 115–126.

Chen, D. R., Pui, D. Y. H., Hummes, D., Fissan, H., Quant, F. R. & Sem, G. J. 1998 Design and evaluation of a nanometer aerosol differential mobility analyzer (nano-DMA). *J. Aerosol Sci.* **29**, 497–509.

Clarke, A. D. 1993 Atmospheric nuclei in the Pacific midtroposphere—their nature, concentration, and evolution. *J. Geophys. Res. Atmos.* **98**, 20 633–20 647.

Clarke, A. D. (and 14 others) 1998 Particle nucleation in the tropical boundary layer and its coupling to marine sulfur sources. *Science* **282**, 89–92.

Clarke, A. D., Varner, J. L., Eisele, F., Mauldin, R. L., Tanner, D. & Litchy, M. 1998 Particle production in the remote marine atmosphere: cloud outflow and subsidence during ACE 1. *J. Geophys. Res. Atmos.* **103**, 16 397–16 409.

Clement, C. F., Kulmala, M. & Vesala, T. 1996 Theoretical consideration on sticking probabilities. *J. Aerosol Sci.* **27**, 869–882.

Covert, A. D., Kapustin, V. N., Quinn, P. K. & Bates, T. S. 1992 New particle formation in the marine boundary layer. *J. Geophys. Res.* **97**, 20 581–20 589.

Covert, D. S., Wiedensohler, A., Aalto, P., Heintzenberg, J., McMurry, P. H. & Leck, C. 1996*a* Aerosol number size distributions from 3 to 500 nm diameter in the Arctic marine boundary layer during summer and autumn. *Tellus* B **48**, 197–212.

Covert, D. S., Kapustin, V. N., Bates, T. S. & Quinn, P. K. 1996*b* Physical properties of marine boundary layer aerosol particles of the mid-Pacific in relation to sources and meteorological transport. *J. Geophys. Res. Atmos.* **101**, 6919–6930.

DeMore, W. B., Sander, S. P., Golden, D. M., Hampson, R. F., Kurylo, M. J., Howard, C. J., Ravishankara, A. R., Kolb, C. E. & Molina, M. J. 1992 Chemical kinetics and photochemical data for use in stratospheric modeling, evaluation no. 10. *Jet Propulsion Laboratory* **92–20**.

Eisele, F. L. & McMurry, P. H. 1997 Recent progress in understanding particle nucleation and growth. *Phil. Trans. R. Soc. Lond.* B **352**, 191–201.

Hegg, D. A., Radke, L. F. & Hobbs, P. V. 1990 Particle production associated with marine clouds. *J. Geophys. Res.* **95**, 13 917–13 926.

Hogan, A. W. 1968 An experiment illustrating that gas conversion by solar radiation is a major influence in the diurnal variation of aitken nucleus concentrations. *Atmos. Environ.* **2**, 599–601.

Hoppel, W. A., Frick, G. M., Fitzgerald, J. & Larson, R. E. 1994 Marine boundary layer measurements of new particle formation and the effects nonprecipitating clouds have on aerosol size distribution. *J. Geophys. Res. Atmos.* **99**, 14 443–14 459.

Jefferson, A., Eisele, F. L., Ziemann, P. J., Weber, R. J., Marti, J. J. & McMurry,

P. H. 1997 Measurements of the H_2SO_4 mass accommodation coefficient onto polydisperse aerosol. *J. Geophys. Res. Atmos.* **102**, 19 021–19 028.

Knutson, E. O. 1976 Extended electric mobility method for measuring aerosol particle size and concentration. In *Fine particles: aerosol generation, measurement, sampling, and analysis* (ed. B. Y. H. Liu), pp. 739–762. Academic Press.

Koutsenogii, P. K. & Jaenicke, R. 1994 Number concentration and size distribution of atmospheric aerosol in Siberia. *J. Aerosol Sci.* **25**, 377–383.

Kulmala, M., Toivonen, A., Makela, J. M. & Laaksonen, A. 1998 Analysis of the growth of nucleation mode particles observed in Boreal forest. *Tellus* B **50**, 449–462.

McGovern, F. M. 1999 An analysis of condensation nuclei levels at Mace Head, Ireland. *Atmos. Environ.* **33**, 1711–1723.

McGovern, F. M., Jennings, S. G. & Oconnor, T. C. 1996 Aerosol and trace gas measurements during the Mace Head experiment. *Atmos. Environ.* **30**, 3891–3902.

McMurry, P. H. 1980 Photochemical aerosol formation from SO_2: a theoretical analysis of smog chamber data. *J. Colloid Interface Sci.* **78**, 513–527.

McMurry, P. H. 1983 New particle formation in the presence of an aerosol: rates, time scales and sub-0.01 μm size distributions. *J. Colloid Interface Sci.* **95**, 72–80.

Mäkelä, J. M., *et al.* 1997 Observations of ultrafine aerosol particle formation and growth in boreal forest. *Geophys. Res. Lett.* **24**, 1219–1222.

Mäkelä, J., Mattila, T. & Hiltunen, V. 1999 *Measurement of the fine and ultrafine particle composition during the particle formation events observed at a boreal forest site.* Tacoma, WA: American Association for Aerosol Research.

Marti, J. 1990 Diurnal variation in the undisturbed continental aerosol: results from a measurement program in Arizona. *Atmos. Res.* **25**, 351–362.

Marti, J. J., Jefferson, A., Cai, X. P., Richert, C., McMurry, P. H. & Eisele, F. 1997 H_2SO_4 vapor pressure of sulfuric acid and ammonium sulfate solutions. *J. Geophys. Res. Atmos.* **102**, 3725–3735.

Perry, K. D. & Hobbs, P. V. 1994 Further evidence for particle nucleation in clear air adjacent to marine cumulus clouds. *J. Geophys. Res. Atmos.* **99**, 22 803–22 818.

Pirjola, L., Laaksonen, A., Aalto, P. & Kulmala, M. 1998 Sulfate aerosol formation in the Arctic boundary layer. *J. Geophys. Res. Atmos.* **103**, 8309–8321.

Radke, L. F. & Hobbs, P. V. 1991 Humidity and particle fields around some small cumulus clouds. *J. Atmos Sci.* **48**, 1190–1193.

Raes, F., Vandingenen, R., Cuevas, E., Vanvelthoven, P. F. J. & Prospero, J. M. 1997 Observations of aerosols in the free troposphere and marine boundary layer of the subtropical Northeast Atlantic: discussion of processes determining their size distribution. *J. Geophys. Res. Atmos.* **102**, 21 315–21 328.

Rao, N. P. & McMurry, P. H. 1989 Nucleation and growth of aerosol in chemically reacting systems: a theoretical study of the near-collision-controlled regime. *Aerosol Sci. Technol.* **11**, 120–132.

Reischl, G. P., Makela, J. M. & Necid, J. 1997 Performance of Vienna type differential mobility analyzer at 1.2–20 nanometer. *Aerosol Sci. Technol.* **27**, 651–672.

Saros, M. T., Weber, R. J., Marti, J. J. & McMurry, P. H. 1996 Ultrafine aerosol measurement using a condensation nucleus counter with pulse height analysis. *Aerosol Sci. Technol.* **25**, 200–213.

Shaw, G. E. 1989 Production of condensation nuclei in clean air by nucleation of H_2SO_4. *Atmos. Environ.* **22**, 2841–2846.

Stolzenburg, M. R. & McMurry, P. H. 1991 An ultrafine aerosol condensation nucleus counter. *Aerosol Sci. Technol.* **14**, 48–65.

Thomson, W. 1871 On the equilibrium of vapour at a curved surface of liquid. *Phil. Mag.* **42**, 448–453.

Weber, R. J., McMurry, P. H., Eisele, F. L. & Tanner, D. J. 1995 Measurement of expected nucleation precursor species and 3–500 nm diameter particles at Mauna Loa Observatory, Hawaii. *J. Atmos. Sci.* **52**, 2242–2257.

Weber, R. J., Marti, J., McMurry, P. H., Eisele, F. L., Tanner, D. J. & Jefferson, A. 1996 Measured atmospheric new particle formation rates: implications for nucleation mechanisms. *Chem. Engng Commun.* **151**, 53–64.

Weber, R. J., Marti, J. J., McMurry, P. H., Eisele, F. L., Tanner, D. J. & Jefferson, A. 1997 Measurements of new particle formation and ultrafine particle growth rates at a clean continental site. *J. Geophys. Res. Atmos.* **102**, 4375–4385.

Weber, R. J., *et al.* 1998*a* A study of new particle formation and growth involving biogenic trace gas species measured during ACE-1. *J. Geophys. Res.* **103**, 16 385–16 396.

Weber, R. J., Stolzenburg, M. R., Pandis, S. N. & McMurry, P. H. 1998*b* Inversion of ultrafine condensation nucleus counter pulse height distributions to obtain nanoparticle (similar to 3–10 nm) size distributions. *J. Aerosol Sci.* **29**, 601–615.

Weber, R. J., McMurry, P. H., Mauldin, L., Tanner, D., Eisele, F., Clarke, A. D. & Kapustin, V. N. 1999 New particle formation in the remote troposphere: a comparison of observations at various sites. *Geophys. Res. Lett. Atmos. Sci.* **26**, 307–310.

Went, F. W. 1964 The nature of Aitken condensation nuclei in the atmosphere. *Proc. Natn. Acad. Sci.* **51**, 1259–1266.

Whitby, K. T. 1978 The physical characteristics of sulfur aerosols. *Atmos. Environ.* **12**, 135–159.

Wiedensohler, A. 1988 An approximation of the bipolar charge distribution for particles in the submicron size range. *J. Aerosol Sci.* **19**, 387–389.

Wiedensohler, A., Covert, D. S., Swietlicki, E., Aalto, P., Heinzenberg., J. & Leck, C. 1996 Occurrence of an ultrafine particle mode less than 20 nm in diameter in the marine boundary layer during Arctic summer and autumn. *Tellus* B **48**, 289–296.

Wiedensohler, A. (and 15 others) 1997 Night-time formation and occurrence of new particles associated with orographic clouds. *Atmos. Environ.* **31**, 2545–2559.

Winklmayr, W., Reischl, G. P., Linder, A. O. & Berner, A. 1991 A new electromobility spectrometer for the measurement of aerosol size distribution in the size range from 1 to 1000 nm. *J. Aerosol Sci.* **22**, 289.

Woo, K. S., Chen, D.-R., Pui, D. Y. H. & McMurry, P. H. 2000 Measurements of Atlanta aerosol size distributrions: observations of ultrafine particle events. *Aerosol Sci. Technol.* (In the press.)

Discussion

R. M. Harrison (*Division of Environmental Health and Risk Management, University of Birmingham, UK*). As some of the molecules forming sulphuric acid clusters in the atmosphere could be small ions, what will be the effect of charge on cluster stability?

P. H. McMurry. Charged clusters are more stable than neutral ones. Therefore, ion-induced nucleation occurs at a higher rate than homogeneous nucleation of neutral species. However, I do not believe the concentration of ions would be high enough to explain the high rates of particle production we observed in the urban Atlanta atmosphere.

C. F. Clement (*Oxon, UK*). What has been used for the evaporation rate in the model described? Particularly with the smaller clusters, it is not obvious that only one molecule could be evaporated.

P. H. McMurry. As you point out, a primary difficulty in nucleation theory is calculating rates at which evaporation occurs from molecular clusters. If nucleation is collision controlled ($E = 0$), then evaporation is negligible relative to condensation and can be neglected. The data for the Atlanta atmosphere appear to be consistent with this hypothesis (i.e. that nucleation is collision controlled). For the theoretical results where evaporation was included ($E > 0$), evaporation rates were calculated by invoking the usual assumptions of classical nucleation theory: molecular clusters are assumed to have the same properties as the bulk liquid, the effect of curvature on vapour pressure is described by the Kelvin equation, and only individual molecules evaporate from clusters. I agree that this is a major area of uncertainty.

CHAPTER 6

PHOTOCHEMICAL GENERATION OF SECONDARY PARTICLES IN THE UNITED KINGDOM

R. G. Derwent and A. L. Malcolm

Climate Research Division, Meteorological Office,
London Road, Bracknell RG12 2SZ, UK

While much of the suspended particulate matter found in the ambient air in urban areas has been emitted directly into the atmosphere, some has been formed there by photochemical reactions from gaseous precursor species. Two major components of this secondary particulate matter have been selected for detailed study in the United Kingdom context. These are particulate sulphate, formed from the precursor, sulphur dioxide, and secondary organic aerosols, formed from oxidation of terpenes and aromatic hydrocarbons. A Lagrangian dispersion model has been used to describe the emissions, transport and transformation of SO_2 into particulate sulphate. The origins of the particulate sulphate are delineated in two separate pollution episodes which occurred during 1996. A photochemical trajectory model is used to describe the formation of secondary organic aerosols and to assess the relative contributions from natural biogenic and man-made precursor sources during conditions typical of photochemical pollution episodes.

Keywords: suspended particulate matter; particulate sulphate; secondary organic aerosols; terpenes; aromatic hydrocarbons; SO_2

1. Introduction

Human health concerns about ambient concentrations of suspended particulate matter, particularly in our cities, are not new. Recently, the application of sophisticated statistical techniques to daily medical records has revealed links between suspended particulate matter and adverse health outcomes at current levels in many cities worldwide (Dockery *et al.* 1993; Pope *et al.* 1995). This has prompted far-reaching reassessments of the potential importance of urban particulate pollution in future air-quality policy.

While much of the suspended particulate matter found in urban areas has been directly injected into the atmosphere from pollution sources such as industrial boilers, furnaces, domestic fires and motor vehicles, some of this material has been formed in the atmosphere by chemical reactions (QUARG 1996). Since most of these chemical reactions are driven by sunlight, they are termed photochemical reactions. The suspended particulate matter formed in the atmosphere is termed secondary particulate matter, or secondary particles, to distinguish it from the primary emitted material. In air-quality policy terms, this distinction is paramount. For emission controls to be effective against secondary particles, they have to operate on the sources of the precursor pollutants that drive the atmospheric chemical production of the secondary particles.

The term 'generation of secondary particulate matter' refers to a rather general and unspecific process which must be split down at the outset into a more specific set of clearly defined atmospheric processes. The term describes primarily the processes whereby gas-phase chemical reactions involving specific precursor gases produce low-volatility products which are capable of homogeneous nucleation to form tiny new particles that can then increase in size by coagulation and capture by pre-existing ambient particles. The term also describes the processes whereby the low-volatility gas-phase reaction products condense onto pre-existing ambient particles, the so-called heterogeneous nucleation process. While homogeneous nucleation may potentially increase both the number of aerosol particles and the mass of the aerosol particles per unit volume in the atmosphere, heterogeneous nucleation can only increase the mass of the aerosol particles per unit volume. Homogeneous nucleation operates in the ultrafine particle size range, and heterogeneous nucleation across the ultrafine and fine particle size ranges.

The main chemical constituents of secondary particulate matter that have been identified generally in urban locations include sulphuric acid and ammonium sulphate, ammonium and other nitrates and organic compounds (Finlayson-Pitts & Pitts 1986). The sulphur- and nitrogen-containing secondary particulate constituents are largely derived from the photochemical oxidation of man-made SO_2 and NO_x precursors. In contrast, the organic constituents appear to have been derived from natural biogenic precursors.

This paper focuses on quality policy for the United Kingdom and aims to address the two following questions.

(1) Is it possible to attribute the origins of the secondary particulate sulphate observed in the UK to UK SO_2 sources alone or are European SO_2 sources also making a contribution?
(2) Is it possible to assess the likely contribution from natural biogenic sources to secondary organic aerosol levels in the UK?

The approach adopted in our study deals with secondary particles in the fine particle size range as a whole and does not deal directly the very smallest of particles in the ultrafine particle size range per se. There are a number of reasons why we have chosen to address fine particles, generally, rather then ultrafine particles in particular. Currently, the modelling tools that we have at our disposal are rudimentary, the gaps in our understanding are wide, and the uncertainties are huge. While there are many years of measurements of fine particles against which our models can be verified, there are few corresponding measurements of ultrafine particles. It is not possible, at present, to quantify accurately how much of the secondary particulate matter in UK urban areas was formed by the homogeneous and heterogeneous nucleation routes. Furthermore, there are internationally accepted air-quality standards and criteria values for fine particles with which to judge public-health significance but none yet exist for ultrafine particles. However, in addressing the above two questions for fine particles, we are necessarily producing answers that are relevant to the special case of ultrafine particles and their importance to public health.

2. Source Attribution of Particulate Sulphate in the UK

Of all the chemical constituents of secondary suspended particulate matter, easily the best quantified are sulphuric acid and ammonium sulphate aerosols, known collectively as particulate sulphate. This situation holds particularly for the United Kingdom (APEG 1999), the focus of this study. The formation mechanisms for sulphate aerosols have been well characterized (Finlayson-Pitts & Pitts 1986), and particulate sulphate observations are available for the United Kingdom (QUARG 1996; APEG 1999) and Europe (Hjellbrekke 1999; Lazaridis *et al.* 1999).

In this study, we address the origins of the particulate sulphate observed in the United Kingdom and ask whether it has been derived from UK

SO_2 sources or whether European SO_2 sources also make a contribution. We have employed the Meteorological Office dispersion model, NAME, to model the formation of particulate sulphate over a European area and provide information on the likely source of the aerosols arriving at particular receptor points within the UK. Previous work (Malcolm *et al.* 2000) studied the year 1996 and in particular two pollution episodes, one in March and the other in July. The model indicated that a high proportion of the particulate sulphate observed during the March episode was due to the import of sulphate aerosols from the rest of Europe, whereas the July episode was dominated by UK sources.

The aqueous phase oxidation scheme has subsequently been revised, and the previously discussed model's underprediction of particulate sulphate in the winter has improved. We have repeated the model run for 1996 for this study and have compared the model results with observations from five rural sulphate measurement sites. Attribution plots during the two episodes are also presented, revealing the likely origins of the observed particulate sulphate.

2.1. *The NAME Model*

NAME is a Lagrangian model in which emissions are simulated by releasing large numbers of particles into a three-dimensional model atmosphere. Detailed descriptions of the model can be found in Physick & Maryon (1995) and Ryall & Maryon (1998).

Meteorological data (such as wind and temperature fields, precipitation and cloud information) are obtained from the Meteorological Office's numerical weather prediction model, the Unified Model (UM) (Cullen 1993). The three-dimensional wind field passively carries the released particles, with turbulent dispersion simulated by random walk techniques. Boundary-layer depth is time varying and is calculated in NAME from wind and temperature profiles. Dry and wet deposition processes act on the pollutant mass carried by each particle. The dry deposition scheme is based on a resistance analogy parametrization to determine the deposition velocity and wet deposition is parametrized by washout and rainout processes using a scavenging coefficient method. Cloud fraction and cloud liquid water output from the UM are used to drive the aqueous phase of the chemistry.

Every particle is labelled with its release location and time of origin, which makes it possible to identify which sources have contributed to a particular receptor area. Each particle is released with an initial mass of pollutant (both sulphur dioxide and ammonia in this study) and exists for as long as it carries mass of any species and it remains inside the model boundaries.

2.2. *Oxidation of Sulphur Dioxide*

In the atmosphere the gas-phase oxidation of sulphur dioxide (SO_2) is dominated by its reaction with the hydroxyl radical, OH. The hydroxyl radical plays an important part in tropospheric chemistry due to both its high reactivity with trace species such as SO_2 and because of its photochemical regeneration in the atmosphere. In the aqueous phase there are two main oxidation pathways, namely those via hydrogen peroxide, H_2O_2, and ozone. These routes are both parametrized in NAME. The reaction with hydrogen peroxide is very rapid and the oxidant can be completely exhausted before there has been time for regeneration of H_2O_2 via the recombination of the hydroperoxy radical, HO_2. The oxidation of SO_2 with O_3 is dependent on the acidity of the cloud droplets and is much more likely to be limited by high acidity (at which point the reaction proceeds very slowly) than low ozone concentrations. In order to parametrize the oxidation of SO_2 by O_3 it is therefore necessary to model the ammonia life cycle so that the concentration of this base species can be included in the calculation of cloud pH.

In the NAME study presented here, both SO_2 and NH_3 are emitted into the model atmosphere using emissions obtained from the EMEP 50 km × 50 km area database (EMEP 1997). The other chemical species required are all obtained from the Meteorological Office global chemistry model, STOCHEM, as monthly average fields. STOCHEM is a three-dimensional Lagrangian tropospheric chemistry model which is driven by global meteorological data from the UM and runs on a much larger scale than NAME (a 5° × 5° grid square is used, which gives a resolution of *ca.* 600 km × 400 km at mid-latitudes), and, hence, is unable to produce the same degree of fine spatial and temporal resolution that can be achieved in NAME. A full description can be found in Collins *et al.* (1997). The fields of OH, O_3 and HO_2 radicals are treated as fixed, their values only changing

Table 1. Statistics for a comparison of modelled versus measured particulate sulphate for 1996.

site	correlation	bias	NMSE	% (within a factor of 2)
Yarner Wood	0.40	−1.16	1.82	51.2
Eskdalemuir	0.39	−1.40	3.23	36.3
High Muffles	0.34	−1.03	1.84	52.5
Strathvaich	0.49	−1.24	6.74	10.8
Lough Navar	0.72	−1.63	5.27	10.1

monthly. The H_2O_2 is initialized using the monthly average field from STOCHEM and thereafter is modelled in NAME as a three-dimensional field.

2.3. *NAME Results for 1996*

Five rural measurement sites (Yarner Wood, Eskdalemuir, High Muffles, Strathvaich and Lough Navar) produce daily values of ambient particulate sulphate, and these data have been compared with output from the NAME model for 1996. The measurement data are obtained from the National Air Quality Information Archive provided by the National Environmental Technology Centre (NETCEN) on behalf of the Department of the Environment, Transport and the Regions (DETR) at

http://www.aeat.co.uk/netcen/airqual/index.html.

The model was run over a domain of longitude 15.0° W to 20.0° E and latitude 43.0° N to 65.0° N. Modelled sulphuric acid has been added to modelled ammonium sulphate to give particulate sulphate in $\mu g\, m^{-3}$ of SO_4. Table 1 shows a set of four standard statistics (correlation, bias, normalized mean square error (NMSE), and percentage within a factor of two) calculated on daily values for the five sites over 1996.

Comparison with a previous model run for this period (Malcolm *et al.* 2000) shows that the average correlation over the five sites for the year remains the same at 0.47, the average bias is less negative by 0.37 (reflecting the improved magnitudes during the winter months), the average NMSE is reduced by 5.41 and the average percentage within a factor of two is increased by 21.6%.

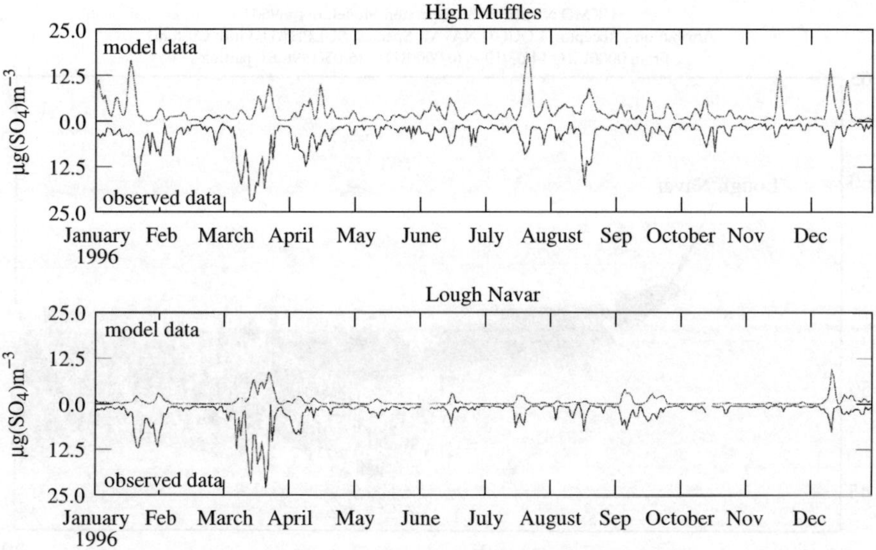

Fig. 1. NAME model daily sulphate aerosols plotted against measured sulphate aerosols at Lough Navar and High Muffles for 1996.

Yearly time-series of daily model particulate sulphate versus observation are presented in figure 1 for Lough Navar and High Muffles. Despite the improved performance of the model in the winter months, the exceptional episode in March is still not fully captured. The negative biases at all sites indicate that the model is generally underpredicting. The obtained correlations are still somewhat low, but given the inherent difficulty of modelling both formation and transport of particulate sulphate, perhaps that is to be expected. To improve model performance significantly, we would need more detailed resolution SO_2 emissions (both spatially and temporally) and also to be able to represent the nonlinear chemical conversion more precisely. It should also be remembered that the meteorology is varying over a 50 km grid scale on a three hourly basis, which means it is unable to resolve sub-grid scale meteorological variations (for example, due to local topography).

2.4. *Source Attribution*

Two periods have been selected from both the March and July episodes in order to demonstrate the origin of the material seen in the modelled data

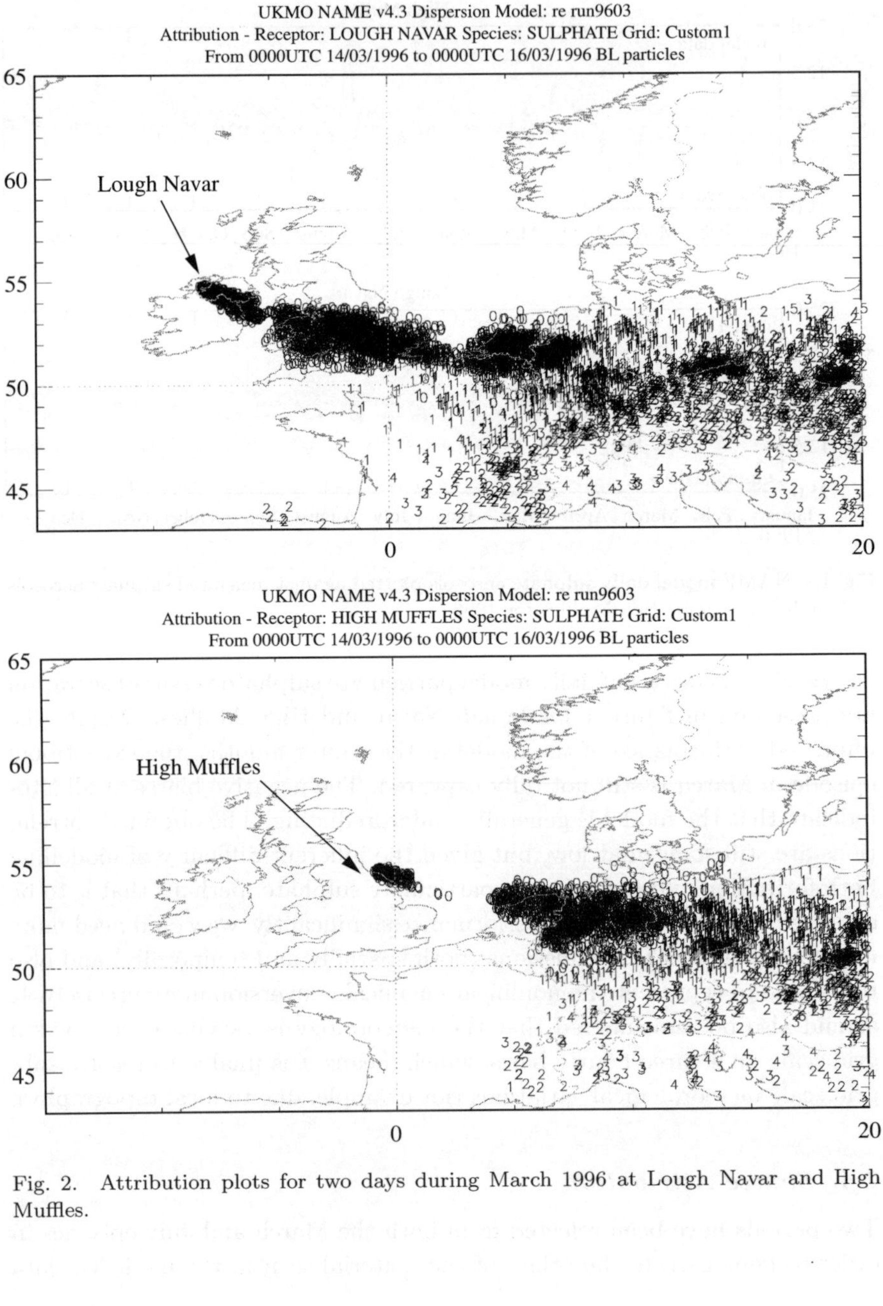

Fig. 2. Attribution plots for two days during March 1996 at Lough Navar and High Muffles.

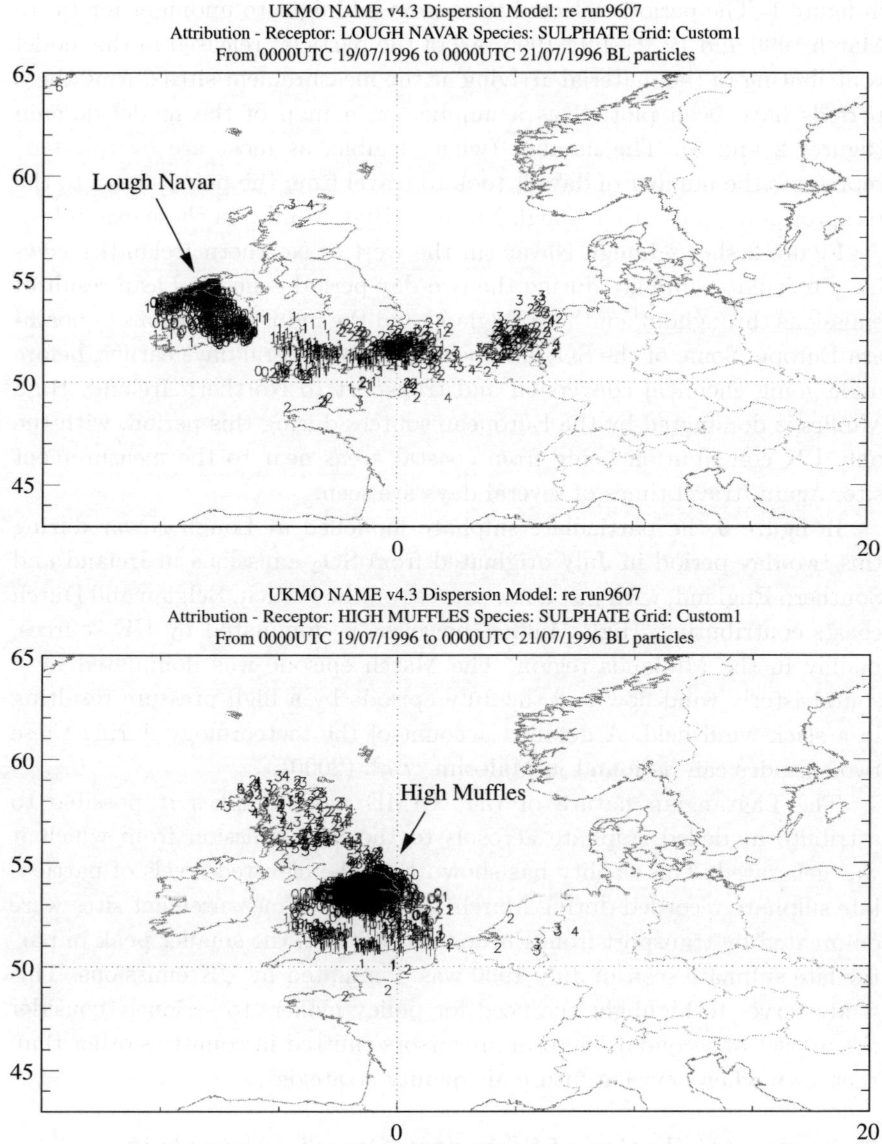

Fig. 3. Attribution plots for three days during July 1996 at Lough Navar and High Muffles.

in figure 1. The periods selected were from midnight to midnight for 14–16 March 1996 and 19–21 July 1996. All of the particles released in the model contributing to the material arriving at the measurement sites during these periods have been plotted as a number on a map of the model domain (figures 2 and 3). The number (where legible, as most are overplotted) represents the number of days it took to travel from the point shown to the receptor point (i.e. either Lough Navar or High Muffles in these examples).

Figure 2 shows Lough Navar (in the west of Northern Ireland) receiving particulate sulphate during the two-day period generated as a result of emissions throughout southern England and the industrial regions of northern Europe. Some of the SO_2 had been emitted several days earlier, before undergoing chemical conversion and transport to Northern Ireland. High Muffles is dominated by the European sources during this period, with the only UK contribution being from coastal areas near to the measurement site. Again, travel times of several days are seen.

In figure 3 the particulate sulphate modelled at Lough Navar during this two-day period in July originated from SO_2 emissions in Ireland and Southern England, with just a few sources on the French, Belgian and Dutch coasts contributing. High Muffles, however, is dominated by UK sources, mainly in the Midlands region. The March episode was dominated by a southeasterly wind flow and the July episode by a high pressure resulting in a slack wind field. A detailed account of the meteorology during these two episodes can be found in Malcolm *et al.* (2000).

The Lagrangian nature of the NAME model makes it possible to attribute modelled sulphate aerosols to the SO_2 emission from which it was generated. This facility has shown that the elevated levels of particulate sulphate recorded during March 1996 at rural measurement sites were dominated by transport from Europe. In contrast, the smaller peak in particulate sulphate seen in July 1996 was dominated by UK emissions. This study serves to highlight the need for policy makers to seriously consider the impact of secondary aerosol precursors emitted in countries other than their own when devising future air-quality strategies.

3. Source Attribution of Secondary Organic Aerosols in the UK

It was noted originally by Went (1960) that natural biogenic hydrocarbons play an important role in the formation of tropospheric aerosols.

The sunlight-driven atmospheric photo-oxidation of high-molecular-weight hydrocarbons has been shown to produce low vapour pressure reaction products that partition between the gas and aerosol phases (Pandis *et al.* 1992). These reaction products are known as semi-volatile organic compounds because of their ability to pass between the gas and aerosol phases (Kamens *et al.* 1999). In the aerosol phase, these reaction products are known as secondary organic aerosols (SOAs). Of the natural biogenic hydrocarbons, terpenes have been found to be effective sources of SOAs (Hoffmann *et al.* 1997), whereas, of the man-made hydrocarbons, aromatics are the most important source (Odum *et al.* 1996).

These considerations have prompted questions about the relative importance of natural biogenic sources as opposed to man-made sources of SOA levels in the United Kingdom. To begin to answer these questions, a photochemical trajectory model has been used to investigate the formation of semi-volatile organic degradation products from the photo-oxidation of both natural biogenic terpene and man-made aromatic hydrocarbon compounds during a summertime regional ozone pollution episode.

3.1. *Application of the UK Photochemical Trajectory Model*

The formation of SOAs during a summertime regional scale pollution episode has been described using the UK Photochemical Trajectory Model (UK PTM). This model addresses the detailed chemical development in an air parcel as it moves across the European emissions grid following a six-day trajectory from Austria through to its arrival point in Wales (Derwent *et al.* 1996). The chemistry is described for a single air parcel whose base is at the surface and whose upper boundary is at the top of the atmospheric boundary layer. Temperatures, humidities, boundary-layer depths, wind speeds and wind directions were all diurnally varying and given values appropriate to the conditions of regional scale pollution episodes.

The UK PTM employs the Master Chemical Mechanism (MCM) to describe the photochemical ozone production from 123 emitted organic compounds that generate 3482 reaction and degradation products and take part in over 10 500 chemical reaction processes (Jenkin *et al.* 1999). The MCM also includes the reactions of the simple atoms and radicals containing oxygen, hydrogen and nitrogen and those of CO, SO_2 and H_2O_2 that together describe the fast photochemistry of the polluted atmospheric boundary layer. The MCM version 2.0 may be downloaded from the World Wide Web at http://chem.leeds.ac.uk/Atmospheric/MCM/mcmproj.html.

The fast photochemistry and regional photochemical ozone production occurring in the UK PTM are driven by the emissions picked up by the air parcel as it traverses Europe. The emissions of NO_x, CO, SO_2, isoprene and volatile organic compounds (VOCs) were employed at 150 km × 150 km scale across Europe based on EMEP emissions (Mylona 1999), at 50 km × 50 km where available from either EMEP or EC CORINAIR (Bouscaren & Cornaert 1995) and at 10 km × 10 km within the United Kingdom from Salway *et al.* (1996). European emission inventories (Mylona 1999) may be downloaded from the World Wide Web at http://www.emep.int.

The emissions of all VOCs were split into the emissions of individual organic compounds using the detailed speciated emission inventory available for the United Kingdom from the NAEI, and this same speciation was assumed to hold across Europe and is given in Derwent *et al.* (1996). The model also treated the dry deposition and surface removal of ozone, nitric acid, hydrogen peroxide and the peroxyacylnitrates.

3.2. *Model Treatment of SOAs*

The formation of SOAs in the UK PTM was driven by the emissions of terpenes from natural biogenic emissions and aromatic hydrocarbons as the air parcel traversed Europe. Emissions of terpenes at a spatial resolution of 1° × 1° and for the month of July for Europe were taken from the Global Emission Inventory Activity emissions database:

http://blueskies.sprl.umich.edu/geia/.

It was assumed that all the terpene emissions occurred into the UK PTM as α-pinene. No explicit temperature or time dependence was assumed for these emissions, and the emissions from a particular grid square were held constant at the monthly average emission rate. Emissions of each of the aromatic hydrocarbons was taken from the EMEP (Mylona 1999), EC CORINAIR (G. McInnes 1994, personal communication) and UK NAEI inventories using the VOC speciation taken from Derwent *et al.* (1996).

The MCM version 2.0 was used to describe the reactions of α-pinene with OH radicals and ozone during daylight and with NO_3 radicals and ozone during nighttime (Jenkin *et al.* 2000). Altogether the α-pinene degradation scheme contained over 329 reactions and formed a number of low-volatility degradation products, which are classed as semi-volatile organic

Table 2. The assumed fractions by mass of each aromatic hydrocarbon oxidized in the UK Photochemical Trajectory Model, which produces SOAs and their percentage contribution to SOA formation. Mass fractions of SOAs produced from each aromatic hydrocarbon oxidized were taken from Odum *et al.* (1997). No SOA was assumed to be formed from the photo-oxidation of benzene, styrene, benzaldehyde, i-propylbenzene, 1,2,3-trimethylbenzene, 1,2,4-trimethylbenzene, 3,5-dimethylethylbenzene and 3,5-diethyltoluene.

aromatic hydrocarbon	fraction by mass of SOA to hydrocarbon oxidized	percentage contribution to SOA formation (%)
toluene	0.089	60.0
o-xylene	0.026	4.0
m-xylene	0.038	4.5
p-xylene	0.025	12.8
ethylbenzene	0.086	9.2
1,3,5-trimethylbenzene	0.031	0.4
m-ethyltoluene	0.065	1.9
p-ethyltoluene	0.054	2.2
o-ethyltoluene	0.062	1.9
n-propylbenzene	0.081	3.0

compounds, including pinonaldehyde, peroxypinonic acid, pinonic acid, norpinonaldehyde and hydroperoxypinonaldehyde. These semi-volatile organic compounds have been scavenged in the UK PTM by pre-existing aerosol species in competition with their subsequent atmospheric degradation. No loss of semi-volatile organic matter from the aerosol back into the gas phase was allowed in order to simulate the upper limit concentrations of SOAs.

The MCM version 2.0 was also used to describe the reactions of aromatic hydrocarbons with OH radicals which generate SOAs. A small fraction of chemical flux through these reactions was assumed to generate low-volatility reaction products, which would be present in the atmosphere as semi-volatile organic compounds. These fractions have been quantified in table 2 for each of the aromatic hydrocarbons where these are available from the literature (Odum *et al.* 1997), otherwise they have been set to zero. Again, these semi-volatile organic compounds have been scavenged in the UK PTM by pre-existing aerosol species. No loss of semi-volatile organic matter from the aerosol back into the gas phase was allowed to simulate an upper limit concentration of SOAs. The semi-volatile organic compounds formed from aromatic hydrocarbon photo-oxidation are thought to be species such as 2,5-furandiones (Forstner *et al.* 1997).

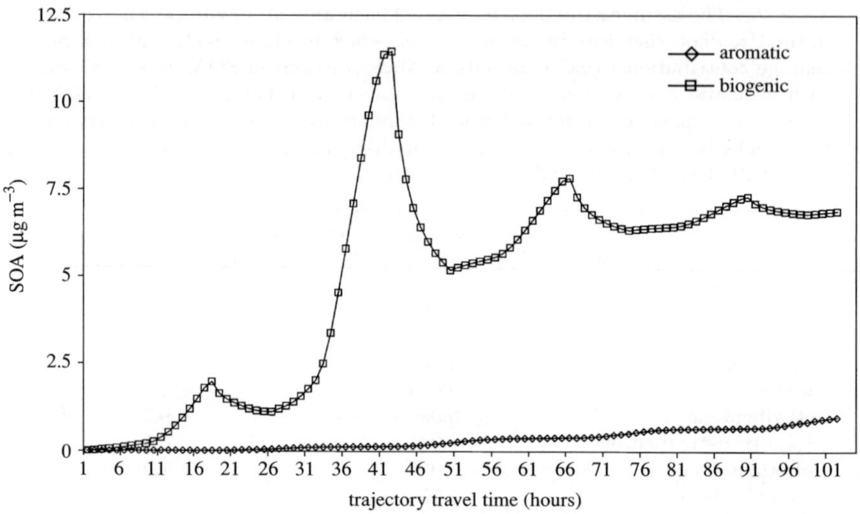

Fig. 4. SOAs from natural biogenic α-pinene and man-made aromatic hydrocarbon photo-oxidation in the UK PTM.

3.3. *Model Results for SOAs*

Figure 4 presents the calculated concentrations of SOAs in the UK PTM as the air parcel traverses Europe from Austria across to Wales. Because no loss of semi-volatile organic matter from the aerosol once scavenged has been allowed, the concentrations of SOAs represent an upper limit to those anticipated in the real atmospheric boundary layer. The figure shows the concentrations of SOAs formed from the photo-oxidation of both natural biogenic hydrocarbons and man-made aromatic hydrocarbons. The conclusion is that the SOA formed from terpene photo-oxidation is several times greater in concentration compared with that from aromatic hydrocarbon photo-oxidation.

By way of comparison, the above model experiment also generated $18\,\mu g\,m^{-3}$ of particulate sulphate from the oxidation of SO_2. The modelled concentrations of $5–10\,\mu g\,m^{-3}$ for SOAs suggest that natural biogenic α-pinene may potentially make a significant contribution to the concentration of secondary particulate matter and, hence, total fine particulate matter during summertime regional pollution episodes. However, significant uncertainties remain concerning the scavenging of the semi-volatile terpene degradation products by the ambient aerosol and the subsequent fate of this aerosol.

The present study in figure 4 shows that the SOA formed from the photo-oxidation of aromatic hydrocarbons produces only *ca.* 10–15% of the total yield of SOAs from both natural biogenic and man-made hydrocarbon photo-oxidation across Europe in the UK PTM. Table 2 provides an analysis of the percentage contributions made by each aromatic hydrocarbon to the overall SOA yield from aromatic hydrocarbons as a class. These calculated contributions reflect the different emissions, OH reactivities and SOA yields for each individual aromatic hydrocarbon. Three species—toluene, *p*-xylene and ethylbenzene—together account for over 80% of the overall SOA yield from aromatic hydrocarbon photo-oxidation under European conditions.

4. Discussion

Particulate sulphate is generally the major observed component of secondary particulate matter in urban areas, and the United Kingdom shows no exception in this regard (APEG 1999). A highly sophisticated Lagrangian dispersion model has been used here to describe the formation of particulate sulphate by the photochemical oxidation of SO_2, its sole precursor species. A comparison of model particulate sulphate with observations for five rural monitoring sites shows good agreement overall, with a close registration of the major pollution episodes, though with a tendency for the model to underestimate the observations somewhat during winter.

The Lagrangian dispersion model has been used to attribute the origins of the particulate sulphate arriving at the measurement sites during two major pollution episodes in March 1996 and July 1996. The origins of the particulate sulphate varied markedly between the different sites for the different episodes. Under some circumstances, particulate sulphate levels are dominated by long-range transport in from the continent of Europe, and this was noticeably the case during March 1996. Under other conditions, United Kingdom SO_2 emissions appear to be the dominant source.

In contrast with the case of particulate sulphate, SOA is much less well understood, and the questions asked are of a much more rudimentary nature. A highly detailed photochemical model has been assembled inside a highly simplistic meteorological model to assess the relative importance of natural biogenic aerosol precursors as opposed to man-made precursors. It is concluded that the formation of SOAs from the photo-oxidation of terpenes is likely to be several times greater in magnitude than that from aromatic hydrocarbon photo-oxidation.

These conclusions from our study necessarily address secondary particles in the fine particle size range as a whole and do not specifically address the very smallest particles in the ultrafine particle size range. There are currently a number of large gaps in understanding which have precluded our focusing on the ultrafine particle size range. There is currently no way of knowing how much of the ultrafine secondary particulate matter in the UK atmosphere has arisen by the homogeneous or heterogeneous nucleation routes. There are so few measurements of ultrafine particles in the United Kingdom that it would be difficult to check model performance against observations in any comprehensive manner. Furthermore, there are no internationally agreed air-quality guidelines with which to assess the public-health significance of ultrafine particle observations.

Whether any of our conclusions concerning the source attribution of particulate sulphate and of SOAs adequately reflect real-world behaviour depends on the adequacy and accuracy of the assumptions and simplifications made in the models and on the accuracy of their input parameters. Without comprehensive monitoring of aerosol composition across the United Kingdom, it will be difficult to make significant progress. However, we have some confidence that our basic conclusions concerning the importance of the long-range trans-boundary transport of particulate sulphate and the importance of natural biogenic precursors for SOAs should be robust.

Acknowledgements

This work was supported as part of the Public Meteorological Service R & D Programme of the Meteorological Office and through the Air Quality Research Programme of the Department of the Environment, Transport and the Regions (contract no. EPG 1/3/128). The authors acknowledge the help and encouragement they have received from Roy Maryon and Derrick Ryall of the Meteorological Office and from Harvey Jeffries of the University of North Carolina. The Master Chemical Mechanism was implemented with the assistance of Michael Jenkin, AEA Technology, and Sandra Saunders and Michael Pilling, University of Leeds.

References

APEG 1999 Source apportionment of airborne particulate matter in the United Kingdom. Report of the Airborne Particles Expert Group. Department of the Environment, Transport and the Regions, London.

Bouscaren, R. & Cornaert, M.-H. 1995 CORINAIR. Technical annexes, vol. 1. Nomenclature and software. European Commission EUR 12586/1, EN, Belgium.

Collins, W. J., Stevenson, D. S., Johnson, C. E. & Derwent, R. G. 1997 Tropospheric ozone in a global-scale three-dimensional Lagrangian model and its response to NO_x emission controls. *J. Atmos. Chem.* **26**, 223–274.

Cullen, M. J. P. 1993 The Unified Forecast/Climate Model. *Meteorolog. Mag. (UK)* **1449**, 81–94.

Derwent, R. G., Jenkin, M. E. & Saunders, S. M. 1996 Photochemical ozone creation potentials for a large number of reactive hydrocarbons under European conditions. *Atmos. Environ.* **30**, 181–199.

Dockery, D. W., Pope, C. A., Xu, X., Spengler, J. D., Ware, J. H., Fay, M. E., Ferris Jr, B. G. & Speizer, F. E. 1993 An association between air pollution and mortality in six US cities. *New England J. Med.* **329**, 1753–1759.

EMEP 1997 Transboundary air pollution in Europe 1997 emissions, dispersion and trends of acidifying and eutrophying agents. Part 1. EMEP/MSC-W report, Norwegian Institute for Air Research, Kjeller, Norway.

Finlayson-Pitts, B. J. & Pitts, J. N. 1986 *Atmospheric chemistry: fundamentals and experimental techniques.* Wiley.

Forstner, H. J. L., Flagan, R. C. & Seinfeld, J. H. 1997 Secondary organic aerosol from the photooxidation of aromatic hydrocarbons: molecular composition. *Environ. Sci. Technol.* **31**, 1345–1358.

Hjellbrekke, A.-G. 1999 Data report 1997. Part 1. Annual summaries. EMEP/CCC report 3/99, Norwegian Institute for Air Research, Kjellet, Norway.

Hoffmann, T., Odum, J. R., Bowman, F., Collins, D., Klockow, D., Flagan, R. C. & Seinfeld, J. H. 1997 Formation of organic aerosols from the oxidation of biogenic hydrocarbons. *J. Atmos. Chem.* **26**, 189–222.

Jenkin, M. E., Hayman, G. D., Derwent, R. G., Saunders, S. M., Carslaw, N., Pascoe, S. & Pilling, M. J. 1999 Tropsopheric chemistry modelling: improvements to current models and application to policy issues. Final report AEAT-4867/20150/R004, AEA Technology, Culham Laboratory, Oxfordshire.

Jenkin, M. E., Shallcross, D. E. & Harvey, J. N. 2000 Development and application of a possible mechanism for the generation of cis-pinic acid from the ozonolysis of α- and β-pinene. *Atmos. Environ.* **34**, 2837–2850.

Kamens, R., Jang, M., Chien, C.-J. & Leach, K. 1999 Aerosol formation from the reaction of α-pinene and ozone using a gas-phase kinetics aerosol partitioning model. *Environ. Sci. Technol.* **33**, 1430–1438.

Lazaridis, M., Semb, A. & Hov, O. 1999 Long-range transport of aerosol particles. EMEP/CCC report 8/99, Norwegian Institute for Air Research, Kjeller, Norway.

Malcolm, A. L., Derwent, R. G. & Maryon, R. H. 2000 Modelling the long-range transport of secondary PM_{10} to the UK. *Atmos. Environ.* **34**, 881–894.

Mylona, S. 1999 EMEP emission data. Status report 1999. EMEP/MSC-W note 1/99, Norwegian Meteorological Institute, Oslo, Norway.

Odum, J. R., Hoffmann, T., Bowman, F., Collins, D., Flagan, R. C. & Seinfeld, J. H. 1996 Gas/particle partitioning and secondary organic aerosol yields. *Environ. Sci. Technol.* **30**, 2580–2585.

Odum, J. R., Jungkamp, T. P. W., Griffin, R. J., Forstner, H. J. L., Flagan, R. C. & Seinfeld, J. H. 1997 Aromatics, reformulated gasoline, and atmospheric organic aerosol formation. *Environ. Sci. Technol.* **31**, 1890–1897.

Pandis, S. N., Harley, R. A., Cass, G. R. & Seinfeld, J. H. 1992 Secondary aerosol formation and transport. *Atmos. Environ.* A **26**, 2269–2282.

Physick, W. L. & Maryon, R. H. 1995 Near-source turbulence parametrization in the NAME model. UK Met Office Turbulence and Diffusion Note 218.

Pope, C. A., Thun, M. J., Namboodiri, M. M., Dockery, D. W., Evans, J. S., Speizer, F. E. & Heath, C. W. 1995 Particulate air pollution as a predictor of mortality in a prospective study of US adults. *Am. J. Resp. Crit. Care Med.* **151**, 669–674.

QUARG 1996 Airborne particulate matter in the United Kingdom. Third report of the Quality of Urban Air Review Group, Department of the Environment, London.

Ryall, D. B. & Maryon, R. H. 1998 Validation of the UK Met Office's NAME model against the ETEX dataset. *Atmos. Environ.* **32**, 4265–4276.

Salway, A. G., Goodwin, J. W. L. & Eggleston, H. S. 1996 UK emissions of air pollutants. AEA Technology Report, Culham Laboratory, Oxfordshire.

Went, F. W. 1960 Blue hazes in the atmosphere. *Nature* **187**, 641–643.

Discussion

N. Rose (*ECRC, University College London, UK*). Does the grid used in your SO_2 model extend to marine areas, and if so is there a significant contribution, to the UK, from shipping sources in the North Sea and English Channel?

R. G. Derwent. The emission inventories used in our modelling work extend over marine areas and included substantial emissions of SO_2 from the North Sea, English Channel and North Atlantic Ocean shipping as well as natural emissions of DMS.

M. Wallis (*FoE Cymru, Cardiff, UK*). I question the correlation between the data and your sulphate meteorological model. The July 1996 episode for Lough Navar shows 93% from the UK and 7% from European sources, as published in the APEG report (APEG 1999) and no significance for validation of your model. The total year 1995 data you presented shows what is well known, that anticyclonic conditions with easterly or southeasterly winds allow accumulation of locally emitted air pollutants. You cannot distinguish this from your 'European' source. APEG para. 4.3.1 says, 'during the winter time, the model clearly underestimates the observations due to the neglect of the ammonic–ozone–SO_2 cloud droplet oxidation route'.

The UK government uses your results to say that we cannot meet PM_{10} standards by UK traffic and industry controls, so the issue is important for policy. The APEG Committee was not convinced. Has your new work been validated by peer review and what confidence can be placed in it?

R. G. DERWENT. The modelling work on particulate sulphate has been validated by comparison with observations and the results have been published by Malcolm *et al.* (2000).

C. N. HEWITT AND H. STEWART (*Institute of Environmental and Natural Sciences, University of Lancaster, UK*). You use the Master Chemical Mechanism to predict the degradation of α-pinene emitted by vegetation in the UK and to describe the formation of semi-volatile organic products that may nucleate or condense onto pre-existing particles. From this, it was shown that biogenic emissions of terpenes have the potential to account for a significant fraction of the secondary organic aerosol in the UK.

In our work on the emissions of volatile organic compounds from the biosphere to the atmosphere, we have shown that relatively few plant species contribute to the emissions of the total flux of VOCs in the UK. In fact, three tree species probably contribute more than 60% of the total biogenic isoprene flux in the UK. These are *Quercus* spp (oak, 27%), *Picea sitchensis* (Sitka spruce, 27%) and *Populus* spp (poplar, 11%). Our current best estimate of the total isoprene emission rate is 88 t h^{-1} at a temperature of 30 °C and a light intensity of 1000 $\mu mol\ m^{-2}\ s^{-1}$ (Stewart *et al.* 2000).

In the case of the C_{10} monoterpene family, our work indicates that 10 plant species probably account for more than 85% of the total monoterpene emission flux in the UK. These are *Picea sitchensis* (Sitka spruce, 35%), *Pinus sylvestris* (Scots pine, 13%), *Calluna vulgaris* (heather, 9%), *Larix* spp (larch, 7%), *Pinus contorta* (beach pine, 6%), *Cirsium arvense* (creeping thistle, 6%), *Picea abies* (Norway spruce, 5%), *Hordeum vulgare* (barley, 2%), *Pisum sativum* (peas, 2%) and *Taraxacum* agg. (dandelion,

2%). The monoterpene compounds known to be emitted from these species are α-pinene, β-pinene, D-limonene, camphene, delta-3 carene, myrcene, β-phellandrene, sabinene and 1,8-cineole. Additionally, there are suggestions that other compounds may be emitted by these species, including α-, β- and γ-terpinene, cymene, α-phellandrene, β-fenchene, tricyclene and α-thujene. Our current best estimate of the total monoterpene emission rate is 68 t h^{-1} at a temperature of 30 °C and a light intensity of 1000 µmol m^{-2} s^{-1}. This is reduced to 11 t h^{-1} or 96 kt yr^{-1} at an average temperature of 10 °C and a light intensity of 500 µmol m^{-2} s^{-1} (Stewart *et al.* 2000).

Interestingly, the commonly held notion that *brassica napus* (oil seed rape) is a prolific emitter of monoterpenes is almost certainly incorrect. It is known to emit α- and β-pinene, δ-limonene, sabinene and α-thujene, but at rates at least an order of magnitude lower (on a per dry weight basis) than the emitting tree species listed above.

Clearly, a quantitive assessment of the role of emissions of VOCs to secondary aerosol formation requires an understanding of the species specific flux rates of the compounds from the biosphere to the atmosphere and of their chemistry in the atmosphere.

R. G. DERWENT. These comments are most helpful, and we will endeavour to use your results in our future work.

Additional reference

Stewart, H., Hewitt, C. N. & Bunce, R. 2000 Emissions of volatile organic compounds from the biosphere to the atmosphere in the United Kingdom. *Atmos. Environ.* (Submitted.)

CHAPTER 7

ULTRAFINE PARTICLES FROM COMBUSTION SOURCES: APPROACHES TO WHAT WE WANT TO KNOW

Henning Bockhorn
Institut für Chemische Technik and Engler-Bunte-Institut/Bereich Verbrennungstechnik, Universität Karlsruhe (TH), Kaiserstraße 12, D-76128 Karlsruhe, Germany

Soot formation and oxidation will be analysed with respect to the most important processes, namely particle inception, coagulation and surface growth. Time-scales of surface growth are estimated for premixed and diffusion flames and compared with time-scales for coagulation. It turns out that characteristic time-scales for soot formation and coagulation are similar and about one order of magnitude larger than the characteristic time-scales for combustion reactions and much smaller than the time-scales of molecular transport.

Coagulation processes will be discussed in detail and a detailed chemistry approach for surface growth will be presented. The detailed information will be put into a soot model that reproduces a number of phenomena in sooting premixed hydrocarbon flames, for example:

(i) the dependence of surface growth and oxidation rates on the chemical 'environment' of soot particles; and

(ii) the fraction of soot formed by particle inception and surface growth reactions and addition of polyacrylic aromatic hydrocarbon (PAH).

The 'fine structure' of soot is not resolved by this approach, and, furthermore, the predictions depend sensitively on information about the kinetics of growth of PAH-like structures, the detailed processes occurring on the surface of soot particles, and, most importantly, the pressure dependence of all these processes.

Keywords: soot formation; soot oxidation; coagulation; surface growth

1. Introduction

Hydrocarbons tend to form soot when burning under fuel-rich conditions. Soot from combustion of hydrocarbons under fuel-rich conditions appears as an ensemble of ultrafine particles in the size range up to a few hundred nanometres. It is this size range of particles that is suspected to exhibit dangerous effects on human health. Particles of this size easily penetrate into the respiratory tracts and are thought either to stimulate the defence mechanisms similar to that against small fibres or act via chemical compounds adsorbed on the surface of the particles.

The formation of soot, i.e. the conversion of a hydrocarbon fuel molecule containing few carbon atoms into a carbonaceous agglomerate containing some millions of carbon atoms, is an extremely complicated process. It is a kind of gaseous–solid phase transition where the solid phase exhibits no unique chemical and physical structure. Therefore, soot formation encompasses chemically and physically different processes, e.g. the formation and growth of large aromatic hydrocarbons and their transition to particles, the coagulation of primary particles to larger aggregates, and the growth of solid particles by picking up growth components from the gas phase. The above-mentioned processes constitute the formation of the bulk of soot. In addition, numerous other processes decide on the 'fine structure' of soot, e.g. the formation of electrically charged soot particles, the formation—charged and neutral—of fullerenes, or the formation of high molecular weight tarry modifications with optical properties quite different from carbon black, and a variety of modifications of soot with different optical and mechanical properties.

While much progress has been achieved in understanding all these processes, numerous problems remain unsolved. In the following, some recent development in mechanisms and models of soot formation will be discussed, focusing on processes of the formation of the bulk of soot and attempting to reduce the gap for a comprehensive understanding of soot formation.

2. Structure of Sooting Flames

The locally resolved structure of laminar and turbulent sooting diffusion flames with respect to soot volume fractions f_V, particle number densities N_V and particle sizes r_m has recently been investigated by Geitlinger *et al.*

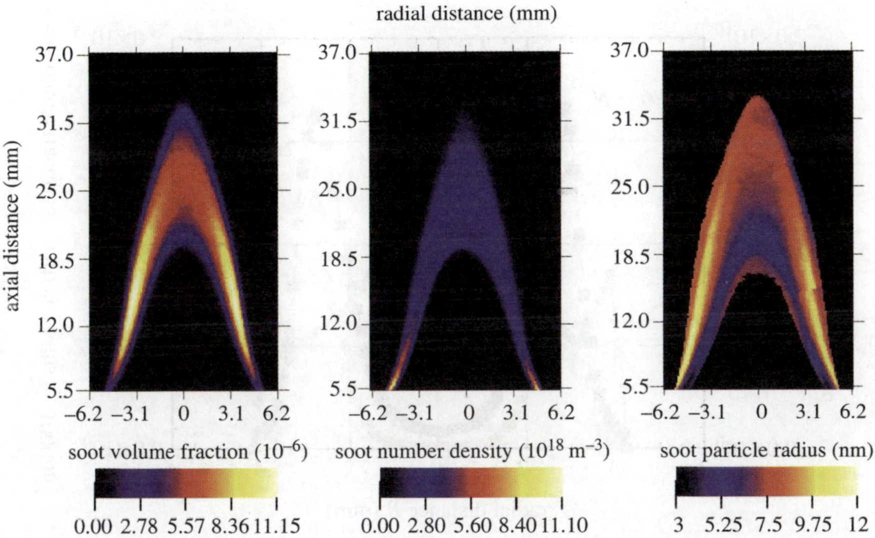

Fig. 1. Two-dimensional maps of particle number density N_V, soot volume fraction f_V, and mean particle radius r_m of a laminar acetylene–air diffusion flame, fuel diluted with nitrogen.

(1998, 1999) by means of a two-dimensional imaging technique employing a combination of Rayleigh scattering and laser-induced incandescence (LII).

Figure 1 gives, as an example from the above-referenced work, two-dimensional maps of soot volume fractions f_V, particle number densities N_V and particle sizes r_m of a bunsen-type, laminar acetylene–air diffusion flame, the fuel of which is diluted with nitrogen. The corresponding profiles at 15 mm height above the burner nozzle are displayed in figure 2.

The figures clearly show that at low heights no soot can be observed in the centre of the flame. At the radial position of the maximum in N_V, a minimum in the particle size appears. The soot formation zone is located at this radial position, where particle inception prevails, generating a large number of small particles. The maximum in the soot volume fraction f_V occurs at somewhat smaller radial distances, indicating that surface growth reactions are taking place in the preheating zone of the fuel, where temperatures are still high enough for this process. Surface growth reactions add mass to the small particles being formed in the particle-inception region. Towards lower radial distances, r_m increases because of surface growth reactions as well as

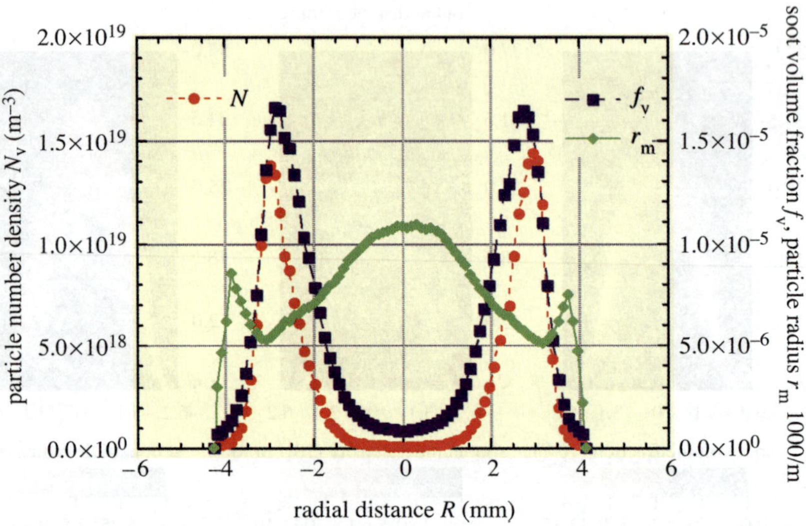

Fig. 2. Profiles of particle number density N_V, soot volume fraction f_V, and mean particle radius r_m of the laminar acetylene–air diffusion flame from figure 1 at 15 mm height above the burner nozzle.

coagulation. The latter process—which adds no mass to the particles but changes their size drastically—is very fast, indicated by the strong decrease in N_V towards lower radial distances. The apparent increase in r_m towards the oxygen-rich zone of the diffusion flame can be explained by coagulation of soot particles as well as the complete oxidation of the smallest soot particles in the reaction zone of the flame. All the profiles exhibit steep gradients when moving towards the oxidation zone of the flame.

At larger heights above the burner in the cone-shaped flame, the profiles are moving towards the centre of the flame. The largest particles are then observed in the centre of the flame. At this position f_V is quite low and the particle radii are dominated by coagulation. At the tip of the sooting region the profiles of f_V and N_V from each side of the flame are fusing together. For f_V no minimum can be observed in the centre of the flame. The maxima in f_V and N_V decrease because of the consumption of soot when reaching the oxidation zone at the flame tip. Particle number density is of the order of 1×10^{18} m^{-3}, whereas mean particle sizes are of the order of 20 nm and soot volume fractions are *ca.* 20 ppm.

Besides the orders of magnitude for N_V, f_V and r_m in diffusion flames of that kind, from the above figures the different main processes leading to the final soot particle ensemble can be identified. These processes can be summarized roughly as follows:

(i) formation of primary soot particles (particle inception),
(ii) surface growth reactions of soot particles, and
(iii) coagulation of soot particles.

These processes are discussed in more detail in the subsequent sections.

3. Processes Leading to Soot

3.1. *Coagulation Processes*

The first and third of the above processes comprise (reactive) coagulation processes, where particles (molecules) of size i collide with those of size j. These processes can be uniquely described by coagulation kinetics.

For a coagulating particle system the change of number density for particles of the size class i with time is given by the Smoluchowsky equation

$$\frac{\mathrm{d}N_i}{\mathrm{d}t} = \frac{1}{2}\sum_{j=1}^{i-1} \beta_{j,i-j} N_j N_{i-j} - N_i \sum_{j=1}^{n_{\max}} \beta_{i,j} N_j, \quad i = 2, \ldots, n_{\max}. \tag{1}$$

In equation (1), N_i and N_j represent the number density of particles in the size class i and j, respectively. The coagulation coefficient $\beta_{i,j}$ for free molecular coagulation is given by

$$\beta_{i,j} = \sqrt{\frac{8\pi k_B T}{\mu_{i,j}}}(r_i + r_j)^2 = C\sqrt{\frac{1}{i} + \frac{1}{j}}(i^{1/3} + j^{1/3})^2, \tag{2}$$

where

$$C = \sqrt{\frac{8\pi k_B T}{m_1}}\left(\frac{3m_1}{4\pi\rho}\right)^{2/3}. \tag{3}$$

The first term on the right-hand side of equation (1) gives the formation rate of particles in the size class i by coagulation of smaller particles, the sizes of which add to the size i, whereas the second term describes the consumption rate of particles in the size class i by collision with other particles. To include addition of large hydrocarbons to the surface of soot particles by sticky collisions, they have to be included in the system of equations (1), while the formation rates of those obey different mechanisms. The smallest particles, the sizes of which are defined so as to consist of two pyrene molecules (Appel & Bockhorn 2000), are balanced by

$$\frac{\mathrm{d}N_1}{\mathrm{d}t} = \tfrac{1}{2}\beta_{0,1}N_0^2 - N_1 \sum_{j=1}^{n_{\max}} \beta_{1,j}N_j, \tag{4}$$

where N_0 is the number density of the last pre-particle species. Again, particle number densities N_0 have to be obtained from different considerations. The Schmoluchowsky equation written for the total particle number density,

$$N = \sum_i N_i,$$

gives

$$\frac{\mathrm{d}N}{\mathrm{d}t} = -\tfrac{1}{2}\beta(i)N^2, \tag{5}$$

where $\beta(i)$ is a weak function of the particle size. Assuming $\beta(i)$ to be independent of particle size, then $\beta \approx 10^{-16}\,\mathrm{m^3\,s^{-1}}$ at 2000 K, and a particle number density of *ca.* $10^{18}\,\mathrm{m^{-3}}$ results in coagulation rates of *ca.* $10^{20}\,\mathrm{m^{-3}\,s^{-1}}$ or characteristic time-scales for coagulation of $\tau_{\mathrm{coag}} \approx$ 10 ms. At incipient soot formation, number densities exceed those in the surface growth region (cf. figure 2), so that characteristic time-scales for coagulation are even smaller and attain similar values to characteristic time-scales for combustion reactions.

With $\beta \approx$ const., the solution of equation (5) results in

$$N = \frac{N_0}{1 + N_0\beta t} = \frac{1}{(1/N_0) + \beta t}. \tag{6}$$

For $\beta t \gg (1/N_0)$, it follows that $N \propto (1/\beta t)$. For comparatively long coagulation times the number density is no longer dependent on initial conditions N_0 and is only given by β and t. The particle ensemble loses its memory and, for typical conditions in flames (temperature 2000 K, $\beta \approx 10^{-16}\,\mathrm{m^3\,s^{-1}}$, coagulation time 100 ms), particle number densities of *ca.* $10^{17}\,\mathrm{m^{-3}}$ are attained. When emitted with the exhaust, particle number density and particle sizes of the soot particle aerosol 'in accumulation mode' exhibit a broad size distribution with low particle number densities, which change only slowly. In contrast, soot particle aerosols in 'nucleation mode' show narrow size distributions with high number densities.

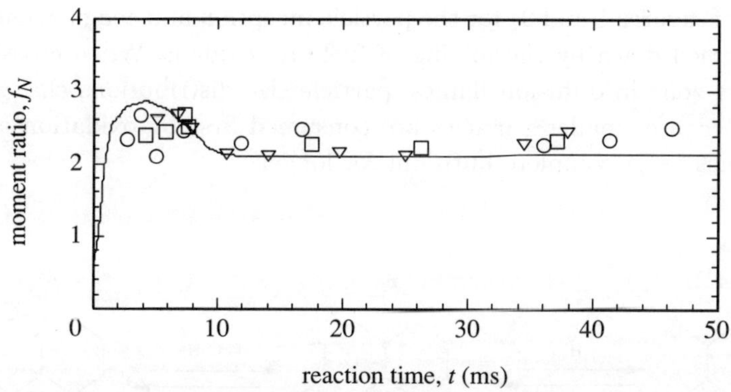

Fig. 3. Evolution of the moment ratio $f_N = \mu_6/\mu_3^2$ for different low-pressure premixed flames (Bockhorn *et al.* 1987); symbols refer to TEM measurements, the solid line denotes the numerical solution of the coagulation equation with particle inception being modelled to produce 5% of the total soot mass within the first 3 ms of soot formation. ∇, propane; $\circ$, benzene; $\square$, acetylene; ——, calculated.

The properties of a coagulating particle system are independent of the initial conditions (after sufficiently large reaction times). Another consequence of this is the evolution of a 'self-preserving' size distribution of the particle ensemble. 'Self-preserving' means that moment ratios of the size distribution $P(r)$ remain constant, e.g. $f_N \equiv (\mu_6/\mu_3^2) = 2.079$, where

$$\mu_i = \int_0^\infty r^i P(r)\,\mathrm{d}r$$

are the moments of the particle size distribution. When increasing the mean particle size by coagulation, the variances of soot particle size distributions, therefore, increase. The evolution of the moment ratio $f_N = \mu_6/\mu_3^2$ for different premixed, low-pressure flames is given in figure 3 (Bockhorn *et al.* 1987) and compared with modelling. For modelling, the appropriate term for the change of number densities in the different size classes by surface growth has been added to equation (1) (cf. Bockhorn *et al.* 1985, 1987). The figure demonstrates that the theoretical value of the moment ratio is quickly attained, and, from the good agreement between measured and simulated values, one can conclude that in flames the largest part of soot is formed by surface growth reactions (more than 95%), rather than by particle inception, and that particle inception occurs to a large extent only during the first few milliseconds of the process. A similar picture is obtained for diffusion flames (cf. figures 1 and 2) for the particle inception and coagulation region that are tied down by the mixing of fuel and oxidant. When crossing the oxidation zone in diffusion flames, particle size distributions change their shape, because smaller particles are consumed first by oxidation and the larger ones resist complete burn-out for longer.

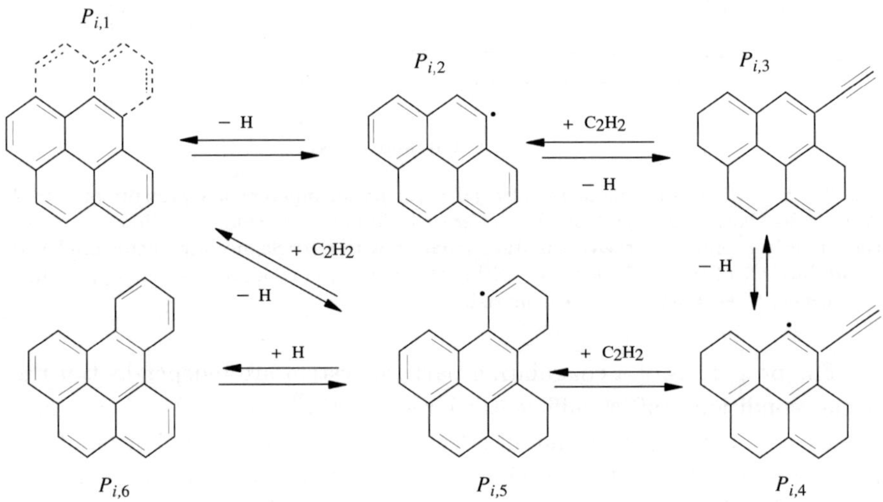

Fig. 4. HACA mechanism for the surface growth of soot (Frenklach & Wang 1994).

3.2. *Surface Growth Processes*

If the major proportion of soot is formed by surface growth reactions, the formation of the bulk of soot is well described via surface growth. Surface growth of soot has been interpreted in terms of the active site model (Woods & Hanyes 1994) as well as the acetylene decomposition model (Harris & Weiner 1990). These explanations provide a chemical interpretation of the appearance rates of soot via the decomposition of acetylene at active sites on the soot particle surface and via the deactivation or thermal stabilization of surface growth sites. The resulting rate expressions are of first order in the partial pressure of acetylene. A mechanistic interpretation of surface growth has been introduced by Frenklach (see, for example, Frenklach & Wang 1994). The basic idea of this approach, which has been adopted meanwhile in numerous works, is the transfer of the H abstraction carbon-addition (HACA) mechanism for the planar growth of polyacrylic aromatic hydrocarbon (PAH) to the heterogeneous surface growth of soot particles. The HACA mechanism provides a linear replication scheme for the planar growth of PAH by a two-step H-abstraction C_2H_2-addition (cf. figure 4). In this approach, PAH growth encompasses reactions between similar classes of particles so that the complex mixture may be described by lumped species classes rather than by single PAH species.

This approach is transferred to the surface growth of soot particles, which represent a weak-interaction cluster of PAH molecules (cf. table 1). In the reaction scheme given in table 1, $C_{soot}H$ represents an armchair site on the soot particle and C^*_{soot} the corresponding radical. S is the surface area of the soot particles and $\chi(C_{soot}H)$ is the number of CH sites per unit surface area accessible for surface growth. This formulation does not necessarily restrict surface growth to the outer surface of soot particles. If 'soot radicals' are replaced by the assumption of quasi-stationarity, the appearance rates of soot can be reproduced by

$$\frac{\mathrm{d}f_{\mathrm{V}}}{\mathrm{d}t} \propto \left(\frac{k_{1a,f}[\mathrm{H}]k_{4a}[\mathrm{O_2}]((k_{\mathrm{w}}[\mathrm{C_2H_2}]/k_{4a}[\mathrm{O_2}]) - 1)}{k_{1a,b}[\mathrm{H_2}]} - k_5[\mathrm{OH}] \right) \chi(\mathrm{C_{soot}H})S. \tag{7}$$

Further assumptions applied to derive equation (7) are that the growth mechanism is mainly initiated by H abstraction from the attack of H via reaction (1a) (see table 1), that the consumption of $C^*_{soot,i}$ is dominated by the reverse of reaction (1a) and that the rate coefficients for C_2H_2 addition ($k_{3a,f}$), C_2H_2 abstraction ($k_{3a,b}$) and ring closure ($k_{3b,f}$) are lumped into k_w.

Equation (7) reveals that only if $k_w[C_2H_2]/k_{4a}[O_2] \gg 1$ are the surface growth rates are of first order in the acetylene concentration. For this case, where surface growth reactions are dominating, the appearance rates are controlled by the ratio of $[H]$:$[H_2]$. The development of this concentration ratio and of the temperature in the soot-formation region is then responsible for the course of the appearance rates of soot. If oxidation is more important, i.e. if $k_w[C_2H_2]/k_{4a}[O_2] \approx 1$ and if $k_5[OH]$ is not negligible, the appearance rates follow a more complicated concentration dependence. For most flame conditions $k_w[C_2H_2]/k_{4a}[O_2] \gg 1$. Therefore, for the conditions in most flames the rate of acetylene addition dominates, so that the soot-formation rate is mostly of first order in the concentration of acetylene.

Table 1. Surface growth reactions for soot particles (after Schäfer *et al.* 1995).

(1a)	$C_{soot,i}H$	$+ H$	$\overset{k_{1a,s}}{\rightleftharpoons}$	$C^{*}_{soot,i}$	$+ H_2$
(1b)	$C_{soot,i}H$	$+ OH$	$\overset{k_{1b,s}}{\rightleftharpoons}$	$C^{*}_{soot,i}$	$+ H_2O$
(2)	$C^{*}_{soot,i}$	$+ H$	$\overset{k_{2,s}}{\rightarrow}$	$C_{soot,i}H$	
(3a)	$C^{*}_{soot,i}$	$+ C_2H_2$	$\overset{k_{3a,s}}{\rightleftharpoons}$	$C^{*}_{soot,i}C_2H_2$	
(3b)	$C^{*}_{soot,i}C_2H_2$		$\overset{k_{3b,s}}{\rightleftharpoons}$	$C_{soot,i+1}H$	$+ H$
(4a)	$C^{*}_{soot,i}$	$+ O_2$	$\overset{k_{4a,s}}{\rightarrow}$	$C^{*}_{soot,i-1}$	$+ 2CO$
(4b)	$C^{*}_{soot,i}C_2H_2$	$+ O_2$	$\overset{k_{4b,s}}{\rightarrow}$	$C^{*}_{soot,i}$	$+ 2CHO$
(5)	$C_{soot,i}H$	$+ OH$	$\overset{k_{5,s}}{\rightarrow}$	$C^{*}_{soot,i-1}$	$+ CH + CHO$

4. Modelling of Soot Formation and Oxidation

When modelling soot formation and oxidation employing the principal processes outlined above, soot formation and oxidation is embedded into the detailed description with the help of the gas-phase chemistry that provides H atom and acetylene concentrations, formation and growth of PAH, and formation and growth of soot particles by particle inception, surface growth and other collision processes. For numerical simulation the mass balances for all of the involved chemical species (about 250 chemical species and 1200 chemical reactions) and the enthalpy balance have to be solved.[a] The soot particle phase is treated as the balance equations of the moments of the size distribution (Frenklach & Wang 1994; Mauß *et al.* 1994; Mauß & Bockhorn 1995), which leads to a closed system of equations. Details of the modelling and numerical methods can be found in Frenklach & Wang (1994), Frenklach & Harris (1987) and Mauß & Bockhorn (1995).

Some results from the application of the above sketched modelling approach are plotted in figures 5 and 6. Figure 5 gives a comparison of the calculated and measured soot volume fractions for a premixed, flat acetylene–oxygen flame. In addition, the different contributions to the soot appearance rates—namely, particle inception, surface growth, PAH

[a]For the formulation of the corresponding balance equations, see, for example, Gardiner (1984) and Warnatz *et al.* (1996).

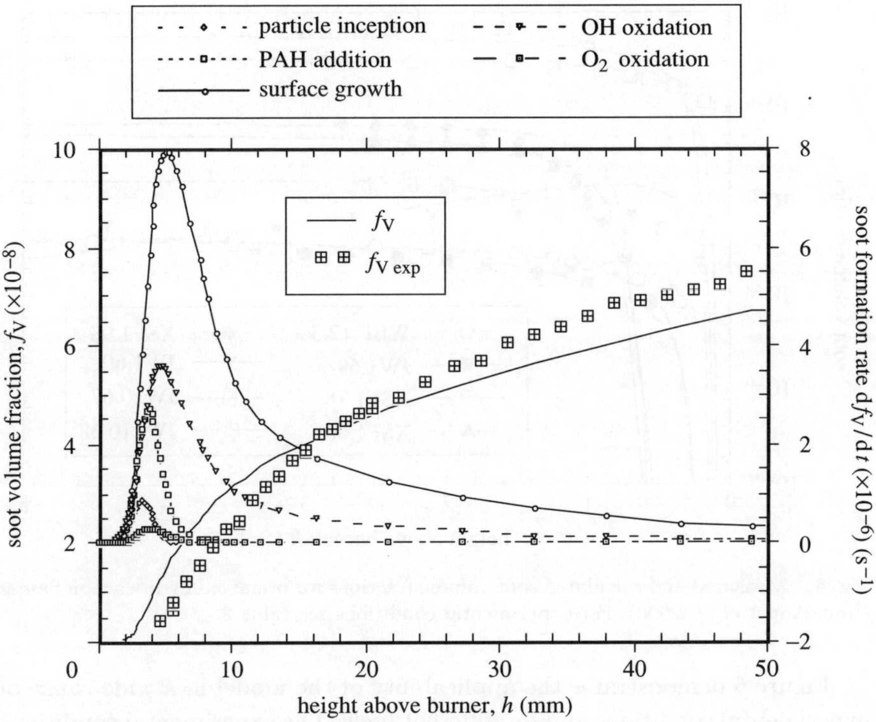

Fig. 5. Measured and calculated soot volume fractions for a premixed acetylene–oxygen–argon flame. Initial conditions: T = 298 K; P = 12 kPa; C:O ratio 1.25; and Ar 60%.

addition, as well as oxidation by oxygen and OH radicals—are indicated. The figure clearly demonstrates that

(i) the experimentally measured soot volume fractions can be predicted well for that flame,
(ii) the most important contribution to soot comes from surface growth, and
(iii) other processes contribute only a little.

Obviously, oxidation by OH takes place simultaneously during the entire soot-formation process, while oxidation by O_2 is of minor importance for the prevailing experimental conditions.

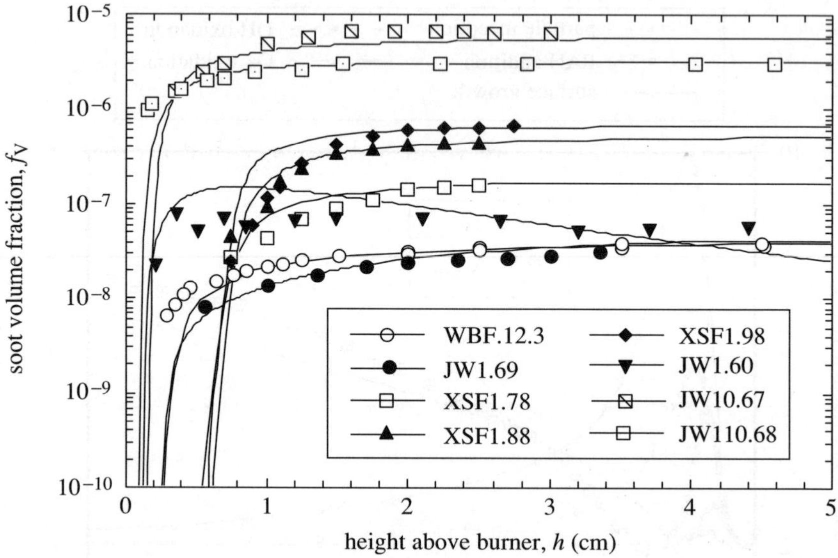

Fig. 6. Measured and calculated soot volume fractions for premixed hydrocarbon flames (from Appel *et al.* 2000). For experimental conditions see table 2.

Figure 6 demonstrates the applicability of the model in a wide range of experimental conditions and for different fuels. The experimental conditions of the flames, the experimentally measured soot volume fraction profiles of which are compared with the corresponding calculations in figure 6, are given in table 2. The model used for this comparison has been modified slightly compared with the concept outlined above (for details see Appel *et al.* (2000)). The figure reveals generally very good agreement between measurements and calculations. Note that the soot volume fractions in the considered flames vary by some orders of magnitude.

Finally, the simulated full particle size distribution is depicted in figure 7 for a premixed, low-pressure propane–oxygen flame from Bockhorn *et al.* (1983). The computations have been performed by solving the coagulation equations of the form

$$\frac{\mathrm{d}N_i}{\mathrm{d}t} = f(N_1, N_2, \ldots, N_m), \qquad i = 1, 2, \ldots, m. \tag{8}$$

In equation (8) N_i is the number density of particles, which are built up from i monomer units. The right-hand side of equation (8) contains all

Table 2. Experimental conditions for the flames given in figure 6.

flame	fuel	fuel (mol %)	O_2 (mol %)	N_2 or Ar (mol %)	v (cm s^{-1})	C/O	T_{max} (K)	P (bar)
WBF.12.3	C_2H_2	22.6	12.4	55.0 (Ar)	20.1	1.3	1992	0.12
JW1.69	C_2H_4	12.66	18.34	69.0 (N_2)	5.9	0.69	1711	1.013
XSF1.78	C_2H_4	14.0	18.0	68.0 (N_2)	4.0	0.78	2104	1.013
XSF1.88	C_2H_4	15.5	17.4	67.1 (N_2)	6.9	0.88	1957	1.013
XSF1.98	C_2H_4	17.0	17.4	65.6 (N_2)	5.3	0.98	1908	1.013
CS1.748	C_2H_6	24.12	32.25	43.36 (Ar)	7.0	0.748	1270	1.013
JW10.60	C_2H_4	11.2	18.65	70.15 (N_2)	6.0	0.60	2017	10
JW10.67	C_2H_4	12.38	18.40	69.22 (N_2)	3.0	0.673	1895	10
JW10.68	C_2H_4	12.5	18.40	69.1 (N_2)	6.0	0.68	1880	10

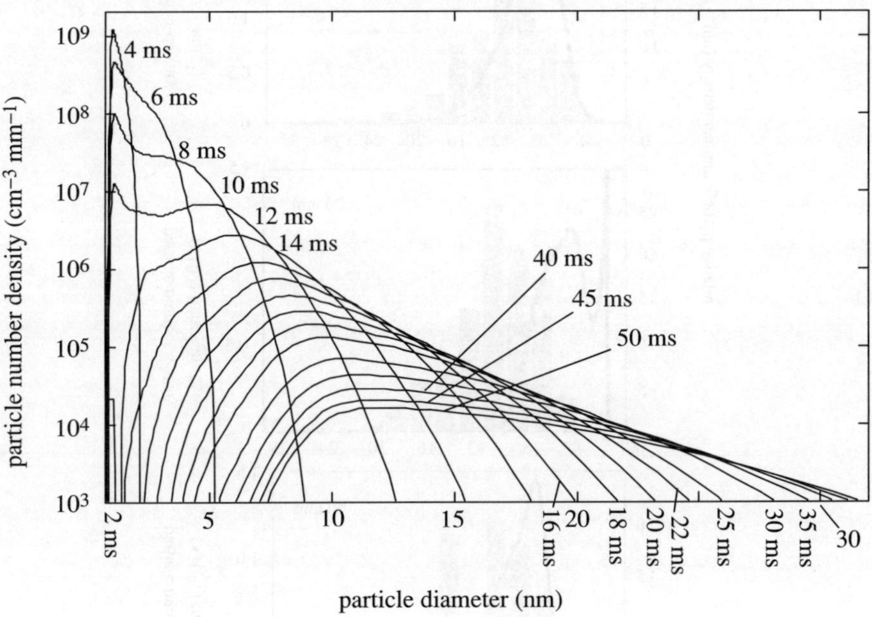

Fig. 7. Evolution of the soot particle size distributions for a premixed, low-pressure propane–oxygen flame (Bockhorn *et al.* 1983).

processes that contribute to the size evolution of the soot particle aerosol, namely particle inception, coagulation, surface growth and deposition of aromatic hydrocarbons at the surface of the soot particles. The algorithm

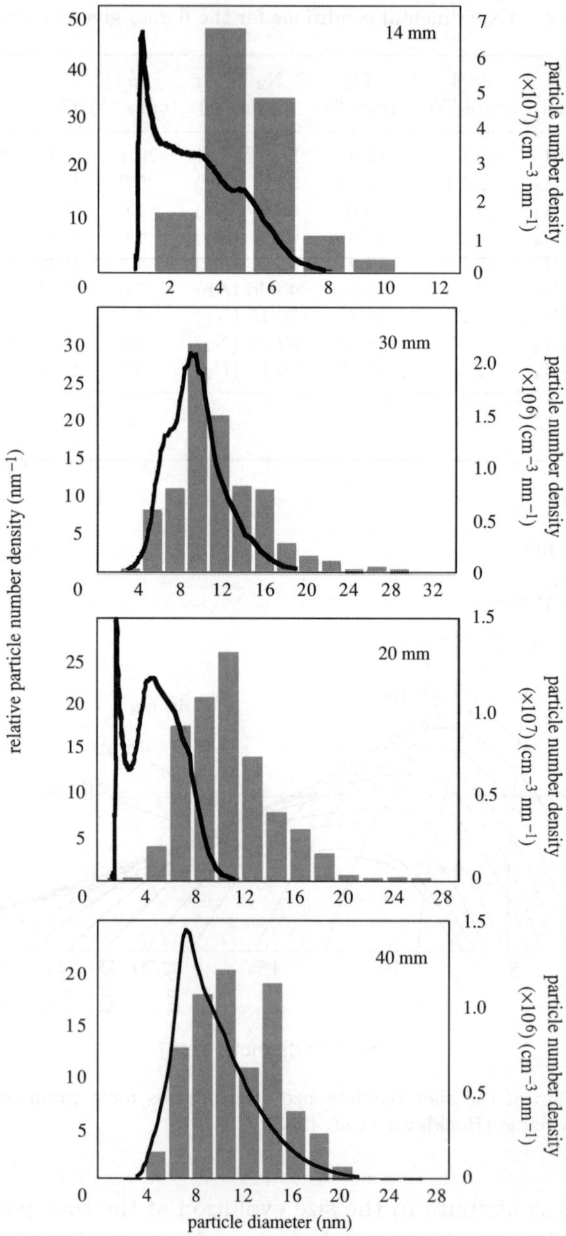

Fig. 8. For description see opposite.

used approximates the size distributions by a multilevel Galerkin h-p-method (Wulkow 1996). The calculation is post-processed after numerical simulation of the complete structure of the premixed flame (Appel & Bockhorn 2000). From the evolution of the soot particle size distributions it can be seen that particle inception and coagulation still dominate the particle dynamics of the system. Surface growth affects the distribution in the main reaction zone. In this region the main amount of soot is added to the solid phase by heterogeneous surface reactions with acetylene. The width of the distribution increases rapidly during this process (cf. figure 7). After the narrow surface growth zone, coagulation is again the dominant source for the evolution of the particle size distribution.

The rate of surface growth is proportional to a fraction of the surface area of the soot particles (see equation (7)). If the particles are assumed to be spherical, the surface area of a particle can be determined by

$$S_i = 4\pi \left(\frac{3m_1}{4\pi \rho_{\text{soot}}} \right)^{2/3} i^{2/3}, \tag{9}$$

and the diameter of the particles is given by

$$d_i = 2 \left(\frac{3m_1}{4\pi \rho_{\text{soot}}} \right)^{1/3} i^{1/3}. \tag{10}$$

The two major processes in soot particle dynamics, namely coagulation and surface growth, have a strong size dependence. For coagulation, high coagulation rates are obtained for the collisions of small particles with large ones, due to the size dependence of the mean velocity of the particles and the collisional cross-sections. This causes a fast consumption of small particles. The size dependence of the rates of surface growth reactions implies that the rate of acetylene addition is high for large particles. However, due to the size dependence of the ratio of surface area to diameter, in regions with many small particles the specific surface area is high and so the overall soot growth rate is also high.

Finally, in figure 8 a comparison of the computed particle size distributions with experimentally determined distributions (Bockhorn *et al.* 1988)

Figure 8. Comparison of experimental and calculated soot particle size distributions. The relative number of particles in the experiment is the number of particles of a certain size divided by the total number of examined particles. Due to the experimental technique used in Bockhorn *et al.* (1988), a comparison of absolute numbers is not possible.

is presented. The experimental results are obtained from molecular beam sampling of the sooting flame and TEM micrographs of the soot particles. The experimental values are relative particle number densities, which indicate the percentage of particles, which were found within a certain range of diameters. The lower detection limit of the experiments was reported to be *ca.* 1 nm. At low heights above the burner, the calculated soot particle size distributions show smaller particles than in the experimental observations. This can be explained by experimental uncertainties, because of the difficulty of extracting the particles from the main reaction zone of the flame. The agreement between the simulated and measured size distributions at 30 and 40 mm above the burner is excellent.

5. Summary

In the preceding sections, soot formation and oxidation have been analysed with respect to the most important processes, namely particle inception, coagulation and surface growth. Time-scales for surface growth, $\tau_{\text{soot}} < 10\,\text{ms}$, can be estimated for premixed flames that are comparable with the time-scales for soot formation in diffusion flames. Time-scales for coagulation are estimated as $\tau_{\text{coag}} \approx 10\,\text{ms}$, so that characteristic time-scales for soot formation and coagulation are similar and about one order of magnitude larger than the characteristic time-scales for combustion reactions. Obviously, the time-scales for soot formation and oxidation for the prevailing conditions in most flames are small compared with the time-scales of molecular transport, so that the soot appearance rates are adjusted to the local flame conditions.

In summary, one can conclude that the approach discussed above for a detailed chemistry soot model reproduces a number of phenomena in sooting premixed hydrocarbon flames, e.g.

(i) the dependence of surface growth and oxidation rates on chemical 'environment' of soot particles, and
(ii) the fraction of soot formed by particle inception and surface growth reactions and addition of PAH.

In addition, the model identifies time-scales of soot formation, soot oxidation and coagulation that are larger than the time-scales of combustion reactions, but are smaller than the time-scales of molecular or turbulent transport. Consequently, 'fast-chemistry' models can be employed when modelling soot formation and oxidation in more complex geometries.

However, the 'fine structure' of soot is not resolved by this approach, e.g.

(i) formation of charged soot particles,
(ii) formation of species such as fullerenes, charged and uncharged, and
(iii) formation of 'non-absorbing soot' that may play a role at incipient soot formation (d'Anna *et al.* 1994).

Furthermore, much more information about the kinetics of growth of PAH-like structures (growth reactions may be size dependent), the detailed processes occurring on the surface of soot particles, and, most importantly, the pressure dependence of all these processes is necessary to extend the validity of this approach to a wider range of experimental conditions.

References

Appel, J. & Bockhorn, H. 2000 *Chemosphere.* (In the press.)

Appel, J., Bockhorn, H. & Frenklach, M. 2000 *Combust. Flame* **121**, 122.

Bockhorn, H., Fetting, F. & Wenz, H. W. 1983 *Ber. Bunsenges. Phys. Chem.* **87**, 1067.

Bockhorn, H., Fetting, F., Heddrich, A. & Wannemacher, G. 1985 In *20th Symp. (Int.) on Combustion*, p. 979. Pittsburgh, PA: The Combustion Institute.

Bockhorn, H., Fetting, F., Heddrich, A. & Wannemacher, G. 1987 *Ber. Bunsenges. Phys. Chem.* **91**, 819.

Bockhorn, H., Fetting, F., Heddrich, A., Meyer, U. & Wannemacher, G. 1988 *J. Aerosol Sci.* **19**, 591.

d'Anna, A., d'Alessio, A. & Minutulo, P. 1994 In *Soot formation in combustion—mechanisms and models* (ed. H. Bockhorn), p. 83. Springer.

Frenklach, M. & Harris, S. J. 1987 *J. Colloid Interface Sci.* **118**, 252.

Frenklach, M. & Wang, H. 1994 In *Soot formation in combustion—mechanisms and models* (ed. H. Bockhorn), p. 165. Springer.

Gardiner, W. C. (ed.) 1984 *Combustion chemistry.* Springer

Geitlinger, H., Streibel, Th., Suntz, R. & Bockhorn, H. 1998 In *27th Symp. (Int.) on Combustion*, p. 1613. Pittsburgh, PA: The Combustion Institute.

Geitlinger, H., Streibel, Th., Suntz, R. & Bockhorn, H. 1999 *Combust. Sci. Technol.* **149**, 115.

Harris, S. J. & Weiner, A. 1990 *Combust. Sci. Technol.* **72**, 67.

Mauß, F. & Bockhorn, H. 1995 *Z. Phys. Chem.* **188**, 45.

Mauß, F., Trilken, B., Breitbach, H. & Peters, N. 1994 In *Soot formation in combustion—mechanisms and models* (ed. H. Bockhorn), p. 325. Springer.

Schäfer, Th., Mauß, F., Bockhorn, H. & Fetting, F. 1995 *Z. Naturf.* **50a**, 1009.

Warnatz, J., Maas, U. & Dibble, R. W. 1996 *Combustion.* Springer.

Woods, I. T. & Haynes, B. S. 1994 In *Soot formation in combustion—mechanisms and models* (ed. H. Bockhorn), p. 275. Springer.

Wulkow, M. 1996 *Macromol. Theory Simul.* **5**, 393.

CHAPTER 8

ULTRAFINE PARTICLES IN WORKPLACE ATMOSPHERES

James H. Vincent[1] and Charles F. Clement[2]
[1]*Department of Environmental Health Sciences, School of Public Health, University of Michigan, 109 S. Observatory, Ann Arbor, MI 48109, USA*
[2]*15 Witan Way, Wantage, Oxon OX12 9EU, UK*

Inhaled ultrafine particles are increasingly being recognized as a potential threat to health. Aerosols in workplace environments may come from a wide variety of sources, depending on the type of activity and processes taking place. Some activities and processes are acknowledged as being 'dusty', where aerosol is generated from the mechanical handling and attrition of solid or liquid material, and are not considered to be plausible sources of ultrafine particles. However, hot processes, involving the vaporization of material, and inevitable subsequent cooling, do have the potential to generate significant number concentrations of ultrafine particles. However, consideration of the physical conditions required for the generation of particles in the range below 100 nm suggests that those conditions are not easily met in workplaces. More generally, the conditions are such that particles grow out of this range, either by continuing condensation (as happens at high vapour concentrations) or by agglomeration between smaller particles (as happens at high number concentrations). Not much is known about ultrafine particles in actual workplaces, mainly because our view has been obscured for the past few decades by the fact that most occupational aerosol standards have been based on the mass concentration of airborne particulate matter. Now that a new awareness has set in, it is expected that new research will address the problem.

Most current aerosol standards are expressed in terms of the mass concentration of particulate matter conforming to a particle size fraction, where the latter is based on knowledge of how particle size relates to where particles deposit in the human respiratory tract and any subsequent effects. At present no such basis exists for ultrafine particles, but one is needed before progress can be achieved towards meaningful standards for occupational ultrafine aerosols. It is expected that, for ultrafine particles, such a standard may, in the future, be expressed in terms of the number concentration of particles less than a certain size, that size to be determined on the basis of the physical and chemical nature of the particle at that size, human physiology and toxicology.

Keywords: workplace; occupational exposure limits; ultrafine particles; occupational hygiene; nucleation; particle-size-selective criteria

1. Introduction

There was a time when, to occupational hygienists and physicians, 'fine' particles were considered to be those which, after inhalation, could penetrate down to the alveolar region of the lung—what is now referred to as *respirable aerosol*. More recently, interest has been drawn towards still finer particle fractions, not least because of their possible association with health effects specific to those sizes of particle. The term *ultrafine* is used broadly to refer to particles whose diameters are less than 100 nm (0.1 μm), and the use of that term seems to suggest that such particles represent the ultimate in 'smallness'. However, that definition, appearing to reflect a boundary that defines what is less harmful (above) and what is more harmful (below), is somewhat arbitrary and needs to be carefully examined.

Interest in still smaller particles in the range a long way below 100 nm is stimulated in large measure by some of the technology-oriented aerosol research which is beginning to yield insights into the physical nature of such particles and their reactivity and how such knowledge can lead to exciting new engineering applications (Pui *et al.* 1998). But there is growing awareness that the very same properties that lie behind such interest might also have a significant bearing on the possible health effects that might arise when ultrafine particles are inhaled by humans and come into contact with living tissue. All discussion about the reactivity of ultrafine particles in such scenarios must begin with the physical nature of the particles themselves. In his review, Preining (1998) described how, as particles become smaller and smaller, their physical state becomes distinctly different from that of larger particles from ultrafine upwards, where the number of molecules or atoms the particle contains becomes so few that a high proportion of them lies at the surface of the particle. For example, a 20 nm particle has 12% of its molecules at the surface. This fraction rises to 25% for a 10 nm particle. Here, therefore, even the concept of a 'surface' has limited meaning because the material structure of the particle can no longer be regarded as a continuum. Thus, according to Professor Preining, the material itself can no longer be thought of simply as 'solid' or 'liquid', and we enter the world of 'cluster physics', where such considerations may become very important in relation to how such particles interact with other molecules and particles. In turn this provides much food for thought in relation to how such particles might interact with biological organisms or cells, and, hence, on their potential toxicological effects.

2. Ultrafine Particles in Living and Working Environments

Much of the current interest in ultrafine particles has stemmed from growing concerns about the role of such fine aerosols in the observed increases in deaths from cardiovascular and respiratory causes (Seaton 1996). Although the net amount of urban particulate air pollution, as expressed in terms of airborne mass concentration, has decreased with reductions in particulate emissions from industry and power stations under clean air regulations that have been enacted in many countries, it is recognized that the nature of urban atmospheric aerosol is now distinctly different. Now, although mass concentrations have fallen, the number concentrations of very small particles, associated for example with emissions from internal combustion engines, has increased. For example, Kittelson (1998) has noted that, although the latest generations of diesel and spark engines are improved in terms of the mass of particulate emissions, the *number* of nano-sized particles emitted has increased sharply. This may also be true for emissions from aircraft engines.

So this changing nature of urban atmospheric aerosol, coupled with the public health epidemiology, is driving much of the discussion on health effects associated with atmospheric aerosol fractions finer than the ones that have been considered in the past. It is also stimulating discussion about workplace aerosols, and an examination of the extent to which occupational health might be similarly impacted by workers' exposures to very small particles. Here it is important to note, firstly, that the occupational environment in general represents a major consideration in public health, since a very high proportion of the population goes to work and, in the process, may be exposed, largely involuntarily, to occupation-related hazards. It was recognized very early on in the 1900s that aerosols in workplace atmospheres contributed to major disease epidemics affecting whole communities (e.g. mining, agriculture, etc.), and so aerosol science became a very important part of the growing field of occupational hygiene science (Walton & Vincent 1998).

Workplace aerosols of interest to occupational hygienists are those that are released into the working environment. Depending on local factors such as workplace layout, the processes going on there and the ventilation, the aerosols to which workers are exposed are likely to be very 'fresh', usually no more than a few minutes old. The range of aerosol types (i.e. chemical composition) and particle sizes can range very widely, and may be roughly categorized as follows.

(i) Dust and sprays from mechanical processes (e.g. mining, textiles, chemical manufacture and transportation, agriculture, etc.), with particle sizes mainly greater than 1 μm ranging all the way up to 100 μm and (even) beyond.
(ii) Fumes from hot processes (e.g. smelting and refining of metals, welding, etc.), with particle sizes usually not much greater than about 1 μm and going down to a few nanometres (i.e. the size of primary particles produced by nucleation).
(iii) Fumes from combustion processes (e.g. transportation, carbon black manufacture, etc.), usually associated with incomplete combustion, again with particle sizes not much greater than about 1 μm but going down to a few nanometres.
(iv) Bioaerosols (e.g. agriculture, biotechnology, etc.), where some particles (e.g. viruses, endotoxins, etc.) may be as small as a few tens of nanometres.

Ultrafine aerosols arising from mechanical processes (e.g. the breaking or fracture of solid or liquid material) are generally unlikely. One possible exception is in the production of very small solid particles from sprays of dilute aqueous solutions like those found in some humidification systems (Vincent 1970). Greater potential for ultrafine particle production exists where there are hot processes or combustion, and where very small particles may be created by the physical process of nucleation from the vapour phase. Such nucleation may take the form of homogeneous nucleation and condensation, by which particles are formed initially by the spontaneous condensation of molecules from the gas to the liquid phase; or heterogeneous nucleation, by which particles are formed by the condensation of molecules onto pre-existing very small nuclei.

3. Ultrafine Particles in Occupational Hygiene

The discipline of 'occupational hygiene' is defined as the anticipation, recognition, evaluation and control of workplace hazards. Yet ultrafine particles have received little attention, either by scholars or practitioners in the field. This lack can be understood from the history of how criteria have evolved for occupational aerosol exposure assessment and standards (Walton & Vincent 1998). In the early 1900s, when it became widely recognized that workplace aerosols were potentially injurious to health, airborne particulate matter was assessed by collecting samples in very rudimentary sampling systems (e.g. with cotton wool filters) and weighing what was collected, from which estimates of mass concentration of the aerosol in question could be made. At that time, therefore, exposure was thought of in terms of particulate mass concentration, and environments were regulated and controlled on that basis. By the 1920s, however, it became recognized that particles below a certain size were the ones primarily associated with mining-related lung disease (e.g. pneumoconiosis) (McCrea 1913). So for many years it was customary in the countries then prominent in occupational hygiene (e.g. the UK, USA, South Africa, Canada, Australia, etc.) to assess occupational aerosol exposures in terms of the number concentrations of particles with diameter less than 5 μm, as evaluated by optical microscopy. Standards for occupational aerosol exposures were correspondingly expressed in this way. It was only in the 1950s that this convention was revisited, leading to the concept that sampling of such aerosols should be carried out in a way that reflects how inhaled particles are aerodynamically transported inside the human respiratory tract and deposited in the various regions of the lung. This led in turn to the development of particle size-selective sampling devices that reflected much more closely the true nature of human exposure and could provide mass concentrations of aerosol in an appropriate particle size range. Epidemiological studies subsequently showed that exposure, expressed in terms of the mass concentration in health-relevant particle size fractions, correlated much better with the prevalence of many lung diseases than the previous index based on particle number concentration. It has remained that way to this day, although the system of conventions for particle size-selective sampling and standards has more recently been expanded (see below). So almost all occupational exposure limits (OELs) for aerosols, with the exception of asbestos and radioactive aerosols, are expressed in terms of mass concentration.

The effect of this history, especially over the past 40 years, has been to focus attention with respect to occupational aerosols on mass concentration as the primary, and most relevant, index of exposure. For ultrafine particles, even though in many instances they might represent the vast majority of the number concentrations of particles in many workplace aerosols, they nearly always represent an infinitesimal proportion of the mass concentration. As a result, ultrafine particles have never appeared in our consciousness. In turn, there is very little to report on how they appear in the overall characterization of workplace aerosols. There are just a few exceptions.

There has long been concern about diesel particulate arising from the use of diesel-powered vehicles in confined industrial spaces (e.g. mines). Here, the numbers of exposed workers around the world are very large. Knight *et al.* (1983), and others since, have examined the particle size distribution of the aerosol in such environments, and have shown significant number concentrations of particles in the range below 100 nm. More recently, Professor Kulmala and his co-workers in Finland (Hämeri *et al.* 1996) reported a study of aerosol generation during the process of ski hot-waxing using fluor powder, where (perhaps surprisingly) over 10 000 people are exposed in Finland alone. They demonstrated that, at certain stages of the hot-waxing process, significant number concentrations of particles are created by nucleation and may occur in the particle size range below 100 nm.

The question of ultrafine particles has been raised in relation to occupational metals exposure, although the available information is somewhat anecdotal. Sandström *et al.* (1989) described incidents that happened in rapid succession early in 1943 at a Swedish smelter, where 13 workers were taken ill. One died of acute respiratory distress syndrome (ARDS), supposedly after exposure to fine particulate nickel; two others died of ARDS supposedly after exposure to fine zinc chloride smoke. The most well-documented, and somewhat similar, case was reported by Rendall *et al.* (1994), describing an occurrence at a South African plant where the arc spraying of nickel was being used to coat metal components. In the process, an arc was created between two nickel wires which were being fed continuously into the jet of a compressed air-driven spray gun. The fine nickel particles were carried away very rapidly by the air jet from the immediate region of the arc and then deposited on the component in question. The first time that the process was used, the operator was soon taken ill, and died a few days later of ARDS. Urine samples, taken before he died, showed a nickel concentration of about 700 $\mu g\, l^{-1}$, a level which was regarded as 'excessively' high. Following the death of the worker, an investigation was carried out that included a simulation of the process he was operating when he became ill (although, of course, the operators this time wore rigorous personal protective equipment). Electron microscope analysis of aerosol samples revealed very large number populations of nickel particles with diameter less than 50 nm.

4. The Physical Scenario

In reality, quite stringent physical conditions need to be met if truly 'ultrafine' particles are to be generated, most notably to ensure that

(a) the primary particles formed by nucleation do not grow to sizes outside the ultrafine range, and
(b) the primary particles undergo minimum coagulation.

The conditions 'favourable' for the generation of ultrafine particles may be summarized as follows:

(i) there must be vaporizable material present;
(ii) the temperature or heat transfer should be sufficient to produce just enough vapour to condense as an independent aerosol, but producing a low number concentration of primary particles;
(iii) there should be rapid cooling of the aerosol that has been formed and a large temperature gradient.

The first of these is obvious. But the other two are more subtle. Below a certain amount of vaporization (as reflected in vapour concentration), the vapour that is present will not nucleate and condense independently, but, rather, will condense on pre-existing and (usually) non-ultrafine particles. This is the case, for example, with involatile radon daughters following the radioactive decay of radon gas. It has recently been found that a condition for the nucleation of initially ultrafine aerosol in Finnish forests is that the concentration of condensing vapour is high enough (Clement *et al.* 2000). This concentration is given by its production rate from solar radiation divided by the removal rate onto existing aerosol. So the observation establishes the need to produce a large enough vapour concentration in the atmosphere, especially when existing aerosol is present. The low concentration requirement ensures that subsequent particle growth by coagulation is minimized. For the third condition, the rapid cooling greatly enhances the probability of nucleation and independent particle production, ensuring that the aerosol is 'frozen' and does not continue to evolve such that the particles continue to increase in size by ongoing condensation.

A typical example where such conditions might be met in an industrial setting might be the heating of a small amount of material by, say, a laser spot. The temperature is very high, but only a very small amount of material is available to be vaporized. So the capacity for growth of the original nucleated particles is limited. Also the number concentration of the newly formed particles is similarly limited at the outset. Further, the very sharp temperature gradient allows the aerosol to be 'frozen' early on in its evolution. With this in mind, therefore, localized heating of volatile material by lasers or arcs might be seen as highly conducive to the production of ultrafine particles, and, hence, might be seen to be potentially the most dangerous. Although the events and subsequent investigation reported by Rendall *et al.* (1994) represent what appears to be an isolated case, and are not documented in a way that allows close inspection in relation to what is currently known about the potential for the generation of ultrafine particles and how they might lead to ill health, it is interesting to note that the scenario is strikingly similar to that described here as being 'favourable' for ultrafine aerosol production.

In general, however, it might be said that the stringent conditions for ultrafine particle formation might be difficult to meet in most of the workplace situations characterized by hot processes. Such conditions are not satisfied, for example, in large-scale accident scenarios associated with overheating in the nuclear industry, where typical particle sizes are of the order of 1 μm or larger (Schikarksi 1988). It is also likely that they will not be met in most large-scale metals smelting and refining processes, one area where concern about ultrafine particles has already been expressed.

5. Health-Related Standards for Aerosols

An ideal standard for any airborne contaminant should include (Vincent 1995):

(i) criteria for exposure, which identify the agent and its specific physical, chemical and/or biological properties relevant to a specific adverse health outcome;
(ii) reference to monitoring instruments and analytical methods with performance characteristics matching the defined exposure criteria;
(iii) reference to a monitoring strategy which sets out to assess exposure in a manner which is representative of the temporal histories and variability of workers' exposures; and
(iv) a health-based OEL derived from considerations of the effects of exposure at various levels, known incidences of the prevalence of the health outcome in question, and what might be an 'acceptable' level of risk.

When the standard is set out in this way, it becomes obvious that, strictly, an OEL cannot be assigned until consideration has been given to the other items listed. This is particularly clear for aerosol exposures, where the evaluation of the health effect includes not only consideration of the intrinsic toxic properties of the particulate material in question but also its physical properties (i.e. particle size, shape, density), which govern how the particles are transported within the respiratory tract and their subsequent fate after deposition. For such exposures, research has delivered a good understanding of the physical nature of how airborne particles are inhaled and, based on knowledge of nasopharynx and lung physiology, how they penetrate to and are deposited in the various parts of the respiratory tract.

From such understanding, particle size-selective criteria have been proposed, identifying:

(i) inhalable aerosol, the fraction of particles that may be inhaled through the nose and mouth during breathing;
(ii) thoracic aerosol, the fraction that may penetrate beyond the larynx and so enter the lung; and
(iii) respirable aerosol, the fraction that may penetrate beyond the airways of the lung and enter the gas-exchange region.

Each of these fractions is described as a curve that expresses the probability of entry or penetration as a function of particle aerodynamic diameter (which mainly governs the physics of particle motion that influences these fractions). In turn, such curves become the yardsticks against which to relate the performances of sampling instruments that will provide scientifically defensible measures of exposure in a way that relates directly to the health effect of interest. For example, respirable aerosol would be chosen as the metric most appropriate to pneumoconiosis, a disease of the gas-exchange region of the lung; while inhalable aerosol would be chosen as the metric most relevant to health effects associated with airborne lead, since all the lead that is inhaled may contribute to lead-related ill health. In turn, therefore, such fractions become sensible bases of standards. Indeed, they underpin the new generation of occupational aerosol standards that is emerging, and on which—over the 20 or so years of their development—a considerable degree of international harmony has been achieved, involving the International Standards Organization (1992), the Comité Européen de Normalization (1992) and the American Conference of Governmental Industrial Hygienists (2000, see also editions from 1993 onwards). Furthermore, of these, the ISO standard is aimed also at aerosol exposures in the ambient atmosphere, and so goes beyond just workplaces.

For ultrafine particles, no such criterion has yet been established upon which to base a standard for the protection of workers. Although 'less than 100 nm' has been identified as a working definition of what is currently referred to as an 'ultrafine' particle, there is at present no physiological or toxicological basis by which to establish a criterion for exposure assessment that, in turn, following the framework outlined above, can form the basis of a scientific standard. Some basic biological research has been conducted to ascertain the specific characteristics of very fine particles, in the size range from about 10 nm (0.01 μm) up to about 500 nm (0.5 μm), in relation to what makes very fine particles more toxic to biological systems than larger particles of the same material (see, for example, Oberdörster *et al.* 1992, 1995; Donaldson *et al.* 1998). The results suggest that the ultrafine particles are much more toxic, although their interpretation is complicated by the fact that, for given particulate mass, particle numbers and particulate surface area increase sharply in this range.

Seaton and co-workers (Seaton *et al.* 1995; Seaton 1996) have considered the possible causative factors linking exposure and the observed mortality in the general population. They proposed a hypothesis in which exposure to ultrafine particles in the size range around 50 nm:

> ...characteristic of air pollution (may) provoke alveolar inflammation leading to acute changes in blood coaguability... [and] result in an increase in the exposed population's susceptibility to acute episodes of cardiovascular disease; the most susceptible suffer the most. This hypothesis, being based on the number, composition and size—rather than on the mass—of particles accounts for the observed epidemiological relations.

However, that there is at present no specific health-related particle size-selective criterion for ultrafine particles reflects the fact that there is insufficient knowledge about cause and effect to permit such a criterion. But, because of the interest stimulated by concerns about health effects in populations exposed to such fine aerosols in workplaces, as well as in the ambient atmospheric environment, it is likely that new research in the years ahead will address the problem. Nevertheless, we may for the time being speculate that a future standard for very, very small particles will be expressed in terms of the concentration of particles less than a certain diameter *according to some metric other than mass* (e.g. number or surface area concentration). That diameter may be in the range of a few nanometres up to a few tens of nanometres, and will be based on new knowledge about the nature of such small particles such that they produce a different—and enhanced—response in exposed people. The concentration that will become the OEL will be determined on the basis of the physical and chemical nature of the particle at that size, human physiology and toxicology. When that criterion eventually emerges, appropriate instrumentation for occupational exposure assessment will be needed. Fortunately, in this regard, aerosol scientists and engineers have already developed a range of sampling and analytical principles and devices that can be used for this purpose (Pui 1996).

6. Conclusions

Ultrafine particles are increasingly being recognized as a potential threat to health. Aerosols in workplace environments may come from a wide variety of sources, depending on the type of activity and processes taking place. Some activities and processes are acknowledged as being 'dusty', where aerosol is generated from the mechanical handling and attrition of solid or liquid material, and are not considered to be plausible sources of ultrafine particles. But hot processes, involving the vaporization of material, and inevitable subsequent cooling, do have the potential to generate significant number concentrations of ultrafine particles. However, consideration of the conditions required for the generation of large concentrations of particles small enough to be regarded as 'ultrafine' suggests that those conditions are not easily met in workplaces. However, very little 'hard' information is available about ultrafine particles in workplaces, mainly because our search for knowledge has been obscured for the past few decades by the focusing of most occupational aerosol standards on the mass concentration of airborne particulate matter. This is clearly an area where further occupational hygiene research is called for.

As mentioned, most current aerosol standards are expressed in terms of the mass concentration of particulate matter conforming to a particle size fraction, where the latter is based on knowledge of how particle size relates to where particles deposit in the human respiratory tract and any subsequent effects. At present, no such basis exists for ultrafine particles, but is needed before progress can be achieved towards meaningful standards.

References

American Conference of Governmental Industrial Hygienists 2000 *Threshold limit values for chemical substances and physical agents and biological exposure indices.* Cincinnati, OH: ACGIH.

Clement, C. F., Pirjola, L., dal Maso, M., Mäkelä, J. M. & Kulmala, M. 2000 Analysis of particle formation bursts observed in Finland. *J. Aerosol Sci.* (Submitted.)

Comité Européen de Normalization 1992 Workplace atmospheres: size fraction definitions for measurement of airborne particles in the workplace. (Standard EN 481.) Brussels: CEN.

Donaldson, K., Li, X.-Y. & MacNee, W. 1998 Ultrafine (nanometre) particle mediated lung injury. *J. Aerosol Sci.* **29**, 553–560.

Hämeri, K., Aalto, P., Kulmala, M., Sammaljärvi, E., Spring, E. & Pihkala, P. 1996 Formation of respirable particles during ski waxing. *J. Aerosol Sci.* **27**, 339–344.

International Standards Organization (ISO) 1992 *Air quality particle size fraction definitions for health-related sampling.* Geneva: International Standards Organization (ISO CD7708).

Kittelson, D. B. 1998 Engines and nanoparticles: a review. *J. Aerosol Sci.* **29**, 575–588.

Knight, G., Bigu, J., Mogan, P. & Stewart, D. B. 1983 Size distribution of airborne dust in mines. In *Aerosols in the mining and industrial work environments* (ed. V. A. Marple & B. Y. H. Liu). Ann Arbor, MI: Ann Arbor Science.

McCrea, J. 1913 *The ash of silicotic lungs.* Johannesburg: South African Institute of Medical Research.

Oberdörster, G., Ferin. J. & Gelein, R. 1992 Role of the alveolar macrophage in lung injury: studies with ultrafine particles, *Environ. Health Perspect.* **97**, 193–199.

Oberdörster, G., Gelein, R. & Ferin, J. 1995 Association of particulate air pollution and acute mortality: involvement of ultrafine particles, *Inhalation Toxicol.* **71**, 111–124.

Preining, O. 1998 The physical nature of very, very small particles and its impact on their behaviour. *J. Aerosol Sci.* **29**, 481–495.

Pui, D. Y. H. 1996 Direct-reading instrumentation for workplace aerosol measurements—a review. *Analyst* **121**, 1215–1224.

Pui, D. Y. H., Brock, J. R. & Chen, D.-R. 1998 Nanoparticles: a new frontier for aerosol research. Guest editorial. *J. Aerosol Sci.* **29**, vi.

Rendall, R. E. G., Phillips, J. I. & Renton, K. A. 1994 Death following exposure to fine particulate nickel from a metal arc process. *Ann. Occupational Hygiene* **38**, 921–930.

Sandström, A. I. M., Wall, S. G. I. & Taube, A. 1989 Cancer incidence and mortality among Swedish smelter workers. *Brit. J. Ind. Med.* **46**, 82–89.

Schikarski, W. O. 1988 Nuclear aerosol science. *Nuclear Technol.* **81**, 137–306.

Seaton, A. 1996 Particles in the air: the enigma of urban air pollution, *J. R. Soc. Med.* **89**, 604–607.

Seaton, A., MacNee, W. & Donaldson, K. 1995 Particulate air pollution and acute health effects. *Lancet* **345**, 176–178.

Vincent, J. H. 1970 Ultrasonic humidifier. Internal technical report no. 400–70-1. Piscatway, NJ: American Standard Corporation.

Vincent, J. H. 1995 *Aerosol science for industrial hygienists.* Oxford: Pergamon.

Walton, W. H. & Vincent, J. H. 1998 Aerosol instrumentation in occupational hygiene: an historical perspective. *Aerosol Sci. Technol.* **28**, 417–438.

Discussion

R. Agius (*University of Edinburgh, UK*). I have a comment regarding nickel toxicity. In the case reported by Rendall *et al.* (1994), the presence of nickel in the urine provides only limited evidence of the time-scale and 'species' of the exposure. Biological monitoring studies have shown that at an individual level the short-term correlation between urinary levels of nickel and exposure is poor, and inhalation of a number of species of nickel or its compounds (as occurs, for example, in welding) can be responsible for elevated urinary nickel. The Australian aqueous refining process which has been remarked upon as a possible source of occupational exposure to particulate elemental nickel is unlikely to have generated ultrafine particulate from an aqueous precipitate. Theoretically, in nickel refining, the likeliest exposure to ultrafine nickel would have arisen in the Mond process after decomposition of nickel tetracarbonyl. This process has resulted in many fatalities. However, the intrinsic toxicity of the tetracarbonyl renders it almost impossible to determine whether or not the ultrafine elemental nickel per se contributed to the toxicity.

CHAPTER 9

THE SURFACE ACTIVITY OF ULTRAFINE PARTICLES

D. A. Jefferson

University Chemical Laboratories, Lensfield Road, Cambridge CB2 1EW, UK

Within the last 20 years, advances in characterization methods, particularly in the field of high-resolution electron microscopy, have made it possible to probe the surface and internal structure of sub-100 nm particles, or nanoparticles. Such studies have indicated conclusively that surface-energy considerations in metal nanoparticles cause these particles to adopt structures which only approximate to close packing but are terminated by close-packed faces. In oxides, where stoichiometry must be maintained, the adoption of low-index crystallographic faces almost invariably necessitates the introduction of cation or anion vacancies, and both have been observed. In such cases, the structure at the edges of the particles differs greatly from that of bulk phases, and it seems highly probable that the physical and chemical properties of these particles are also different. In certain cases it appears that new structural types, found only in nanoparticulate form, may exist. The significance of these findings, particularly as regards their relevance to particulate pollutants in the atmosphere, may be of great interest.

Keywords: particle size; nanoparticle chemistry; particulate pollutants; nanoparticle structure

1. Introduction

The toxicological effects of ultrafine particles present in the atmosphere depend on many factors, both physical and chemical. Possibly the most important factor, apart from their concentration, is the size of the particles, as this determines where they are deposited in the respiratory tract and, hence, the manner in which they may interact with living tissue. Particle size is also crucial, as this is the main factor governing their removal from the atmosphere by filtration. However, the chemistry of such nanoparticles is also a factor to be considered. Although it has been shown that relatively inert materials have enhanced toxicological effects when present as sub-100 nm particles, implying that physical size is the main factor determining interaction with lung tissue (Amdur *et al.* 1988; Gilmour *et al.* 1997), it has also been suggested that the greater specific surface of such particles may raise kinetically or thermodynamically unfavourable reactions to significant levels. However, the relevant chemical properties of these particles have always been assumed to be those of the bulk material: for this reason, chemical action by sub-100 nm particles of inert oxides such as titania and alumina has generally been disregarded.

The high surface area—or, more correctly, the enhanced ratio of surface to bulk atoms—is the dominant factor in all nanoparticle properties. This has long been realized in the field of heterogeneous catalysis, where economic factors frequently necessitate the use of catalytic species in a finely divided particulate form. The importance of surface effects can be readily seen when metal nanoparticles are exposed to an environment where stable surface adsorbate layers are formed: formation of the adsorbate can frequently provide enough energy for the complete reconstruction of the particles into a new morphology (Harris 1986; Jefferson & Harris 1988; Gribelyuk *et al.* 1994). As part of the search for increased catalyst efficiency, many structural investigations of nanoparticulate species have been performed, and these indicate that the small size and high proportion of surface atoms may have a profound effect on the internal structures of such particles, so much so that nanoparticles may differ appreciably from bulk material. The toxicological consequences of these differences may well be significant.

2. Metal Nanoparticles

The simplest examples come from metals with the face-centred cubic structure. In order to minimize surface energy, it can be expected that nanoparticles will try to assume a shape which approximates to a sphere, as shown in figure 1*a*. In a close-packed metal structure, this is relatively easy, but for small particles the surface is marked by numerous steps, where the local coordination number is low and the energy is markedly raised from that of a bulk atom, which is 12-coordinated. A more stable surface can be created by using low-index crystallographic planes, such as $\{111\}$ and

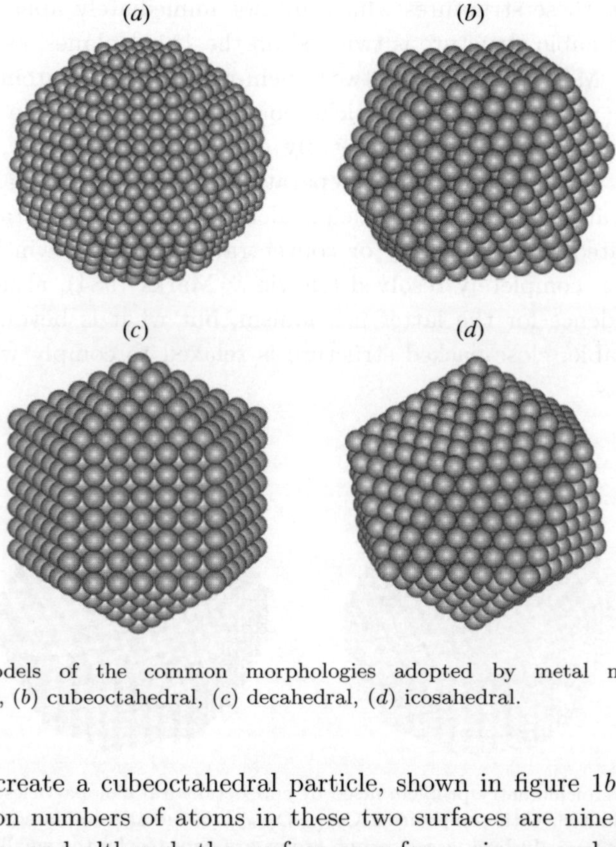

Fig. 1. Models of the common morphologies adopted by metal nanoparticles: (*a*) spherical, (*b*) cubeoctahedral, (*c*) decahedral, (*d*) icosahedral.

{100}, to create a cubeoctahedral particle, shown in figure 1*b*. Here, the coordination numbers of atoms in these two surfaces are nine and eight, respectively, and although the surface area for a given number of atoms is increased, the overall energy is greatly reduced. Such faceted particles have been observed using high-resolution electron microscopy (Heinemann *et al.* 1979), but other more complex shapes have also been noted, particularly the so-called multiply twinned particles (MTPs) (Marks & Smith 1981, 1983). These, which comprise either decahedral or icosahedral particles (figure 1*c*, *d*) increase the ratio of the higher-coordinated {111} surfaces relative to the {100} type by twinning the structure such that each particle is made up of a number of smaller regions, numbering five in the decahedron and 20 in an icosahedron, the latter having only {111} surfaces.

It is relatively easy to show that decahedral and icosahedral configurations are much more stable than a simple spherical particle in both metal and non-metallic systems (Uppenbrink *et al.* 1992). However, there are subtleties in these structures which are not immediately apparent. If the face-centred cubic structure is twinned on the {111} planes, as must happen in these MTPs, the angle between twin-related rows of atoms is 70.52°, whereas the geometry of the particle requires a 72° angle. The MTPs are, therefore, not compatible with a truly close-packed structure, and some strain must exist, as atom–atom separations parallel to the particle faces must be greater than their radial equivalents. Whether this strain is either accommodated homogeneously or concentrated near the twin boundaries has not been completely resolved (Howie & Marks 1984), although there is more evidence for the latter mechanism, but what is beyond doubt is that the stable, close-packed structure is relaxed to comply with surface requirements.

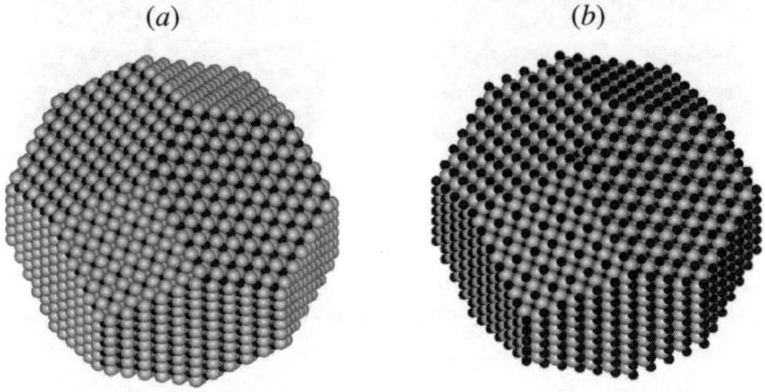

Fig. 2. The stoichiometry problem faced by cubeoctahedral nanoparticles of CeO_2: (*a*) oxygen terminated, with composition $Ce_{2735}O_{5688}$; (*b*) metal terminated, with composition $Ce_{2735}O_{4600}$. In both cases cerium atoms are depicted by the small dark circles, with oxygen being the larger, lighter circles.

3. Oxide Nanoparticles with Anion Vacancies

Many oxides are based upon approximately close-packed arrangements of oxygen anions, and similar behaviour might be expected in oxide nanoparticles, but when more than one type of atom is involved, a more important consideration becomes paramount, namely that of maintaining the oxide stoichiometry. A very simple example is given by the case of ceria, CeO_2, which is of considerable commercial importance as a catalyst support and oxygen storage medium, and is particularly easy to prepare in sub-10 nm form (Brinker & Scherer 1990). Ceria has the fluorite structure, and in nanoparticle form adopts a cubeoctahedral morphology, showing {111} and {100} surfaces, presumably to minimize surface-energy effects. However, the atomic arrangement on crystallographic planes with either of these sets of indices alternates between metal atoms and oxygen atoms, but no single plane contains both. A nanoparticle of ceria is therefore faced with an impossible dilemma, in that if it is terminated with planes of metals atoms, there is an excess of metal in the particle, but if oxygen termination is selected, there is an equal excess of oxygen. These two arrangements are shown schematically in figure 2.

The simplest way to overcome the problem of the surface excess of metal or oxygen atoms is to introduce vacancies of the opposite species within the bulk. High-resolution electron micrographs of ceria (figure 3*a*) do not appear to indicate any significant metal vacancies, as a regular array of metal atoms is clearly visible, but cerium does form a series of reduced oxides, which are based on regular arrangements of oxygen vacancies within the fluorite structure (Brauer 1964; Bevan 1973), so the presence of the latter is most likely, inferring a metal atom termination of the particles, although this cannot be substantiated by high-resolution electron microscopic studies, as the scattering from the oxygen is minimal at current resolution limits. Consequently, no conclusions can be made concerning the location of such vacancies, if they are present.

Ceria is known to form solid solutions with many other metal oxides, particularly if they possess similar structures, and an excellent example of this is given by the solid solution with lanthana, La_2O_3. Lanthana normally adopts a hexagonal structure, but a cubic form is also known (Gschneider & Eyring 1979) and is based on an oxygen-deficient fluorite arrangement. Depending on the temperature of preparation, solid solubility of lanthana in ceria may extend up to more than 50% (Bevan 1955; Morris *et al.* 1993), and similar behaviour has also been found in mixed nanoparticles prepared by sol-gel methods (Tilley 1997). In the latter case, however, the limits of solid solubility are extended considerably, with two-thirds replacement of cerium by lanthanum being confirmed by microanalysis, although electron microscopic images indicate apparently normal ceria particles (figure 3*b*). Analysis of the surface composition of a specimen of uniformly sized particles using X-ray photoelectron spectroscopy, however, indicates a large preponderance of lanthanum atoms at the surface, although a high oxygen signal suggests that the surfaces are by no means metal terminated. These

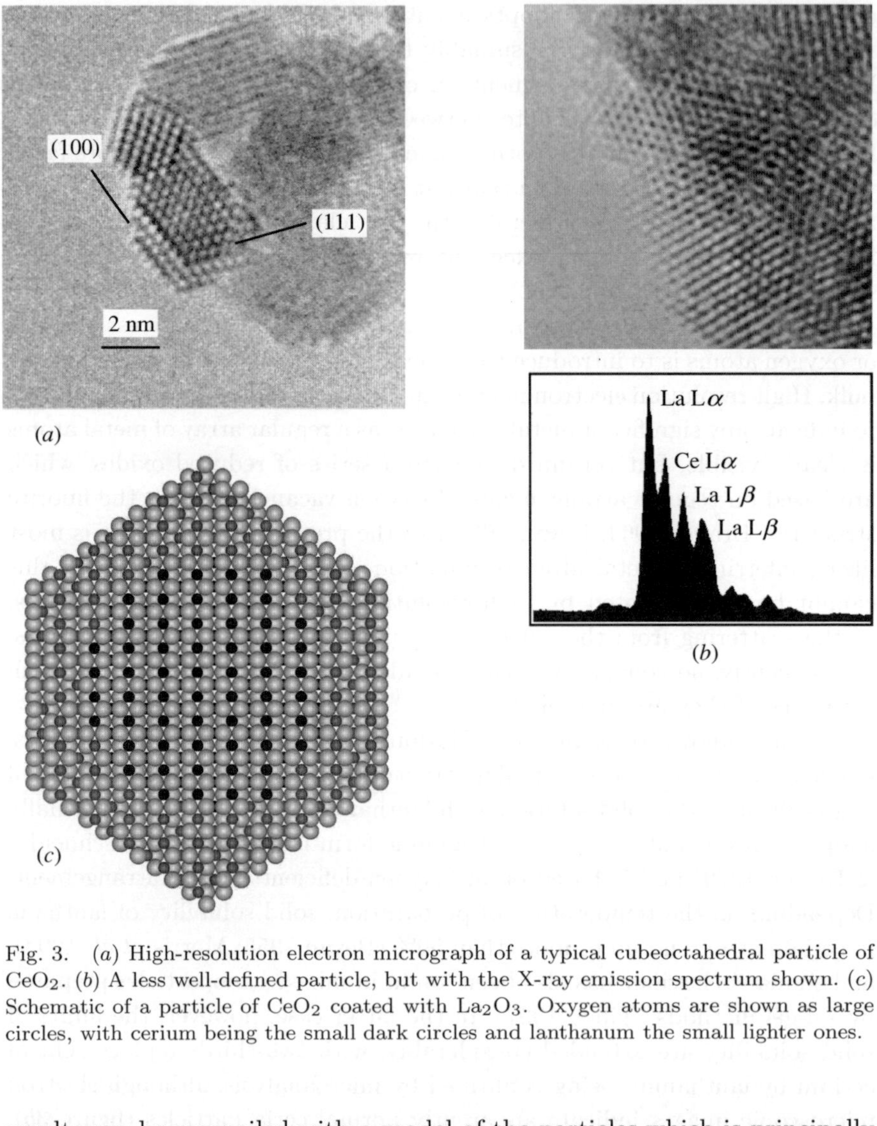

Fig. 3. (*a*) High-resolution electron micrograph of a typical cubeoctahedral particle of CeO_2. (*b*) A less well-defined particle, but with the X-ray emission spectrum shown. (*c*) Schematic of a particle of CeO_2 coated with La_2O_3. Oxygen atoms are shown as large circles, with cerium being the small dark circles and lanthanum the small lighter ones.

results can be reconciled with a model of the particles which is principally normal ceria in the interior, but then accommodates an increasing number of lanthanum atoms at or near the surfaces, with ordered oxygen vacancies (as found in cubic La_2O_3) located at the particle surfaces. This is illustrated schematically in figure 3*c*. This implies that particles of pure ceria may well behave in a similar manner, and consequently the surfaces of such particles might possess a reactivity not normally associated with the bulk oxide. This could explain the apparent ease with which ceria nanoparticles seem to dissolve other metals, as has been observed in electron microscopic studies (Hutchison 1990).

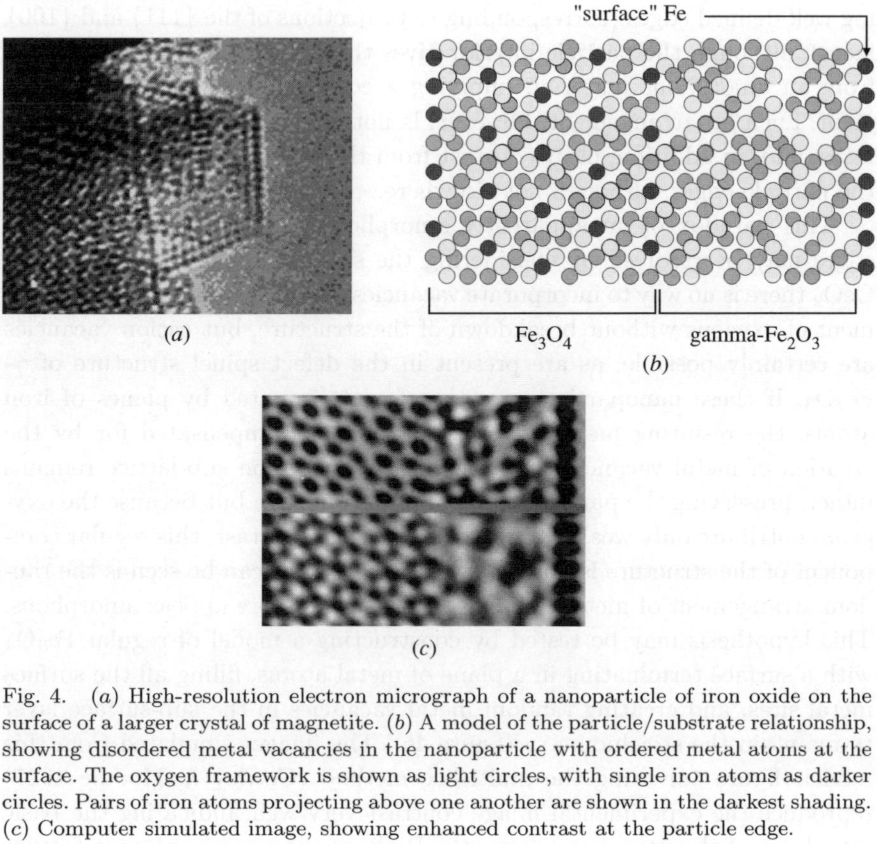

Fig. 4. (*a*) High-resolution electron micrograph of a nanoparticle of iron oxide on the surface of a larger crystal of magnetite. (*b*) A model of the particle/substrate relationship, showing disordered metal vacancies in the nanoparticle with ordered metal atoms at the surface. The oxygen framework is shown as light circles, with single iron atoms as darker circles. Pairs of iron atoms projecting above one another are shown in the darkest shading. (*c*) Computer simulated image, showing enhanced contrast at the particle edge.

4. Oxide Nanoparticles with Cation Vacancies

Oxides with the spinel structure, notably γ-alumina and Fe_3O_4, have exactly the same problems as ceria when produced in nanoparticle form, as they adopt either octahedral or cubeoctahedral morphologies, and although the structure is different, low index planes still contain either metal or oxygen atoms. In their case, however, the solution to the problem is very different. Nanoparticles of Fe_3O_4, which form the precursor of several types of iron catalyst, show remarkable features when observed in the electron microscope. One such image is shown in figure 4*a*, where a particle is observed at the margins of a much larger crystal, and it is notable in having well-defined edges (corresponding to projections of the $\{111\}$ and $\{100\}$ faces), although the particle interior gives the contrast normally expected from an amorphous material, implying a completely disordered arrangement. Such an arrangement, however, is not compatible with well-defined faces, and, in addition, clear contrast from the metal atoms is observed at the particle edges, although not elsewhere.

This paradox of an apparently amorphous particle with well-defined edges may be resolved by considering the stoichiometry problem. Unlike CeO_2 there is no way to incorporate vacancies into the close-packed arrangement of oxygens without breakdown of the structure, but cation vacancies are certainly possible, as are present in the defect spinel structure of γ-Fe_2O_3. If these nanoparticles are therefore terminated by planes of iron atoms, the resulting metal atom excess can be compensated for by the creation of metal vacancies in the interior. The anion sub-lattice remains intact, preserving the particle shape and morphology, but because the oxygens contribute only weakly to the overall image contrast, this regular component of the structure is not observed, and all that can be seen is the random arrangement of metal atoms, which will therefore appear amorphous. This hypothesis may be tested by constructing a model of regular Fe_3O_4 with a surface terminating in a plane of metal atoms, filling all the surface metal sites, and creating random metal vacancies in the sub-surface layer to maintain the stoichiometry (figure 4*b*). The images simulated from this model (figure 4*c*) using the multislice method (Cowley & Moodie 1957) reproduce the experimental image contrast very well, indicating the basic soundness of this structural principle. Perhaps the most important feature is that to obtain sufficient contrast at the surface layer, it is necessary to fill all the octahedrally coordinated sites in the surface layer with metal atoms, as in the manner of stoichiometric FeO. The surface regions of these nanoparticles are, thus, very different from the structure of bulk Fe_3O_4.

Similar images have also been observed in other spinel-based oxides. Electron-beam induced recrystallization of α-Al_2O_3 into the γ-form, which

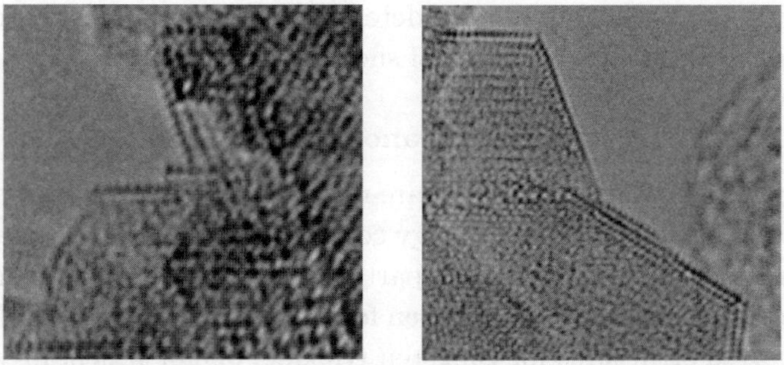

(a) (b)

Fig. 5. Octahedral and cubeoctahedral particles of γ-Al_2O_3, showing the same enhanced surface contrast as that observed in the iron oxides.

has the defect spinel structure, has been noted (Smith *et al.* 1986), and although γ-Al_2O_3 is difficult to prepare in the pure state, there is strong thermodynamic evidence that, as the particle size decreases, it becomes the thermodynamically stable structure (McHale *et al.* 1997). Al_2O_3 is widely used as a support for metal catalysts, and it is believed that the 'active' support, which facilitates the monodispersion of metals, is in fact the γ-form. Images of particles of γ-Al_2O_3 are shown in figure 5*a*, *b*. That in figure 5*a* is at the higher end of the nanoparticle size regime, but still shows strong contrast at the edges, with only weak fringe contrast in the interior regions. The only difference from Fe_3O_4 is the truncation of the {100} faces so that the overall particle shape is octahedral. In the smaller particle shown in figure 5*b*, the central fringe contrast is almost entirely absent and the interior appears to be amorphous. These images can be interpreted using the same model as Fe_3O_4, using metal atom terminations and an excess of metal vacancies in the interior (Jefferson *et al.* 1992). In addition, because of the reduced difference in the scattering powers of oxygen and aluminium, the enhanced contrast at the particle edges can only be explained if the surface is truly metal terminated, with no outer oxygen atoms. Bearing in mind the reactivity of aluminium, this is chemically very surprising, but it may explain the ease with which metals such as platinum and rhenium disperse when supported on γ-Al_2O_3, as when these metals are added to a specimen of γ-Al_2O_3 nanoparticles they can 'dissolve' in the surface metal layer and release aluminium ions that migrate to the particle interior, further stabilizing the particle. The exact valence of metal atoms added in this way has not yet been determined, but it is unlikely to be zero, explaining the extreme reactivity of such catalysts.

5. New Oxide Structures in Nanoparticle Form

The nanoparticles described above are modified variants of bulk structures. Given the influence of surface-energy considerations, however, the possibility exists of new structures in nanoparticles that have no bulk counterpart. A phase of this type has recently been found in tungsten trioxide. There are three reported structures for tungsten trioxide, one of which, m-WO_3, is a perovskite network of corner-sharing WO_6 (Wells 1984), and two further structures which have been prepared using 'wet' methods, namely a simple hexagonal form, h_1-WO_3, and a pyrochlore-like form, p-WO_3 (Figlarz 1989). Both of the latter contain tunnels formed by six WO_6 octahedra, and convert irreversibly to m-WO_3 at temperatures above 700 K (Gerand *et al.* 1979). The hexagonal form is basically a pure oxide equivalent of some alkali tungsten bronzes (Ekstrom & Tilley 1980), although the thermodynamic stability of h_1-WO_3 and p-WO_3 is open to question.

Nanoparticles of WO_3 may be prepared using sol-gel techniques from acidified sodium tungstate followed by refluxing with either 30% H_2O_2 or NH_4Cl solution at a higher pH until a fine yellow precipitate forms (Tilley 1997). Specimens produced in this way show particles with both the m-WO_3 and h_1-WO_3 structures, but also nanoparticles of a new phase, also hexagonal, but with a much larger unit cell than that of h_1-WO_3 (Tilley & Jefferson 1999). A micrograph of a particle of this phase is shown in figure 6*a*, and a schematic diagram of the structure, which has been confirmed from image simulations in figure 6*b*. This phase, which has been designated h_2-WO_3, is intermediate between the known monoclinic and hexagonal forms, in that it contains the hexagonal tunnels of the latter separated by groups of four octahedra from the former. A similar configuration has been observed in bulk specimens of $Sb_{0.2}WO_3$, although in the latter the separation of the hexagonal tunnels by elements of the m-WO_3 structure is only in one direction (Dobson *et al.* 1987). It is believed that the tunnels of the h_1-WO_3 structure form around H_3O^+ ions which are present at low pH: raising the pH effectively reduces their concentration and ensures that the monoclinic structure begins to form. At intermediate pH values, however, the hexagonal tunnels will still form but their overall density is reduced, and the space between them is filled with elements

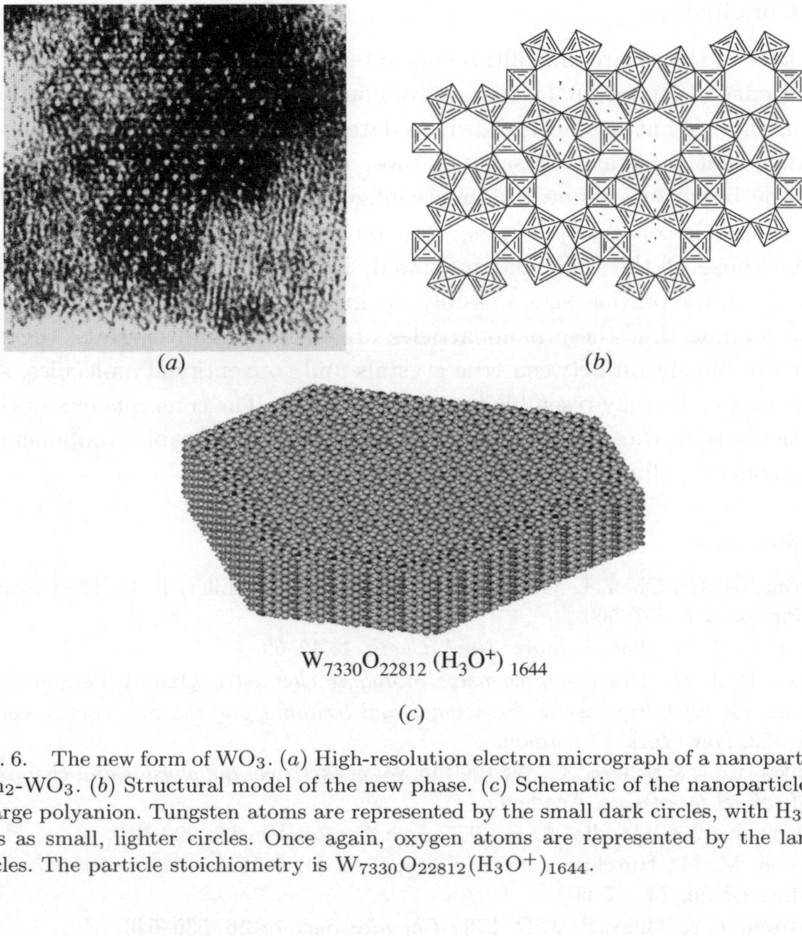

Fig. 6. The new form of WO_3. (*a*) High-resolution electron micrograph of a nanoparticle of h_2-WO_3. (*b*) Structural model of the new phase. (*c*) Schematic of the nanoparticle as a large polyanion. Tungsten atoms are represented by the small dark circles, with H_3O^+ ions as small, lighter circles. Once again, oxygen atoms are represented by the larger circles. The particle stoichiometry is $W_{7330}O_{22812}(H_3O^+)_{1644}$.

of the monoclinic structure. A whole series of intermediates is, therefore, possible, but although disordered nanoparticles have been noted, only the h_2-structure has been observed in a perfect arrangement. That part of the new arrangement derived from m-WO_3 is heavily distorted and extremely strained, and it is probable that in bulk specimens such strain could not be accommodated. In the original solution, these nanoparticles are almost certainly gigantic large polyanions of the type shown in figure 6*c*, and it is therefore quite likely that other hitherto unknown structural variants can exist.

6. Conclusions

Because of the severe difficulties encountered in their characterization, our knowledge of the internal structures of non-metallic nanoparticles is only in its infancy. What has been shown to date, however, is that it is unwise to assume that these are the same as those of bulk materials, although they may be based on a known atomic configuration. It therefore follows that the properties of such particles, both physical and chemical, are unlikely to be those of the bulk and may well, like the structure itself, depend heavily on the particle size. Possibly the greatest mistake that can be made is to assume that these nanoparticles are merely small crystals: they lie in a size dimension between true crystals and conventional molecules, and their properties may resemble those of the latter. The consequences of this, particularly as these particles form a potentially intractable component of atmospheric pollution, may well be significant.

References

Amdur, M. O., Chen, L. C., Guty, J., Lam, H. F. & Miller, P. D. 1988 *Atmos. Environ.* **22**, 557–560.

Bevan, D. J. M. 1955 *J. Inorg. Nucl. Chem.* **1**, 49–59.

Bevan, D. J. M. 1973 *Comprehensive inorganic chemistry.* Oxford: Pergamon.

Brauer, G. 1964 *Progress in the science and technology of the rare earths*, vol. 1, p. 152. New York: Pergamon.

Brinker, C. J. & Scherer, G. W. 1990 In *Sol-gel science: the physics and chemistry of sol-gel processing.* Academic.

Cowley, J. M. & Moodie, A. F. 1957 *Acta Crystallogr.* **10**, 609–619.

Dobson, M. M., Hutchison, J. L., Tilley, R. J. D. & Watts, K. A. 1987 *J. Solid State Chem.* **71**, 47–60.

Ekstrom, T. & Tilley, R. J. D. 1980 *Chemica Scripta* **26**, 535–546.

Figlarz, M. 1989 *Progr. Solid State Chem.* **19**, 1–46.

Gerand, B., Nowogrocki, G., Guenot, J. & Figlarz, M. 1979 *J. Solid State Chem.* **29**, 429–434.

Gilmour, P., Brown, D. M., Beswick, P. H., Benton, E., MacNee, W. & Donaldson, K. 1997 *Ann. Occup. Hygiene* **41**, 32–38.

Gribelyuk, M. A., Harris, P. J. F. & Hutchison, J. L. 1994 *Phil. Mag.* **69**, 655–669.

Gschneider, K. A. & Eyring, L. 1979 *Handbook on the physics and chmistry of rare earths*, vol. 3. Amsterdam: North Holland.

Harris, P. J. F. 1986 *Nature* **323**, 792–794.

Heinemann, K., Yacaman, M. J., Yang, C. Y. & Poppa, H. 1979 *J. Cryst. Growth* **47**, 177–183.

Howie, A. & Marks, L. D. 1984 *Phil. Mag.* A **49**, 95.

Hutchison, J. L. 1990 *Proc. 12th Int. Congr. Electron Microscopy*, vol. 1, pp. 478–479. San Francisco Press.

Jefferson, D. A. & Harris, P. J. F. 1988 *Nature* **332**, 617–620.

Jefferson, D. A., Kirkland, A. I., Reller, A., Tang, D., Williams, T. B. & Zhou, W. 1992 *Electron microscopy 1992*, vol. 2, pp. 611–614. Universidad de Granada.

McHale, J. M., Auroux, A., Perotta, A. J. & Navrotsky, A. 1997 *Science* **277**, 788–791.

Marks, L. D. & Smith, D. J. 1981 *J. Cryst. Growth* **54**, 425.

Marks, L. D. & Smith, D. J. 1983 *J. Microscopy* **130**, 249–261.

Morris, B. C., Flavell, W. R., Mackrodt, W. C. & Morris, M. A. 1993 *J. Mater. Chem.* **3**, 1007–1013.

Smith, D. J., Bursill, L. A. & Jefferson, D. A. 1986 *Surf. Sci.* **175**, 673–683.

Tilley, E. E. M. 1997 Synthesis and characterisation of nanocrystalline metal oxides. PhD thesis, University of Cambridge, UK.

Tilley, E. E. M. & Jefferson, D. A. 1999 *Particulate matter*, pp. 63–84. Oxford: BIOS Scientific.

Uppenbrink, J., Kirkland, A. I., Wales, D., Jefferson, D. A. & Urban, J. 1992 *Phil. Mag.* B **65**, 1079–1096.

Wells, A. F. 1984 *Structural inorganic chemistry*, 4th edn, pp. 516–573. Oxford University Press.

CHAPTER 10

RESPIRATORY DOSE OF INHALED ULTRAFINE PARTICLES IN HEALTHY ADULTS

Chong S. Kim[1] and Peter A. Jaques[2]

[1] *Human Studies Division, National Health and Environmental Effects Research Laboratory, US Environmental Protection Agency, Research Triangle Park, NC 27711, USA* (kim.chong@epamail.epa.gov)

[2] *Center for Environmental Medicine and Lung Biology, University of North Carolina, Chapel Hill, NC 27599, USA*

Ultrafine particles (less than 0.10 µm in diameter) are ubiquitous in the atmosphere and possess unique physicochemical characteristics that may pose a potential health risk. To help elucidate the potential health risk, we measured respiratory dose of ultrafine particles (0.04, 0.06, 0.08 and 0.10 µm in diameter) in healthy young adults using a novel serial bolus-delivery method. Under normal breathing conditions (i.e. tidal volume of 500 ml and respiratory flow rate of 250 ml s^{-1}), bolus aerosols were delivered sequentially to a lung depth ranging from 50–500 ml in 50 ml increments and deposition was measured for each of ten equal-volume compartments.

Results show that regional deposition varies widely along the depth of the lung regardless of the particle sizes used. Peak deposition was found in the lung regions situated between 150 and 200 ml from the mouth. Sites of peak deposition shifted proximally with a decrease in particle size. Deposition dose per unit surface area was largest in the proximal lung regions and decreased rapidly with an increase in lung depth. Peak surface dose was 5–7 times greater than the average lung dose. The results indicate that local enhancement of dose occurs in normal lungs, and such a dose enhancement may play an important role in the potential health effects of ultrafine aerosols.

Keywords: ultrafine aerosol; regional lung deposition; respiratory dose; particulate matter; ambient aerosol

1. Introduction

Although the mass fraction of ultrafine particles in ambient particulate matter is small, their presence in great number and surface area has been a source of concern as a potential health hazard. In a recent epidemiological study, a decrement of lung function measured in asthmatic adults has been shown to correlate better with the number of ultrafine particles than with the mass of fine particles (Peters *et al.* 1997). Animal studies have shown that ultrafine particles were capable of causing acute toxic effects and even death after short-term exposure in rats and that the observed toxic effects were correlated better with the surface area than with the mass of particles (Oberdörster *et al.* 1992, 1995). However, most epidemiological studies consistently reported a good correlation between relative health risk and mass concentration of presumably fine particles (Schwartz 1994; Pope *et al.* 1995). At present, there is no clear explanation for how ambient particles can cause adverse health effects at low concentrations. As such, it is unclear whether there are differential roles for fine and ultrafine particles on health effects at ambient conditions. However, from the dosimetric point of view, a greater deposition dose poses a greater risk to health. Previous studies have shown that total lung deposition of ultrafine particles increases with a decrease in particle size, i.e. the smaller the particle size, the greater the lung deposition (Tu & Knutson 1984; Wilson *et al.* 1985; Schiller *et al.* 1986; Jaques & Kim 2000). Although the size-dependent deposition characteristics are different from those of fine and coarse particles for which lung deposition increases with an increase in particle size, total lung deposition values are generally comparable for ultrafine versus fine and coarse particles (Stahlhofen *et al.* 1989). However, inhaled particles deposit variably in different regions of the lung and this may result in a marked enhancement of dose in local regions, while overall lung dose may be considered to be safe. Because local regions receiving greater doses are likely to be affected more severely and may become initiating points for subsequent adverse health effects, assessment of local dose would be of great interest in evaluating potential health risk of inhaled particles. Previously, we have shown that local deposition dose can be many times greater than the average lung dose in healthy subjects for fine and coarse particles (Kim *et al.* 1996; Kim & Hu 1998). These results may not be applied directly to ultrafine particles because particles with different sizes deposit in the lung by different deposition mechanisms. Ultrafine particles deposit in the lung by diffusion, whereas fine and coarse particles deposit by gravitational sedimentation and inertial impaction. Therefore, it is important to know if there is any uniqueness in deposition patterns of ultrafine particles that can be related to detrimental health effects. In the present study, we measured total as well as detailed regional lung deposition for four different sizes of ultrafine particles under normal breathing conditions and compared the results with those obtained previously for fine and coarse particles. The purpose of the study was to obtain a detailed site–dose relationship for ultrafine particles in healthy lungs, which may be used for evaluating the potential health risk of ambient particulate matter.

2. Experimental Methods

2.1. *Subjects*

Twenty-two healthy adults (11 men and 11 women) ranging in age from 20 to 40 years old were studied. The subjects either had no history of smoking or had not smoked in the past five years. All subjects underwent a screening procedure that included a complete medical history, physical examination, SMA-20 blood chemistry screen, and complete differential blood count. Those who passed the initial screening had their basic lung function measured by both spirometry and body plethysmography. Subject characteristics and lung function test results are shown in table 1.

Table 1. Summary of subject characteristics and lung function test results. All values are mean ± SD of $n = 11$ each. FVC denotes forced vital capacity; FEV_1 denotes forced expired volume at 1 s; R_{aw} denotes airway resistance; FRC denotes functional residual capacity; TLC denotes total lung capacity.

sex	age (yr)	height (cm)	FVC (ml)	FEV_1 (ml)
men	31 ± 4	173 ± 7	5388 ± 847	4404 ± 708
women	31 ± 4	165 ± 6	4278 ± 587	3467 ± 540

sex	R_{aw} (cm H_2O l^{-1} s^{-1})	FRC (ml)	TLC (ml)
men	1.00 ± 0.6	3911 ± 892	6598 ± 980
women	1.24 ± 0.6	3314 ± 547	5282 ± 599

2.2. *Generation of Ultrafine Aerosols*

Ultrafine aerosols were generated by condensing sebacate oil (di-2-ethylhexyl sebacate) vapour on non-hygroscopic metallic nuclei particles. The aerosol generator consisted of a monodisperse condensation aerosol generator (model 3470, TSI Inc., St Paul, MN) and a nuclei aerosol generator using a nickel–chromium heating wire (80% Ni and 20% Cr and *ca.* 0.5 mm in diameter; Omega Engineering, Stamford, CT). The TSI aerosol generator uses NaCl aerosols as a source of condensation nuclei. However, ultrafine sebacate oil particles generated with NaCl nuclei were found to be somewhat hygroscopic. Therefore, NaCl nuclei were replaced with non-hygroscopic metallic nuclei. Briefly, metallic nuclei are produced by heating a coiled Ni–Cr wire (*ca.* 3–4 Ω) at low electric voltage (*ca.* 1.1–1.6 V AC). The nuclei aerosol (*ca.* $3\,\mathrm{l\,min^{-1}}$) is then passed through a *boiler* in which sebacate oil is heated and vaporized at 70–100 °C. The mixture of nuclei and oil vapour from the boiler is passed through a *reheater* that is maintained at 190 °C and subsequently through an unheated vertical column designed to induce condensation of oil vapour on the surface of nuclei particles. The aerosols emerging from the generator are diluted with filtered air (*ca.* $100\,\mathrm{l\,min^{-1}}$) and supplied to the inhalation system. In the present study, ultrafine aerosols with four different particle sizes were generated; 0.04, 0.06, 0.08 and 0.1 μm in number median diameter (NMD) with a geometric standard deviation (σ_g) in the range 1.27–1.34. The size distribution was measured using a scanning mobility particle sizer (SMPS) (model 3934, TSI Inc., St Paul, MN).

2.2.1. *Inhalation System*

The core of the system consists of an ultrafine condensation particle counter (UCPC), an aerosol bolus-injection module, and an on-line data-acquisition system (see figure 1). In the bolus-injection module, test aerosols are introduced into the inspiratory line as a small bolus (half width of *ca.* 45 ml) by activating a solenoid valve. The duration of valve opening is initially set to 100 ms and adjusted to an appropriate value depending on flow and pressure conditions upstream. The aerosol chamber upstream of the solenoid valve is maintained at a positive pressure (1–5 cm H_20) slightly above room conditions to help inject the aerosol. During inhalation, the aerosol is sampled continuously into a UCPC (model 3025A, TSI Inc., St Paul, MN)

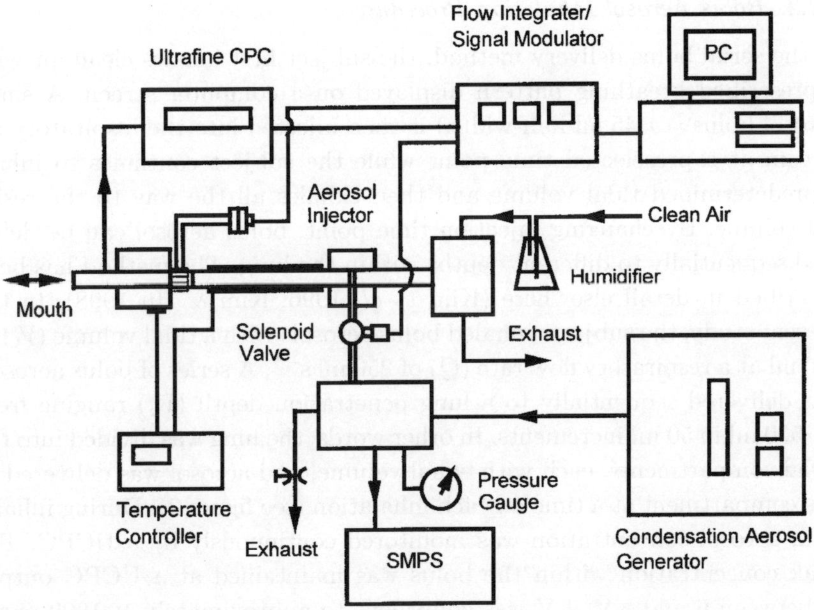

Fig. 1. Experimental system used for determining regional lung deposition of ultrafine particles. CPC denotes condensation particle counter; PC, personal computer.

at a rate of 25 ml s^{-1} via the sidearm port attached to the mouthpiece. In the UCPC, ultrafine particles pass through an alcohol vapour chamber (38 °C), and the mixture of the aerosol and vapour is introduced into a tube cooled to 4 °C in which alcohol vapour condenses on the surface of particles. As a result, ultrafine particles grow to a super-micrometre size, and the enlarged particles are detected by a laser sensor. The TSI UCPC outputs an aerosol signal averaged over a 2 s period. In the present system, the averaging circuitry was bypassed and aerosol signals were taken directly from the sensor for continuous output. Respiratory flow rates are measured by a pneumotachograph (Fleisch Size no. 1, Linton Instrumentation, Norfolk, UK) in conjunction with a pressure transducer (model 239, ±1.27 cm H_2O range, Setra Systems Inc., Acton, MA) that is connected to the mouthpiece in-line. Both flow and aerosol signals are supplied to an on-line data acquisition system at a rate of 200 Hz and subsequently analysed breath by breath.

2.2.2. *Bolus Aerosol Inhalation Procedure*

In the serial bolus-delivery method, the subject first inhales clean air with a prescribed breathing pattern displayed on a computer screen. A small aerosol bolus (*ca.* 45 ml half-width) is then injected into the inspiratory air stream at a preselected time point while the subject continues to inhale a predetermined tidal volume and then exhales all the way to the residual volume. By changing injection time point, bolus aerosol can be delivered sequentially to different depths within the lung. The method has been described in detail elsewhere (Kim *et al.* 1996; Kim & Hu 1998). In the present study, the subjects inhaled bolus aerosols with a tidal volume (V_t) of 500 ml at a respiratory flow rate (Q) of 250 ml s^{-1}. A series of bolus aerosols was delivered sequentially to a lung penetration depth (V_p) ranging from 50–500 ml in 50 ml increments. In other words, the lung was divided into ten serial compartments, each with equal volume, and aerosol was delivered to one compartment at a time on each inhalation (see figure 2). During inhalation, aerosol concentration was monitored continuously by a UCPC. The peak concentration within the bolus was maintained at a UCPC output of between 6 and 8 V; 1 V was equivalent to approximately 100 000 particles cm^{-3}. For a given inhalation condition, at least five repeated measurements were obtained. The procedure was repeated for each of four different aerosols (d_p = 0.04, 0.06, 0.08 and 0.1 μm; d_p refers to number median diameter here and elsewhere). The total number of particles inhaled (N_{in}) and subsequently exhaled (N_{ex}) was calculated for each bolus inhalation, and the recovery (RC = N_{ex}/N_{in}) of bolus was obtained from each of ten volumetric compartments. Using a series of simultaneous mathematical formulae, local deposition efficiency (X) and subsequently local deposition fraction (LDF) were determined for each volumetric compartment (see figure 2). LDF was defined by the fraction of total aerosol inhaled that was deposited in each compartment.

3. Results and Discussion

3.1. *Deposition Distribution in Sequential Volumetric Lung Regions*

The values of LDF of ultrafine aerosols (d_p = 0.04–0.1 μm) in sequential lung regions, each consisting of a 50 ml volume compartment, are shown in figure 3 for both men and women. All subjects inhaled ultrafine aerosols

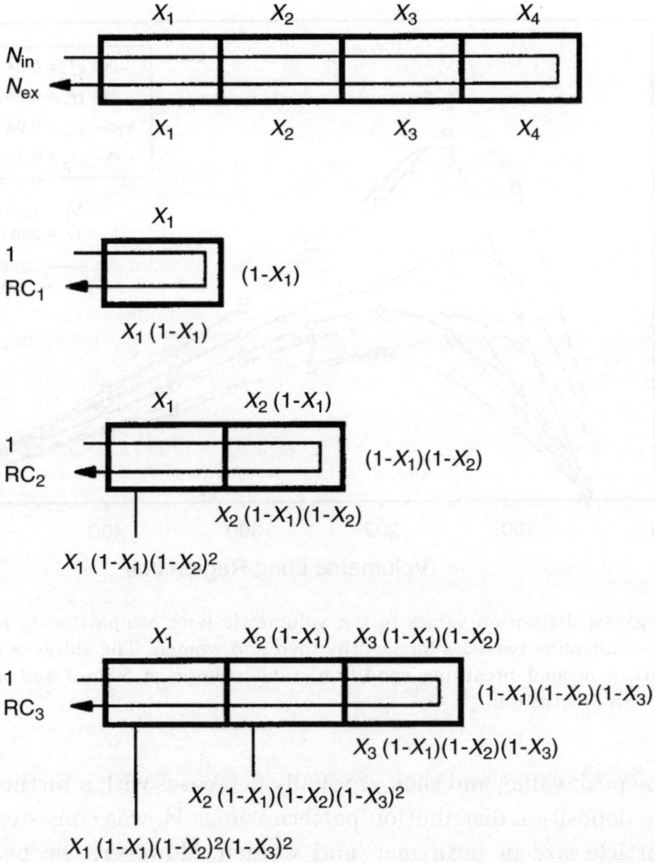

Fig. 2. Calculation procedures for determining regional deposition efficiencies (X_i) and deposition fraction values for serial lung compartments. Bolus aerosol recovery (RC) is defined by the ratio of the total number of particles exhaled (N_{ex}) to the total number inhaled (N_{in}). Deposition efficiencies are assumed to be the same for inspiratory and expiratory flow in each compartment. Deposition fractions for inspiratory and expiratory phases are shown on the top and bottom of each compartment, respectively. Aerosol fractions remaining at end inspiration are as follows: $\mathrm{RC} = N_{\mathrm{ex}}/N_{\mathrm{in}}$; $\mathrm{RC}_1 = (1 - X_1)^2$; $\mathrm{RC}_2 = (1 - X_1)^2(1 - X_2)^2$; $\mathrm{RC}_3 = (1 - X_1)^2(1 - X_2)^2(1 - X_3)^2$; $\mathrm{RC}_n = \prod_{m=1}^{n}(1 - X_m)^2$; $\mathrm{RC}_n/\mathrm{RC}_{n-1} = (1 - X_n)^2$; $X_n = 1 - \sqrt{(\mathrm{RC}_n/\mathrm{RC}_{n-1})}$.

at a fixed breathing pattern consisting of a tidal volume of 500 ml and breathing frequency of 15 breaths min^{-1}. Mean respiratory flow rate was 250 ml s^{-1}. Figure 3 shows that LDF increases with V_{p} from the mouth,

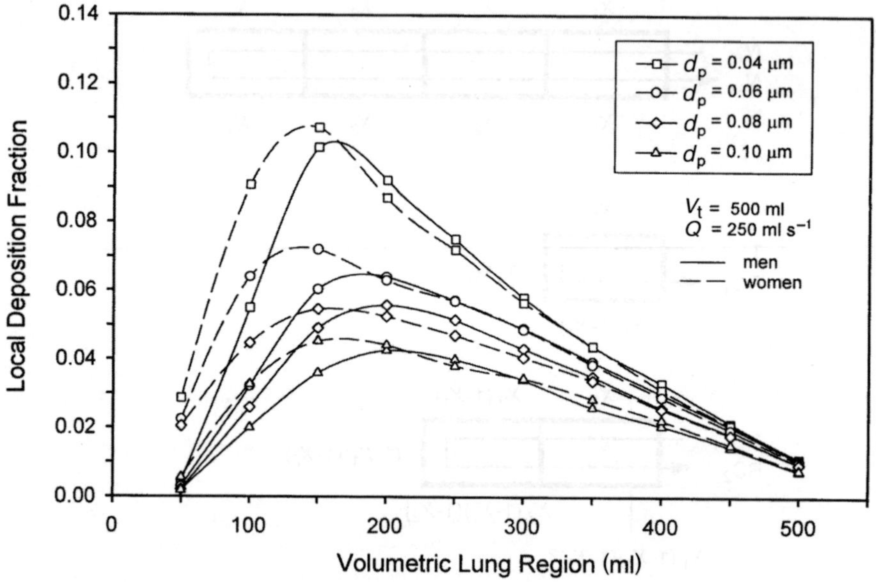

Fig. 3. Regional deposition values in ten volumetric lung compartments for four different sizes of ultrafine particles for healthy men and women. The subjects inhaled the aerosols with a normal breathing condition: tidal volume of 500 ml and a breathing frequency of 15 breaths min^{-1}.

reaches the peak value, and then gradually decreases with a further increase in V_p. The deposition distribution pattern versus V_p was consistent regardless of particle size in both men and women. However, the peak height and position varied depending on particle size and gender of subjects. In men, the peak deposition was found in the lung region $V_p = 150$–200 ml for $d_p = 0.1$ μm. The peak position gradually shifted towards the mouth with decreasing particle size and was found in the lung region $V_p = 100$–150 ml for $d_p = 0.04$ μm. LDF was greater with smaller d_p throughout the entire lung regions. The increase in deposition was particularly prominent in the peak deposition regions. The peak deposition was nearly 2.5 times greater for $d_p = 0.04$ μm than for $d_p = 0.1$ μm. In women, deposition patterns were similar to those of men, but peak deposition regions shifted closer to the mouth and peak heights were slightly elevated for all d_p compared with those of men. LDF was consistently greater in shallow lung regions ($V_p < 150$ ml), particularly for regions of $V_p = 0$–50 ml and $V_p = 50$–100 ml.

In deeper lung regions (i.e. $V_p > 200$ ml), deposition was comparable for men and women.

These results clearly show that regional deposition values vary widely in normal lungs and that local deposition dose can be many times greater than the average dose of the entire lung.

Peak deposition occurs in lung regions between 150 and 200 ml depth that encompasses the transition zone between the conducting airways and alveolar region. It should be noted that deposition efficiency in local lung regions increases monotonically with an increase in lung depth (Kim *et al.* 1996) because airway dimensions are smaller and particle residence time is longer in deeper lung regions. Therefore, deposition enhancement in the transition zone is not related to any unique structural features in the region, but is, rather, a logical outcome of a sequential filtration process in the respiratory airways. Deposition increases initially with an increase in lung depth and then decreases with a further increase in lung depth, because air reaching the deeper lung regions contains fewer particles. Longitudinal variation of lung deposition is an inevitable consequence of human lung anatomy and sequential respiratory airflow. Figure 3 shows that the longitudinal variation is more pronounced for smaller ultrafine particles (i.e. $d_p = 0.04\,\mu m$). This can be expected because the deposition efficiency of these small particles is very high (i.e. high diffusivity), resulting in a rapid increase in deposition in shallow lung regions followed by a rapid decrease in the deeper regions. Therefore, deposition tends to be concentrated over a small volumetric region of the lung. On the other hand, particles with low deposition efficiency (i.e. $d_p = 0.1\,\mu m$) can easily penetrate into deep lung regions, and deposition spreads out over a large area of the lung. The results also show that regional deposition is more pronounced in women than in men. Deposition enhancement is particularly noted in the proximal airway regions for women versus men. Similar findings have been reported previously for coarse particles (i.e. $d_p = 3$ and $5\,\mu m$; see Kim & Hu (1998)), and enhanced proximal deposition in women was attributed to small dimensions of the upper airways (i.e. pharynx and larynx), which, in turn, could result in an increase in inertial impaction. Inertial impaction is not relevant to deposition of ultrafine particles. However, airflow conditions in the upper airways are usually turbulent because of complex airway geometry, and enhanced turbulence in the smaller upper airways could result in an increase in diffusive deposition of ultrafine particles.

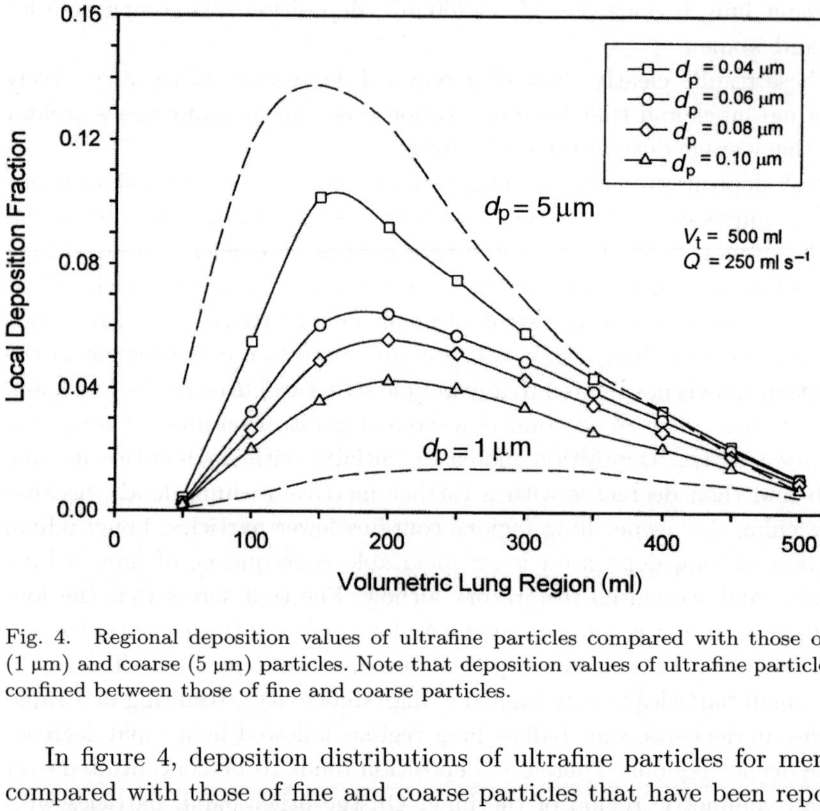

Fig. 4. Regional deposition values of ultrafine particles compared with those of fine (1 μm) and coarse (5 μm) particles. Note that deposition values of ultrafine particles are confined between those of fine and coarse particles.

In figure 4, deposition distributions of ultrafine particles for men are compared with those of fine and coarse particles that have been reported in earlier studies (Kim *et al.* 1996; Kim & Hu 1998). In the figure, it can be seen that deposition distributions of ultrafine particles are confined between those of fine ($d_p = 1$ μm) and coarse ($d_p = 5$ μm) particles, and that for particles of smaller size deposition patterns become more like those of coarse particles. In other words, very small ultrafine particles deposit in the lung more like large coarse particles. It should be noted that all of the present results are based on a typical breathing pattern (i.e. $V_t = 500$ ml and $Q = 250$ ml s^{-1}), and as such, the results may not be applied freely to different breathing conditions.

3.2. *Three-Compartment Regional Lung Deposition*

Conventionally, regional lung deposition is expressed for three anatomic regions: head (larynx and above), tracheobronchial (TB) and alveolar

Table 2. Three-compartment regional lung deposition values (%) for men and women. All values (mean ± SD) are percentage of total aerosol inhaled via the mouth. Breathing pattern was 500 ml tidal volume and 250 ml s^{-1} flow rate (i.e. 15 breaths per min).

lung regions	particle diameter (μm)			
	0.04	0.06	0.08	0.10
men ($n = 11$)				
head	0.4 ± 0.7	0.3 ± 0.5	1.0 ± 1.9	0.2 ± 0.5
tracheobronchial	15.6 ± 4.6	9.2 ± 3.8	8.2 ± 3.7	5.7 ± 3.2
alveolar	33.1 ± 2.7	27.2 ± 3.8	23.9 ± 5.6	18.2 ± 6.2
total	49.2 ± 6.6	36.7 ± 7.2	33.1 ± 9.2	24.1 ± 8.9
women ($n = 11$)				
head	2.9 ± 2.5	2.2 ± 2.3	2.0 ± 2.2	0.6 ± 0.7
tracheobronchial	19.8 ± 3.4	13.6 ± 2.9	9.9 ± 2.7	7.8 ± 1.8
alveolar	32.2 ± 3.9	26.5 ± 4.1	22.7 ± 4.7	19.0 ± 2.9
total	54.9 ± 5.9	42.3 ± 6.9	34.7 ± 7.8	27.4 ± 4.1

region. Because these regions can be defined approximately by $V_p < 50$ ml for head, $V_p = 50$–150 ml for TB, and $V_p > 150$ ml for alveolar (Kim & Hu 1998), deposition in each of the regions can be obtained from the present sequential compartment results. For both men and women, deposition values in three regions are summarized in table 2 for a breathing pattern with $V_t = 500$ ml and $Q = 250$ ml s^{-1}. Total lung deposition values also are shown in table 2. All deposition values (mean ± SD) are a percentage of total aerosol inhaled via the mouth. Results show that deposition decreases consistently in all regions with an increase in particle size. This is consistent with the theory of particle deposition by diffusion: a greater deposition is expected with smaller ultrafine particles having greater diffusivity. Deposition in the head regions (mainly oropharynx and larynx) was very small (less than 3%). TB and alveolar deposition ranged from 5.7 to 15.6% and 18.2 to 33.1%, respectively, depending on particle size. Of the total deposition in the lung, 23–32% was deposited in TB and 68–77% was deposited in the alveolar region. These values are in general agreement with predictions by a mathematical lung deposition model adopted by the International Commission on Radiological Protection (ICRP 1994) at a similar breathing condition. In table 1, it is noted that, compared with men, deposition in women is consistently greater in the TB region (21–47%), but was comparable or slightly smaller in the alveolar region. As a result, total lung deposition was greater in women than in men (5–15%). These results are consistent with those obtained by conventional non-bolus inhalation methods (Jaques & Kim 2000).

3.3. *Surface Dose in the Regional Lung Compartment*

LDF values in sequential volume compartments of the lung are essential for deriving deposition values at specific anatomic regions, e.g. tracheobronchial versus alveolar region, as discussed above. However, such data are less useful for evaluating toxicological effects that may result from particle dose at a tissue level. Therefore, surface dose in each volumetric compartment was calculated and the result was plotted in figure 5 for the men's data. The surface dose was defined by LDF divided by surface area of each volumetric compartment. The surface area was calculated from Weibel's symmetric lung model at a lung volume of 3500 ml (Weibel (1963); see also table 3). The figure shows that surface dose is largest in the most proximal

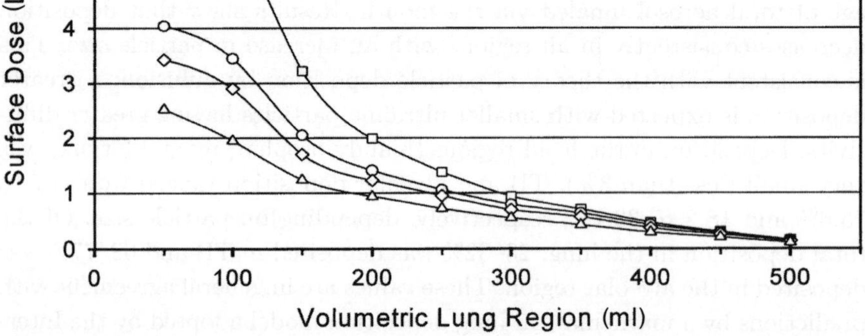

Fig. 5. Surface dose of ultrafine particles in sequential lung-volume compartments. Surface dose is defined by LDF divided by the surface area (S in cm^2) of the same local compartment.

Table 3. Regional deposition values of ultrafine particles ($d_p = 0.04$ μm) in sequential volumetric lung regions for men inhaling aerosols at a fixed concentration of 10 μg m^{-3} with a normal breathing pattern[a]. MDR denotes deposition rate of particle mass (density of particles equals 1 g cm^{-3}); NDR denotes deposition rate of particle number; S_pDR denotes deposition rate of particle surface area; and pS_pDR denotes deposition rate of projected surface area of particles.

lung regions[b] (ml)	airway surface[c] (cm^2)	LDF	MDR (μg h^{-1})	NDR (particles h^{-1})	S_pDR (mm^2 h^{-1})	MDR/S (μg h^{-1} mm^{-2})	NDR/S (particles h^{-1} mm^{-2})	S_pDR/S ($μm^2$ h^{-1} mm^{-2})	pS_pDR/S ($μm^2$ h^{-1} mm^{-2})
50–100	927	0.055	0.247	7.37×10^9	37.0	2.66×10^{-6}	7.95×10^4	399.5	99.9
100–150	3 149	0.102	0.457	1.37×10^{10}	68.6	1.45×10^{-6}	4.34×10^4	217.8	54.5
150–200	4 540	0.090	0.406	1.21×10^{10}	61.0	8.95×10^{-7}	2.68×10^4	134.4	33.6
200–250	5 411	0.075	0.335	1.00×10^{10}	50.3	6.20×10^{-7}	1.85×10^4	93.0	23.3
250–350	11 953	0.101	0.453	1.35×10^{10}	68.0	3.79×10^{-7}	1.13×10^4	56.9	14.2
350–500	20 343	0.066	0.295	8.82×10^9	44.3	1.45×10^{-7}	4.34×10^3	21.8	5.5
50–150	4 076	0.156	0.703	2.10×10^{10}	105.5	1.73×10^{-6}	5.15×10^4	258.8	64.7
150–500	42 247	0.331	1.49	4.45×10^{10}	223.6	3.53×10^{-7}	1.04×10^4	52.9	13.2
50–500	46 323	0.487	2.19	6.55×10^{10}	329.3	4.73×10^{-7}	1.41×10^4	71.1	17.8

[a]Tidal volume of 500 ml and breathing frequency of 15 breaths min^{-1}.
[b]Lung regions between specified volumetric depths from the mouth.
[c]Airway surface area (S) calculated from Weibel's symmetric lung model at a lung volume of 3500 ml.

lung region and decreases rapidly with an increase in V_p. This was to be expected, because the surface area of the lung increases rapidly as the airways branch out into the deeper lung regions. Peak surface dose was 3–6 times (men; 5–7 times in women) greater than average lung dose, depending on particle size used.

Within each volumetric compartment, deposition distribution has been shown to be highly uneven, and a large portion (greater than *ca.* 80%) of deposition is focalized at specific anatomic sites, particularly in the conducting airway regions (Schlesinger *et al.* 1982; Kim & Fisher 1999). Therefore, a local tissue dose can be much greater than the peak surface dose shown in figure 5. Because adverse effects are likely to be initiated at the local site receiving greater tissue dose, this suggests that a risk assessment based on the average lung dose may substantially underestimate the potential health hazard of inhaled particles. The results also indicate that conducting airway regions take major insults of inhaled particles. This may be considered to be good because sensitive pulmonary regions may be protected from major insults of particles. However, the airway itself could be subject to serious injuries and may become a source of aetiology. This is particularly relevant to patients with obstructive airway disease, in whom particle deposition is greatly concentrated in the airway regions (Taplin *et al.* 1977; Kim & Kang 1997).

3.4. *Exposure-Dose Considerations*

The deposition values presented above are based on a fraction of the total amount of aerosol inhaled and are independent of dose metrics. However, the actual practice of risk or toxicological assessment requires a specific dose metric, e.g. mass, number or surface area. Ultrafine particles contribute very little to mass concentrations of typical ambient aerosols, but they may constitute a large portion of the number and surface area of the aerosols (Whitby *et al.* 1974). To elucidate a potential role of ultrafine particles on adverse health effects, deposition dose needs to be analysed for appropriate dose metrics easily applicable to toxicological assessment. Table 3 shows the regional deposition dose of ultrafine particles ($d_p = 0.04\ \mu m$) in eight different dose metrics for men inhaling the aerosol at a fixed concentration of $10\ \mu g\ m^{-3}$ with a normal breathing pattern at rest (i.e. $V_t = 500$ ml, $Q = 250\ ml\ s^{-1}$). The ambient concentration of $10\ \mu g\ m^{-3}$ is rarely expected for ultrafine particles, even in heavily polluted areas, but it was used as a reference concentration in the present calculation because, in epidemiological studies, the health risk of exposure to particulate pollutant has been routinely analysed for an increment of $10\,\mu g\,m^{-3}$ (Pope *et al.* 1995). In table 3, it can be seen that the whole lung dose of $0.04\,\mu m$ particles may accumulate at a rate of $2\,\mu g$ in mass, 6.6×10^{10} particles in number, and $329\,mm^2$ in surface area per hour under the exposure conditions described above (see bottom row). If these values are normalized by the surface area of the lung, the deposition rate will be $4.7 \times 10^{-7}\,\mu g$ in mass, 1.4×10^4 particles in number, $71\,\mu m^2$ in surface area, and $18\,\mu m^2$ in projected surface area per hour in a $1\,mm^2$ area of the surface of the lung. Regional surface doses vary depending on values of local LDF and surface area, but it can be 5–6 times greater than the average surface dose (see the first row of table 3). These results are useful for estimating microscopic cellular or tissue doses if the number of specific cells or tissue volumes are known at the local lung regions. It should be noted that local particle burdens do not necessarily accumulate at the same rate as the deposition rate discussed above because particles are constantly cleared out of the deposition site by mucociliary or other transport mechanisms. Dose accumulation depends on the net balance between deposition and clearance. The results shown in table 3 did not consider clearance of particles, and, therefore, may be considered as a worst-case scenario for local dose accumulation. Although table 3 provides data for only one particle size, the data may be used as a guide for estimating deposition dose of particles of different sizes. LDF values of different ultrafine particles are shown in figure 3, and those of fine and coarse particles can be obtained from our earlier reports (Kim *et al.* 1996; Kim & Hu 1998).

4. Conclusions

Detailed regional deposition of ultrafine particles was measured in healthy men and women under normal breathing conditions at rest by a novel bolus-delivery method. From the results, the following conclusions can be drawn.

(1) Deposition of ultrafine particles in serial compartments of the lung varies widely along the volumetric depth of the lung. Peak deposition occurs in the transition zone between the conducting airways and alveolar region.

(2) Proximal airway regions receive the largest surface dose that amounts to a value several times greater than the average lung dose.
(3) Women receive a greater dose than men in the head and tracheobronchial regions.

Because adverse health effects are more likely to develop from local sites that are subject to excessive particle dose, the present results for local peak dose and dose distribution may prove to be useful for understanding the potential health risk of exposure to ultrafine particles.

Disclaimer

Although the research described in this article has been supported by the United States Environmental Protection Agency, it has not been subjected to Agency review and therefore does not necessarily reflect the views of the Agency and no official endorsement should be inferred. Mention of trade names or commercial products does not constitute endorsement or recommendation for use.

Acknowledgements

The authors thank Paulette DeWitt of the US EPA for skilful performance of pulmonary function tests on volunteer subjects, and the medical staff of the US EPA Human Studies Facility for a careful screening of study subjects.

References

ICRP (International Commission on Radiological Protection) 1994 Human respiratory tract model for radiological protection. *Ann. ICRP* **66**, 50, 423.

Jaques, P. A. & Kim, C. S. 2000 Measurement of total lung deposition of inhaled ultrafine particles in healthy men and women. *Inhal. Toxicol.* **12**, 715–731.

Kim, C. S. & Fisher, D. M. 1999 Deposition characteristics of aerosol particles in sequentially bifurcating airway models. *Aerosol Sci. Technol.* **31**, 198–220.

Kim, C. S. & Hu, S. C. 1998 Regional deposition of inhaled particles in human lungs: comparison between men and women. *J. Appl. Physiol.* **84**, 1834–1844.

Kim, C. S. & Kang, T. C. 1997 Comparative measurement of lung deposition of inhaled fine particles in normal subjects and patients with obstructive airway disease. *Am. J. Respir. Crit Care Med.* **155**, 899–905.

Kim, C. S., Hu, S. C., DeWitt, P. & Gerrity, T. R. 1996 Assessment of regional deposition of inhaled particles in human lungs by serial bolus delivery method. *J. Appl. Physiol.* **81**, 2203–2213.

Oberdörster, G., Ferin, J., Gelein, R., Sonderholm, S. C. & Finkelstein, J. 1992 Role of the alveolar macrophage during lung injury: studies with ultrafine particles. *Environ. Health Perspect.* **97**, 193–199.

Oberdörster, G., Gelein, R. M., Ferin, J. & Weiss, B. 1995 Association of particulate air pollution and acute mortality: involvement of ultrafine particles? *Inhal. Toxicol.* **7**, 111–124.

Peters, A., Wichmann, H. E., Tuch, T., Heinrich, J. & Heyder, J. 1997 Respiratory effects are associated with the number of ultrafine particles. *Am. J. Respir. Crit. Care Med.* **155**, 1376–1383.

Pope, C. A., Dockery D. W. & Schwartz, J. 1995 Review of epidemiological evidence of health effects of particulate air pollution. *Inhal. Toxicol.* **7**, 1–18.

Schiller, C. F., Gebhart, J., Heyder, J., Rudolf, G. & Stahlhofen, W. 1986 Factors influencing total deposition of ultrafine aerosol particles in the human respiratory tract. *J. Aerosol Sci.* **17**, 328–332.

Schlesinger, R. B., Gurman, J. L. & Lippmann, M. 1982 Particle deposition within bronchial airways: comparison using constant and cyclic inspiratory flows. *Ann. Occup. Hyg.* **26**, 47–64.

Schwartz, J. 1994 Air pollution and daily mortality: a review and meta-analysis. *Environ. Res.* **64**, 36–52.

Stahlhofen, W., Rudolf, G. & James, A. C. 1989 Intercomparison of experimental regional aerosol deposition data. *J. Aerosol Med.* **2**, 285–308.

Taplin, G. W., Tashkin, D. P., Chopra, S. K., Anselmi, O. E., Elam, D., Calvarese, B., Coulson, A., Detels, R. & Rokaw, S. N. 1977 Early detection of chronic obstructive pulmonary disease using radionuclide lung imaging procedures. *Chest* **71**, 567–575.

Tu, K. W. & Knutson, E. O. 1984 Total deposition of ultrafine hydrophobic and hygroscopic aerosols in the human respiratory system. *Aerosol Sci. Technol.* **3**, 453–465.

Weibel, E. R. 1963 *Morphometry of the human lung.* Academic.

Whitby, K. T., Clark, W. E., Marple, V. A., Sverdrup, G. M., Sem, G. J., Willeke, K., Liu, B. Y. H. & Pui, D. Y. H. 1974 Characterization of California aerosols. I. Size distribution of freeway aerosol. *Atmos. Environ.* **9**, 463–482.

Wilson Jr, F. J., Hiller, H. C., Wilson, J. D. & Bone, R. C. 1985 Quantitative deposition of ultrafine stable particles in the human respiratory tract. *J. Appl. Physiol.* **58**, 223–229.

CHAPTER 11

SURFACTANT–ULTRAFINE PARTICLE INTERACTIONS: WHAT WE CAN LEARN FROM PM_{10} STUDIES

Peter Gehr[1], Marianne Geiser[1], Vinzenz Im Hof[1] and Samuel Schürch[2]

[1]*Institute of Anatomy, University of Bern, Bühlstrasse 26, PO Box, CH-3000 Bern 9, Switzerland*

[2]*Health Sciences Centre, The University of Calgary, 3330 Hospital Drive NW, Calgary, Alberta, Canada T2N 4N1*

There is increased concern about the associations between particulate air pollution and human health. Inhaled and deposited particles play a crucial role in the aetiology of a range of pulmonary diseases. A variety of pulmonary diseases develop from the inhalation and deposition of pathogenic organisms or noxious particles (e.g. viruses, bacteria, spores, pollen, etc.). The inhalation of soot, burned tobacco and paper leads to common pulmonary diseases: chronic bronchitis and lung cancer.

It has been suggested that ultrafine particles might be taken up by cells, including by airway epithelial cells, through a process related to the surface forces exerted on them at the cell membrane–particle interfacial region.

Keywords: airways; surfactant; surface forces; primary defence barrier; particle–surfactant interaction; particle–cell interaction

1. Introduction

With each breath millions of particles enter the lungs, where they may land on the surface of the conducting airways or the alveoli in the gas-exchange region. Upon making contact with the wall (deposition), the processes of retention and clearance begin. These processes depend on many factors, including

(1) particle size, shape, solubility, surface chemistry and elastic properties of both the particles and the lung surfaces;

(2) the anatomical location of the deposition (alveoli, conducting airways), which is important for the route and distance of particle clearance;
(3) the histological structures the particles interact with at the site of deposition, including cells and the surfactant film at the air–liquid interface.

All these factors determine the fate of deposited particles and, hence, they are important for the therapeutic or pathogenic potential of the retained particles.

Clearance of particles deposited in the gas-exchange region is slow; it may last for months or years and involves phagocytosis by cells of the defence system, cell migration and, eventually, mucociliary transport. Particles deposited in the conducting airways, on the other hand, may generally be cleared within 24 h. However, data from the GSF–National Research Centre for Environment and Health in Munich (Scheuch *et al.* 1996) and from our laboratories (Geiser *et al.* 1990*a*) have shown that particles of 6 μm in diameter and smaller are not entirely cleared within this period. Inhalation studies in humans have shown that the fraction cleared within 24 h decreases with decreasing particle size from greater than 90% for particles greater than 6 μm in diameter to less than 30% for particles less than 1 μm in diameter (Scheuch *et al.* 1996).

There are reports on ultrafine particles (less than 0.1 μm in diameter) which show that these tiny particles are more toxic than larger particles (Oberdörster *et al.* 1994). The enhanced toxicity may be related to their greater surface area relative to the particle mass, and so may depend on the provision of more sites to interact with cell membranes and a greater capacity to adsorb and transport toxic substances such as acids (Chen *et al.* 1992). Acidic dusts are powerful irritants and stimulate mucus secretion. A single intratracheal instillation of a sample of airborne dust collected in Ottawa (3–5 μm mass median aerodynamic diameter, 1.9 μm geometric standard deviation) was found to cause a marked increase in thickness of the mucus layer from 2 to greater than 30 μm within *ca.* 5 min (Green *et al.* 1995).

Could the adverse health effects of small particles in the lungs be due to their prolonged retention time? With regards to the particles deposited on the airway wall and their clearance, the air–liquid interface is of high

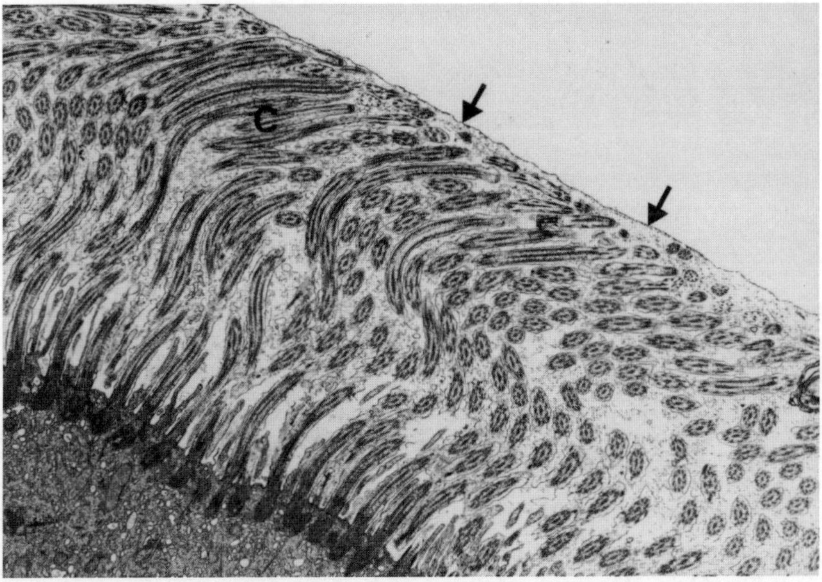

Fig. 1. Transmission electron micrograph of tracheal epithelium of horse. It shows the apical part of the epithelium with cilia (C) in the extracellular aqueous liquid and a surfactant film at the air–liquid interface (arrows) (magnification 6.500×).

significance. Like the alveolar surfaces, the airway surfaces are also covered by a surfactant film (figure 1). This surface layer consists of phospholipids and specific proteins. It is surface active and, hence, results in a reduction of the surface tension at the airway surface (Geiser *et al.* 2000; Gehr *et al.* 1990; Schürch *et al.* 1990).

The alveolar surfactant film stabilizes the gas-exchange area of the lung by reducing the surface tension at the air–liquid interface. In contrast, the mechanical functions of the airway surfactant are not well established. One suggested function is particle transport from the alveoli to the ciliated airways by a surface tension gradient (Horn & Davis 1975; Podgorski & Gradon 1993), another property of airway surfactant is particle displacement toward the epithelium. In the following we will focus on this latter experimentally established property of airway surfactant.

After their deposition on the surfactant film of the airway wall, particles are wetted and displaced toward the epithelium by the surfactant film during the retention process. This translocation of the particles by

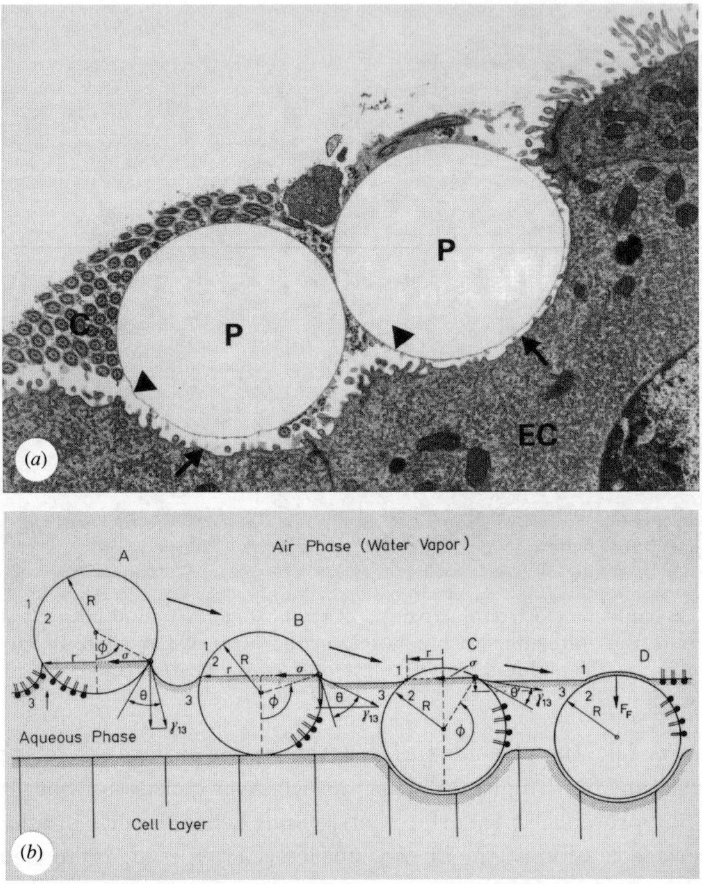

Fig. 2. Particle displacement. (*a*) Transmission electron micrograph of the airway epithelium of a hamster. It shows two polystyrene particles (P) of a test aerosol, 6 µm in diameter, deposited on the airway wall and subsequently displaced into the aqueous liquid lining layer. Note the indentations (arrows) into the epithelial cells (EC) and the cilia (C) surrounding the particles. Arrow heads point to surfactant film (magnification 9.000×; see Geiser *et al.* (1990*b*)). (*b*) Schematic of the immersion of a particle. In addition to surface tension, line tension is considered. Line tension is the one-dimensional analogue of surface tension or the excess free energy density associated with the linear phase where the phases vapour, fluid+film and solid join. A, situation immediately after deposition. B, particle is further displaced. C, surface tension in conjunction with line tension promotes further particle displacement. D, particle is below surfactant film, which may be considered as an elastic skin keeping the particle submerged. θ is the contact angle, R the particle radius, r the radius of the three-phase line, σ the line tension, γ the surface tension, and ϕ indicates position of the three-phase line (Gehr *et al.* 1996).

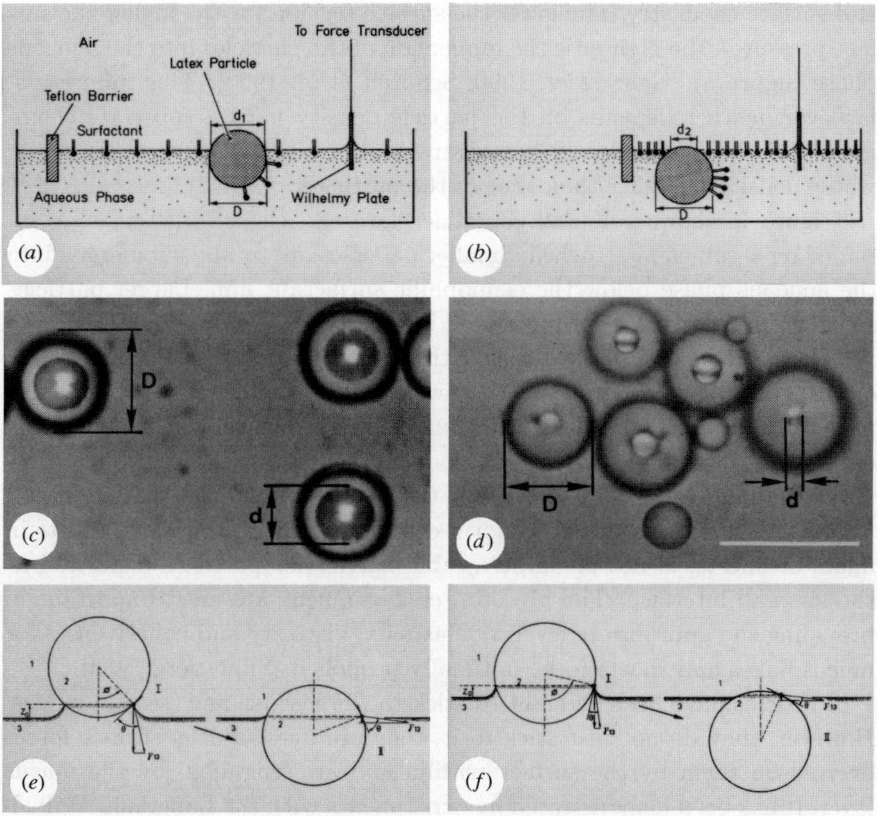

Fig. 3. *In vitro* experiment with particles in a Langmuir–Wilhelmy surface balance. Two mechanically enforced stationary states of polystyrene spheres. Schematic of sphere that had been deposited on a DPPC monolayer at higher (a) and at lower (b) surface tension. (c) and (d) show polystyrene particles placed onto a DPPC monolayer on saline–sucrose subphase; the density of the subphase was higher than the density of the particles. The surface tension was higher in (c) than in (d). (e) and (f) are schematic diagrams of the two stationary states of particles. Higher contact angle (e); lower contact angle (f), characteristic for greater wettability by the fluid phase (Gehr *et al.* 1996).

capillary forces is one of the mechanical properties of the surfactant film (figure 2). *In vitro* experiments using a modified Langmuir–Wilhelmy balance have demonstrated that the extent of particle immersion depends on the surface tension of the surfactant film (e.g. dipalmitoyl phosphatidylcholine (DPPC) of lipid extract surfactant) for a particular particle shape and surface chemistry. The lower the surface tension (or the higher the surface pressure), the greater is the immersion of the particles into the aqueous phase (figure 3) (Gehr *et al.* 1990; Schürch *et al.* 1990). This immersion, however, neither depends on the particle density nor on substrate chemistry. This has been clearly demonstrated by choosing a liquid substrate whose density is higher than that of the particles.

Figure 4*a* shows polymethylmethacrylate particles of different sizes retained on a guinea pig trachea. Smaller particles are totally submerged into the aqueous phase below the osmiophilic surfactant film. Larger particles show various degrees of displacement into the liquid phase.

The rheological characteristics of the aqueous liquid phase, which partly consists of mucus, are well matched with the characteristics of the ciliary propulsion system, and facilitate mucociliary transportation. It depends on the interplay between three components, the cilia, the periciliary (less viscous) fluid, and the mucus, with the surfactant film at the liquid–air interface. The mucus, which is not a continuous layer, but rather exists in flakes or sheets, is not an impermeable barrier for particles deposited at the air–wall interface. The glycoproteins of mucus are most important in providing the appropriate levels of elasticity, viscosity and cohesiveness for mucus flakes and sheets to be optimally propelled (Silberberg 1983).

Particles have been thought to stick to the viscous and 'sticky' mucus. However, they do not just stick to it, they are displaced by surface forces exerted on them by the surfactant film and are 'engulfed' by the mucus layer. It has been demonstrated by experiments with the Langmuir–Wilhelmy balance that, for particle immersion, the aqueous substrate on the airway wall may be mimicked *in vitro* by a viscous Newtonian fluid whose density is higher than that of the particles, provided the surface tension of the surfactant film in the balance is the same as that at the airway wall–air surface.

The *in vitro* experiment also showed that smaller particles are wetted by the substrate to a substantially greater extent than larger ones (figure 4*b*). One may also conclude from these experiments that small particles are more effectively displaced by a fluid phase covered with a surfactant film than larger particles, and that this effect is not related to the thickness of the aqueous liquid phase. The exact displacement mechanisms, especially the initial wetting process of particles of differing sizes, are not yet understood and require further experiments.

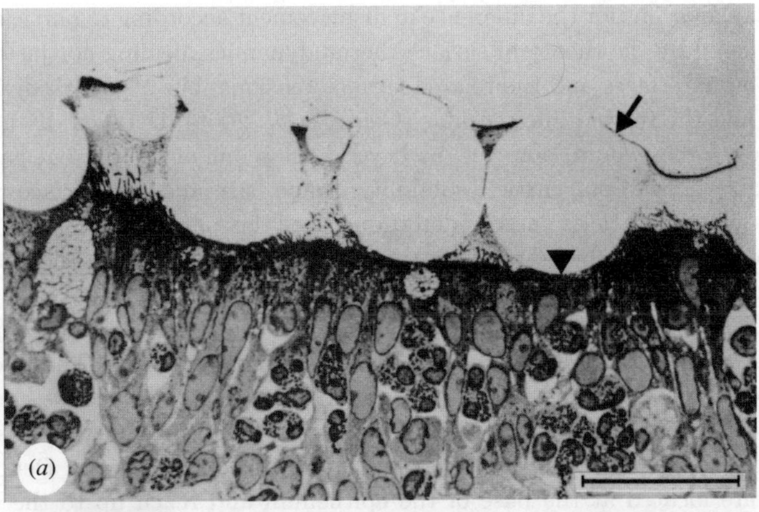

Fig. 4. Displacement of particles of different sizes. (*a*) Light micrograph of tracheal wall of guinea pig, which had been exposed to a polydispersed aerosol of polymethylmethacrylate particles. The particles are submerged beneath an osmiophilic film (arrow) and indent the underlying epithelium (arrow head). Bar equals 30 μm (Gehr *et al.* 1996). (*b*) Light micrograph of polymethylmethacrylate particles of different size on a DPPC film, supported by a saline–sucrose subphase. D is the total diameter, d is the diameter of the segment exposed to air. The ratio d/D for the small particles is much smaller than that for the larger particles. This indicates greater immersion into the subphase of the small particles compared with the large particles. Bar equals 50 μm (Geiser *et al.* 2000).

Explanations for the difference in displacement according to particle size are offered by the theory of surface thermodynamics, dealing not just with dividing interfaces and the related surface tensions, but also with dividing lines and the related line tensions (Geiser *et al.* 2000). Detailed discussion of these forces would, however, be beyond the scope of the present paper.

After deposition, particle solubility, shape, size and the surface properties of the particles largely determine their fate and, hence, also their residence time in the respiratory tract, with consequences for the generation of lung disease. As shown by the model studies outlined above, wetting and displacement of particles at the air–aqueous substrate interface depend on surface forcesa and also, probably, on line tension effects. These forces also determine whether or not the particle is brought into close contact with the epithelial cells and, in particular, with cells of the defence system, like macrophages on the epithelial layer, or with dendritic cells. Dendritic cells are located at the base of the epithelium and reach up to the tight junctions between the epithelial cells with long fine cytoplasmic processes. Surface and line tension forces depend on the interfacial properties of the interacting systems, including the particles themselves and the surrounding aqueous medium, with the interfacial film between medium and particle. Particles with a low surface free energy, such as Teflon, will generally be immersed less readily than high-energy particles, such as glass. Hydrated particles, such as bacteria, with a relatively high surface free energy would be substantially more readily wetted and displaced than hydrophobic particles.

The forces associated with the free energy of interfaces and dividing lines might also contribute substantially to particle–cell interactions. These interactions are considered non-specific, in contrast with specific receptor–ligand interactions. Thus, non-specific interactions may contribute to the uptake of particles by cells such as epithelial cells that do not generally function as phagocytic cells. These findings may have implications for particle pathogenicity and persistence of small particles in the submicrometre range (ultrafine particles, nanoparticles) (Oberdörster *et al.* 1994).

The process of dislocation of particles into the aqueous liquid phase is important, as only particles in proximity to the cells of the defence system can interact with these cells, that is with macrophages and eventually with dendritic cells and epithelial cells. We postulate that the initial step of retention, the particle–surfactant film interaction, determines whether the

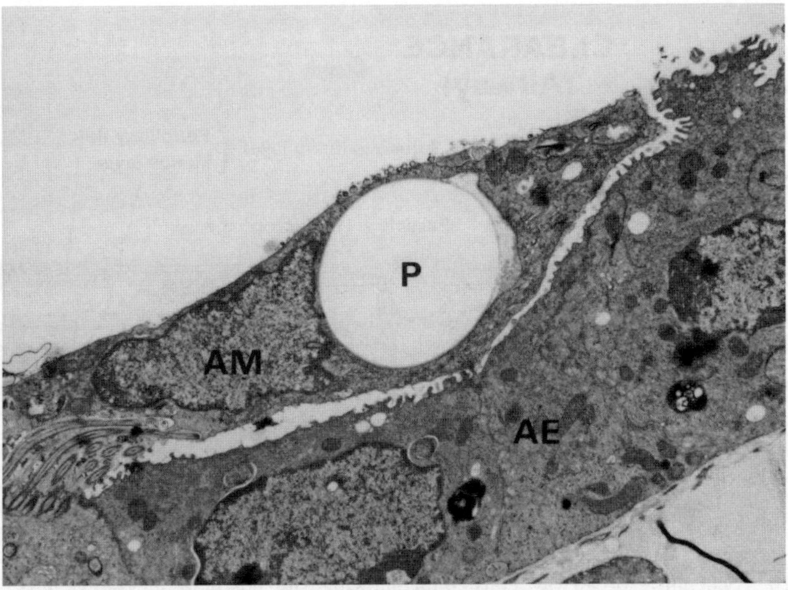

Fig. 5. Transmission electron micrograph of airway macrophage (AM) which had phagocytized a polystyrene particle (P), closely associated with airway epithelium (AE) (magnification 6.000×; see Geiser *et al.* (1990*b*)).

particles are carried away free or in macrophages (figure 5), i.e. by professional phagocytes via the airways (mucociliary clearance), or whether they are transported into the tissue via dendritic cells, i.e. professional antigen-presenting cells. These cells may process the particles and carry them to the specific immunological defence system, present them to T-lymphocytes and eventually initiate an immune reaction (figure 6). Because of this function and the crucial role they play in the pulmonary defence system, dendritic cells in the lungs are called 'sentinels' of the pulmonary immune system (McWilliam *et al.* 2000; Holt & Schon-Hegrad 1987).

In the primary defence against noxious particles, the macrophages play the leading role. They engulf as much material as possible and produce and secrete mediators, causing an inflammatory or immune reaction. In order to be able to trigger an immune reaction, T-lymphocytes must first be activated. As T-lymphocytes cannot react independently to an antigen, they need the help of accessory cells absorbing and processing the antigen and presenting it to them in a proper form.

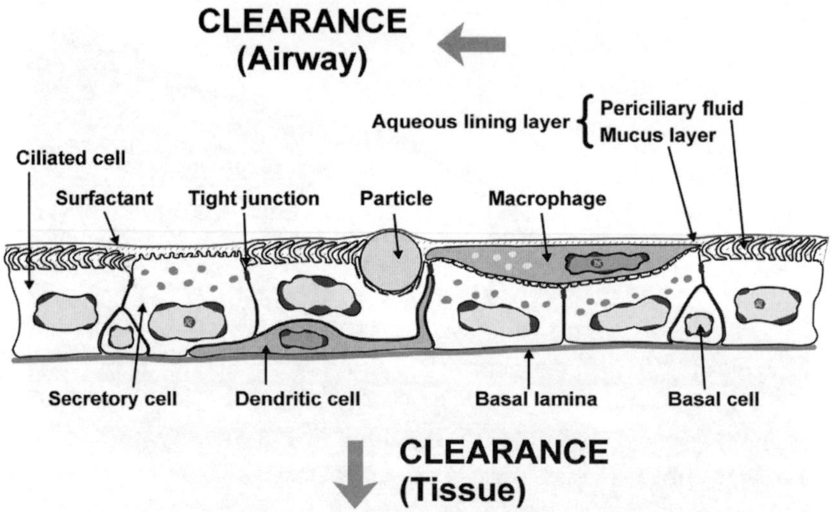

Fig. 6. Schematic of airway epithelium showing the postulated close vicinity of a retained particle with epithelial cells, airway macrophages and dendritic cells (McWilliam *et al.* 2000).

Dendritic cells build a three-dimensional network in the airways, and seem to stay in contact with each other via long cytoplasmic processes reaching out between the epithelial cells and oriented parallel to the surface of the airway epithelium (figure 7*a*). The body of the cells is situated at the epithelial basis but on the epithelial side of the basal membrane. By means of cytoplasmic processes, dendritic cells push through inter-epithelial spaces perpendicular to the epithelial surface, almost reaching the airway lumen and separated from it only by the tight junctions (figure 7*b*) (McWilliam *et al.* 2000; Holt *et al.* 1990). Hence, they are in close association with particulate antigens and airway macrophages, which are in the aqueous layer below the surfactant (figure 6).

In contrast to the macrophages, which engulf as much material as possible for efficient clearance, dendritic cells probably take up only as much antigen material as necessary to stimulate an immune reaction when presented to the T-cells.

The mechanisms by which dendritic cells reach the particles or how the particles get to the cells have not yet been determined. On the one hand,

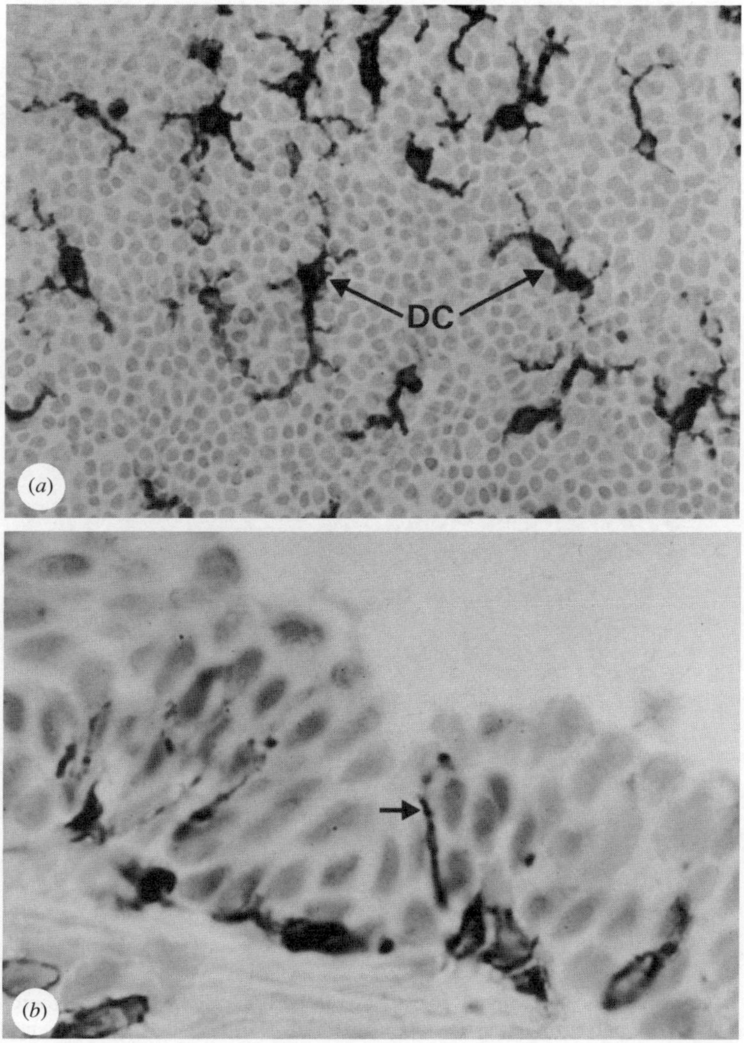

Fig. 7. Light micrographs of dendritic cells (DCs), stained with Ox6 against rat MHC class II antigen, in rat tracheal epithelium. (*a*) Horizontal section through the epithelium, showing the DCs eventually communicating with each other with their long cytoplasmic processes (magnification 500×). (*b*) Section perpendicular to epithelial surface, showing the cell bodies at the base of the epithelium and long cytoplasmic processes reaching up close to the airway lumen (arrow) (magnification 800×). (Courtesy of P. G. Holt, Institute for Child Health Research, University of Western Australia, Perth, Western Australia).

particles phagocytized by macrophages are probably carried out of the lungs by mucociliary transport via airways. Particles engulfed by dendritic cells, on the other hand, are transported into the tissue, from where they reach the lymph nodes via the lymphatic drainage system (figure 6) (McWilliam *et al.* 2000; Gehr *et al.* 1996). However, a cooperation of the dendritic cells with macrophages on the luminal side of the epithelium and with the epithelial cells themselves may be a possible mechanism for the trans-epithelial transport of particles.

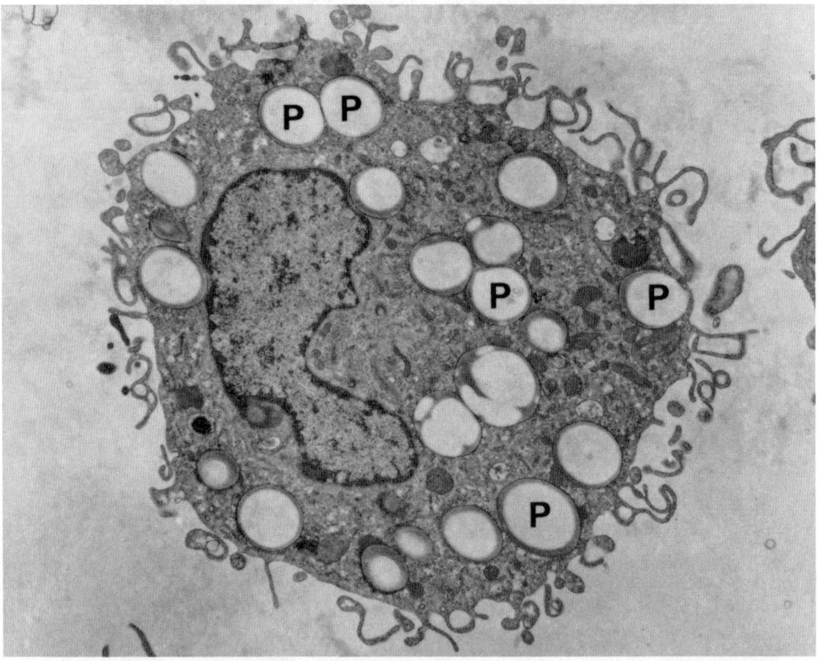

Fig. 8. Transmission electron micrograph of dendritic cell in culture, where it was exposed to 1.5 μm polystyrene particles (P) (magnification 11.000×). (Specimen courtesy of L. P. Nicod, Cantonal Hospital, University of Geneva, Geneva; micrograph courtesy of S. G. Kiama, Institute of Anatomy, University of Bern, Bern, Switzerland).

In vitro experiments have indicated that dendritic cells are less capable of phagocytosis than macrophages, but the phagocytic capacity of dendritic cells is much higher than expected, even though this is only true during their immature stage (Kiama *et al.* 1999) (figure 8). While the cells mature, their capability of antigen presentation is developed and the capability of phagocytosis disappears almost completely. During this process the dendritic cells leave the epithelium and migrate to the lymph nodes (McWilliam *et al.* 2000; Kiama *et al.* 1999).

The displacement of particles deposited on the surfactant film by surface and possibly line tension forces exerted on them by this film is probably the initial step of a complex cascade of defence processes in the lungs. Surfactant, or more precisely its film, may, therefore, be called a primary defence barrier. Through opsonization by surfactant or surfactant components during displacement by wetting, particles are probably rendered less toxic and more attractive to phagocytic cells. It is proposed that the opsonization helps guiding particles along that clearance pathway which is most beneficial for our health, namely up the airways and out of the organism, or into the tissue to be presented to the specific defence system (figure 6) (Gehr *et al.* 1996).

It is of great importance and remains to be determined whether ultrafine, sub-micron particles experience the same fate as the micrometric size particles considered above. It has already been suggested that ultrafine particles might be taken up by cells, including by airway epithelial cells, through a process related to the surface forces exerted on them at the cell membrane–particle interfacial region.

References

Chen, C. C., Miller, P. D., Amdur, M. O. & Gordon, T. 1992 Airway hyperresponsiveness in guinea pigs exposed to acid-coated ultrafine particles. *J. Toxicol. Environ. Health.* **35**, 165–174.

Gehr, P., Schürch, S., Berthiaume, Y., Im Hof, V. & Geiser, M. 1990 Particle retention in airways by surfactant. *J. Aerosol Med.* **3**, 27–43.

Gehr, P., Green, F. H. Y., Geiser, M., Im Hof, V., Lee, M. M. & Schürch, S. 1996 Airway surfactant, a primary defense barrier: mechanical and immunological aspects. *J. Aerosol Med.* **9**, 163–181.

Geiser, M., Cruz-Orive, L. M., Im Hof, V. & Gehr, P. 1990*a* Assessment of particle retention and clearance in the intrapulmonary conducting airways of hamster lungs with the fractionator. *J. Microsc.* **169**, 75–88.

Geiser, M., Im Hof, V., Gehr, P. & Cruz-Orive, L. M. 1990*b* Histological and stereological analysis of particle retention in the conducting airways of hamster lungs. *J. Aerosol Med.* **3**, 131–145.

Geiser, M., Schürch, S., Im Hof, V. & Gehr, P. 2000 Retention of particles: structural and interfacial aspects. In *Particle–lung interaction* (ed. P. Gehr & J.

Heyder), vol. 143. In *Lung biology in health and disease* (exec. ed. C. Lenfant), pp. 291–322. New York: Marcel Dekker.

Green, F. H. Y., Lee, M. M., Roth, S. H., Karkhanis, A., Schürch, S., Schürch, D., Bjarnason, S. G., Vincent, R. & Gehr, P. 1995 Effects of solid and liquid acid particles on airway mucus. *J. Aerosol Med.* **8**, 111.

Holt, P. G. & Schon-Hegrad, M. A. 1987 Localization of T-cells, macrophages and dendritic cells in respiratory tract tissue: implications for immune function studies. *Immunology* **62**, 349–356.

Holt, P. G., Schon-Hegrad, M. A., Oliver, J., Holt, B. J. & McMenamin, P. G. 1990 A contiguous network of dendritic antigen-presenting cells within the respiratory epithelium. *Arch. Allergy Appl. Immunol.* **91**, 155–159.

Horn, L. W. & Davis, S. H. 1975 Apparent surface tension hysteresis in dynamical systems. *J. Colloid Interface Sci.* **51**, 459–478.

Kiama, S. G., Cochand, L., Nicod, L. P. & Gehr, P. 1999 Stereological assessment of phagocytosis by monocytes, alveolar macrophages and dendritic cells. *J. Aerosol Med.* **12**, 49 (abstract).

McWilliam, A., Holt, P. G. & Gehr, P. 2000 Dendritic cells as sentinels of immune surveillance in the airways. In *Particle–lung interaction* (ed. P. Gehr & J. Heyder), vol. 143. In *Lung biology in health and disease* (exec. ed. C. Lenfant), pp. 291–322. New York: Marcel Dekker.

Oberdörster, G., Ferin, J. & Lehnert, B. E. 1994 Correlation between particle size, *in vivo* particle persistence, and lung injury. *Environ. Health Perspect.* **102** (Suppl. 5), 173–179.

Podgorski, A. & Gradon, L. 1993 An improved mathematical model of hydrodynamical self-cleansing of pulmonary alveoli. *Ann. Occup. Hygiene* **37**, 347–365.

Scheuch, G., Stahlhofen, W. & Heyder, J. 1996 An approach to deposition and clearance measurements in human airways. *J. Aerosol Med.* **9**, 35–41.

Schürch, S., Gehr, P., Im Hof, V., Geiser, M. & Green, F. H. Y. 1990 Surfactant displaces particles toward the epithelium in airways and alveoli. *Respir. Physiol.* **80**, 17–32.

Silberberg, A. 1983 Biorheological matching: mucociliary interaction and epithelial clearance. *Biorheology* **20**, 215–222.

Discussion

R. L. Maynard (*Department of Health, Skipton House, 80 London Road, London*). You pointed out that as a surfactant film contracts and its surface tension falls, particles are drawn through the surface. Surfactants have a range of roles including stabilizing alveoli of different sizes and reducing the movement of water into alveoli: these effects are dependent on the low and variable surface tension of the surfactant film. Is the effect you described an undesirable side-effect of these functions?

P. Gehr. I wouldn't call the displacement of particles by surface forces exerted on them by surfactants a side-effect. It is just another effect which was unknown until particle retention was investigated, i.e. '[s]ince the second half of the eighties'. This effect is more pronounced in the alveoli since the surface tension becomes very low (less than 1 mN/m) at expiration. However, as recently found by my colleagues Marianne Geiser and Samuel Schürch, even with the much higher surface tension in the airways (but lower than that of mucus and other biopolymers), all particles smaller than 10 μm were found to be displaced.

C. Kim (*National Health and Environmental Effects Research Laboratory, US EPA, Research Triangle Park, NC, USA*). Particles deposited in the airways are cleared out of the lung by mucociliary escalator.

It is believed that mucociliary transport is efficient only if particles are positioned on the mucus layer. If particles are forced to sink down through the mucus layer, or they are landed in the areas where mucus layer is absent, those particles may not be cleaned by mucociliary escalator. Do you have any suggestions concerning how much of the particles deposited in the airways may be found on the epithelial surface?

P. Gehr. But these particles may eventually be cleared by ciliary activity (as we have observed but which needs to be defined). The displacement may get particles into (not through) the mucus; that is, the mucus will act as a containing space for transportation of the particles.

The number of particles found on the epithelial surface probably depends on the size of the particles, it might be most of them or only a few. We only have observations of this process, this needs to be studied in more detail. From studies of the GSF (W. Stahlhofen, G. Scheuch and J. Heyeder) in Munich it is known that the smaller the deposited particles the smaller the fast clearing fraction is.

H. Rees (*Therapeutics and Toxicology Centre, UWCM Academic Centre, Llandough Hospital, Cardiff, UK*). You presented information on the interactions between fine polystyrene particles and respiratory epithelium. Have you studied the behaviour of other materials? Have you found that polytetrafluorethylene (Teflon) particles and asbestos fibres behave much like polystyrene, and that the behaviour of different materials can be predicated by size?

P. Gehr. Our inhalation studies with different types of fine particles like polystyrene, Teflon, puff ball spores and glass fibres showed that they are all displaced into the liquid layer by surface forces if deposited on the surfactant film at the air–liquid interface. The displacement process depends on the size rather than on the shape and type of the particles, i.e. all inhalable particles will be displaced. I would predict that asbestos fibres also behave in the same way.

CHAPTER 12

TOXICOLOGY OF ULTRAFINE PARTICLES: *IN VIVO* STUDIES

Günter Oberdörster

Department of Environmental Medicine, School of Medicine and Dentistry, University of Rochester, Rochester, NY 14642, USA

Ultrafine particles (less than 100 nm in diameter) are encountered in ambient air and at the workplace. Normal background levels in the urban atmosphere for ultrafine particles are in the range $1\text{–}4 \times 10^4$ cm^{-3}; however, their mass concentration is normally not greater than 2 µg m^{-3}. At the workplace, ultrafine particles occur regularly in metal fumes and polymer fumes, both of which can induce acute inflammatory responses in the lung upon inhalation. Although ultrafine particles occurring at the workplace are not representative, and, therefore, are not relevant for urban atmospheric particles, their use in toxicological studies can give valuable information on principles of the toxicity of ultrafine particles. Studies in rats using ultrafine polymer fumes of polytetrafluoroethylene (PTFE) (count median diameter *ca.* 18 nm) showed that (i) they induced very high pulmonary toxicity and lethality in rats after 15 min of inhalation at 50 µg m^{-3}; (ii) ageing of PTFE fumes resulted in agglomeration to larger particles and loss of toxicity; (iii) repeated pre-exposure for very short periods protected against the toxic and lethal effects of a subsequent 15 min exposure; (iv) rapid translocation of PTFE particles occurred to epithelial, interstitial and endothelial sites. Since one characteristic of urban ultrafine particles is their carbonaceous nature, exposure of rats to laboratory-generated ultrafine carbonaceous (elemental, and organic, carbon) particles was carried out at a concentration of *ca.* 100 µg m^{-3} for 6 h. Modulating factors of responses were prior low-dose inhalation of endotoxin in order to mimic early respiratory tract infections, old age (22-month old rats versus 10-week old rats) and ozone co-exposure. Analysis of results showed that (i) ultrafine carbon particles can induce slight inflammatory responses; (ii) LPS priming and ozone co-exposure increase the responses to ultrafine carbon; (iii) the aged lung is at increased risk for ultrafine particle-induced oxidative stress. Other studies with ultrafine and fine TiO_2 showed that the same mass dose of ultrafine particles has a significantly greater inflammatory potential than fine particles. The increased surface area of ultrafine particles is apparently a most important determinant for their greater biological activity. In addition, the propensity of ultrafine particles to translocate may result in systemic distribution to extrapulmonary tissues.

Keywords: ultrafine particles; carbon; pulmonary inflammation; dosimetry; adaptation; LPS; age

1. Introduction

Epidemiological studies have shown an association between increased particulate air pollution and adverse health effects in susceptible parts of the population, in particular the elderly with respiratory and cardiovascular diseases (EPA 1996). Urban airborne particles consist of three modes: ultrafine particles, accumulation-mode particles (which together with the ultrafines form the fine particle mode), and coarse-mode particles. Ultrafine particles (particles less than 0.1 μm in diameter) contribute very little to the overall mass, but they contribute most to the number concentration of airborne urban particles. Any of the three modes could causally be associated with adverse health effects, although epidemiological studies suggest that fine-mode particles are better correlated with such effects than coarse particles (EPA 1996). We are testing the hypothesis that urban ultrafine particles consisting of a carbonaceous core with attached inorganic and organic materials can cause adverse health effects in compromised subjects during episodic high increases in concentration.

Ultrafine particles are also encountered in the workplace as fumes generated from smelting processes of metals and heating of polymers. Resulting acute effects in exposed workers have been well documented as metal-fume fever and polymer-fume fever, consisting of acute pulmonary and inflammatory responses, which can also be accompanied by systemic effects with symptoms of fever, nausea and headaches (Drinker *et al.* 1927; Gordon *et al.* 1992; Goldstein *et al.* 1987; Rosenstock & Cullen 1986). Concentrations of such ultrafine particles at the workplace are very high compared with those in the urban atmosphere, and often the high concentrations encountered at the workplace result in particles which have agglomerated, by classical coagulation phenomena, to larger sized particles.

Although workplace exposures to ultrafine fumes are known to result in potentially severe acute responses, a direct extrapolation to effects of urban ultrafine particles is not possible. The toxicity of carbonaceous urban ultrafine particles is apparently much lower compared with that of freshly generated metal or polymer fumes. However, an important difference between exposures to ultrafine particles at the workplace and in the general environment lies in the exposed population. These are healthy adults at the workplace, and all of the population, including the most sensitive members, in the urban setting. Despite this, it is likely that certain principles and concepts of ultrafine particle toxicology are common between workplace ultrafine particles and urban ultrafine particles. Therefore, it could be very useful to assess the toxicity and behaviour of ultrafine particles encountered at the workplace, which might be helpful for designing studies with ambient ultrafine particles. This paper will first summarize some of our studies with polymer fumes, discuss certain principles of ultrafine particle dosimetry and behaviour, and finally describe results of toxicological studies with carbonaceous and other ultrafine particles as surrogates for ambient particles.

2. Studies with Ultrafine Particles from Polymer Thermodegradation

2.1. *Characterization and Toxicity of Inhaled Polytetrafluoroethylene (PTFE)*

Heating of PTFE (Teflon) to *ca.* 480 °C results in the generation of fumes that consist of ultrafine particles and some gas-phase products, mainly HF. Heating to temperatures above 500 °C will generate additional gas-phase products like perfluoroisobutylene and others, which are highly toxic (Lee & Seidel 1991; Waritz & Kwon 1968; Makulova 1965). Studies in our laboratory were performed at heating temperatures below 500 °C to avoid confounding of the responses due to the toxic gas-phase components. The ultrafine particle size of the PTFE fumes ranges between 10 and 50 nm. Exposure of rats to a concentration of 5×10^5 particles cm^{-3} (equivalent to *ca.* 50 μg m^{-3}) for 10–20 min results in the development of severe pulmonary inflammation and hemorrhage within 4 h post exposure (Oberdörster *et al.* 1995). High mortality due to the severe pulmonary oedematous response also occurs. Lung lavage fluid shows significantly increased neutrophils, up to 80% of total cells, highly increased protein levels, and increases in lysosomal and cytoplasmic enzymes. Histologically, a high degree of alveolar and interstitial oedema is present and both epithelial and endothelial cell layers are severely disrupted. These ultrafine particles are thought to be the cause of the toxicity of PTFE fumes. Using electron energy-loss spectroscopy (EELS), ultrafine fluorine-containing particles can be found shortly after the exposure in epithelial, interstitial and endothelial sites of both the conducting airways and alveolar regions of the lung (Oberdörster *et al.* 2000). Analysis of lung mRNA shows significant upregulation of proinflammatory cytokines such as *IL*-6, *IL*-1b, *IL*-1a, TNFα, and of antioxidants such as MnSOD, and metallothionein and the chemokine MIP-2. These responses are consistent with injuries due to severe oxidative lung damage.

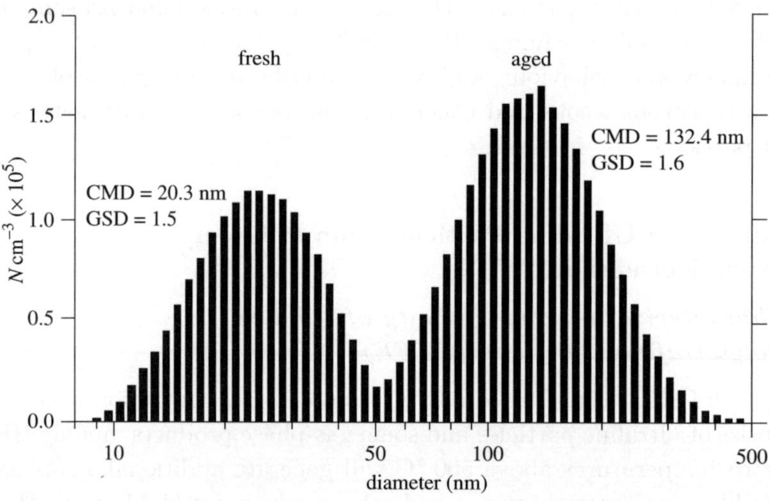

Fig. 1. Particle size distribution of freshly generated and aged PTFE fumes showing the shift from ultrafine particle distribution to a size distribution with a median greater than 100 nm due to coagulation after 3.5 min of ageing.

2.2. *Effect of Ageing on PTFE Toxicity*

Particles present in the air at high number concentrations tend to coagulate and thereby form larger particles over time. This coagulation process can occur within fractions of a second if number concentrations exceed 10^8 particles cm^{-3}, and will be progressively slower at lower concentrations

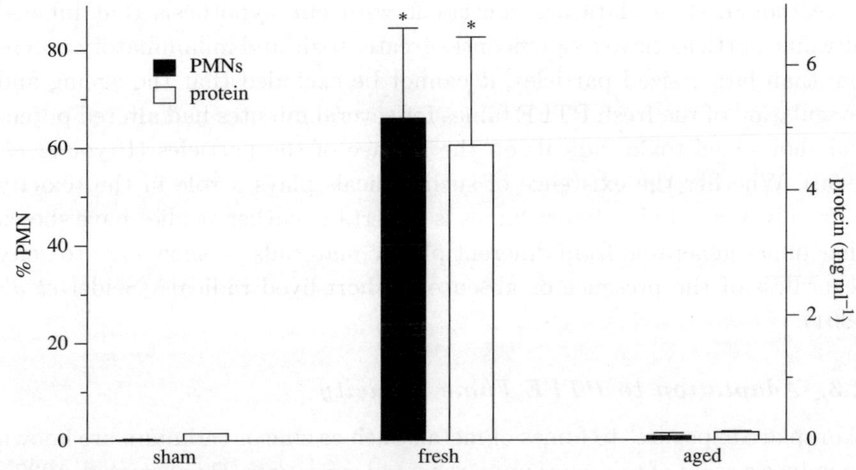

Fig. 2. Lung lavage parameters of rats 4 h after a 50 min exposure to fresh and aged PTFE fumes. The percentage of neutrophils of the total lavage cells and the lavage protein content are shown. ($N = 4$ per group; mean ±SD; *, significantly different from sham and aged (ANOVA, $p < 0.05$).)

(Hinds 1982). Ageing of the freshly generated PTFE fumes for 3–5 min results in a shift of the particle size distribution from a count median diameter (CMD) of 15–20 nm to one above 100 nm (figure 1). Gas-phase products do not seem to change in this ageing process. If the toxicity of PTFE fumes is indeed caused by the presence of ultrafine particles and if larger particles are less toxic, one would predict that exposure to the aged PTFE fumes with the larger particles will result in significantly less toxicity. Figure 2 shows the results of a study in which rats were exposed to either the freshly generated PTFE fumes at a concentration of 5×10^5 particles cm^{-3} or to aged PTFE fumes at a concentration of 1.5×10^5 particles cm^{-3}. The mass concentrations of the particles were 50 and 70 μg m^{-3}, respectively. Control rats were separately exposed to filtered air, and inflammatory lung lavage parameters were determined 4 h post exposure. As can be seen from figure 2, the rats exposed to the freshly generated fumes showed the expected high increase in lavage neutrophils and protein, indicating the presence of severe inflammatory pulmonary oedema. In contrast, sham-exposed control rats and rats exposed to the aged fumes for 15 min did not show significant changes in the lavage parameters.

Although these data are consistent with our hypothesis that inhaled ultrafine particles have a significantly greater toxic and inflammatory potential than larger sized particles, it cannot be excluded that the ageing and coagulation of the fresh PTFE fumes for several minutes had altered potential short-lived toxic radicals on the surface of the particles (Pryor *et al.* 1990). Whether the existence of such radicals plays a role in the toxicity of freshly generated polymer fumes is uncertain; earlier studies have shown that fumes generated from different plastic materials all show high toxicity regardless of the presence or absence of short-lived radicals (Seidel *et al.* 1991).

2.3. *Adaptation to PTFE Fume Toxicity*

Workplace exposures to fumes of metals such as zinc or cadmium are known to induce a state of tolerance upon repeated exposures (Drinker *et al.* 1927). Such adaptation is associated with increases of antioxidant proteins including metallothionein (Hart *et al.* 1989), which protect the lung from the toxic effects of subsequent exposure to high concentrations of these fumes. Such adaptive responses have not been demonstrated for polymer fumes. However, since the mechanism for the high acute toxicity of these metal fumes is most likely to be due to oxidative stress, one might expect that similar adaptive changes occur after exposures to polymer fumes consisting of ultrafine particles.

In order to investigate the induction of PTFE fume tolerance, we exposed rats to PTFE fumes containing 5×10^5 particles cm^{-3} (equivalent to *ca.* 50 $\mu g\, m^{-3}$) for 5 min each on three consecutive days, followed by a 15 min exposure to the same concentration on day four. A second group of rats was sham exposed for 5 min on three consecutive days and on day four received the 15 min PTFE fume exposure together with animals from group one. A third group was sham exposed to filtered air on all four days. Four hours after the 15 min exposure, surviving animals were euthanized and their lungs lavaged for measurement of inflammatory parameters. Figure 3 shows the result of this study with respect to lavage neutrophils and lavage protein content. During the three day adaptation, animals of the adapted group did not show any respiratory symptoms of being affected by these short, 5 min, exposures. Neither did the rats of this group show any clinical signs of respiratory effects after the 15 min exposure of PTFE fumes on day four. In contrast, rats of the non-adapted group were severely

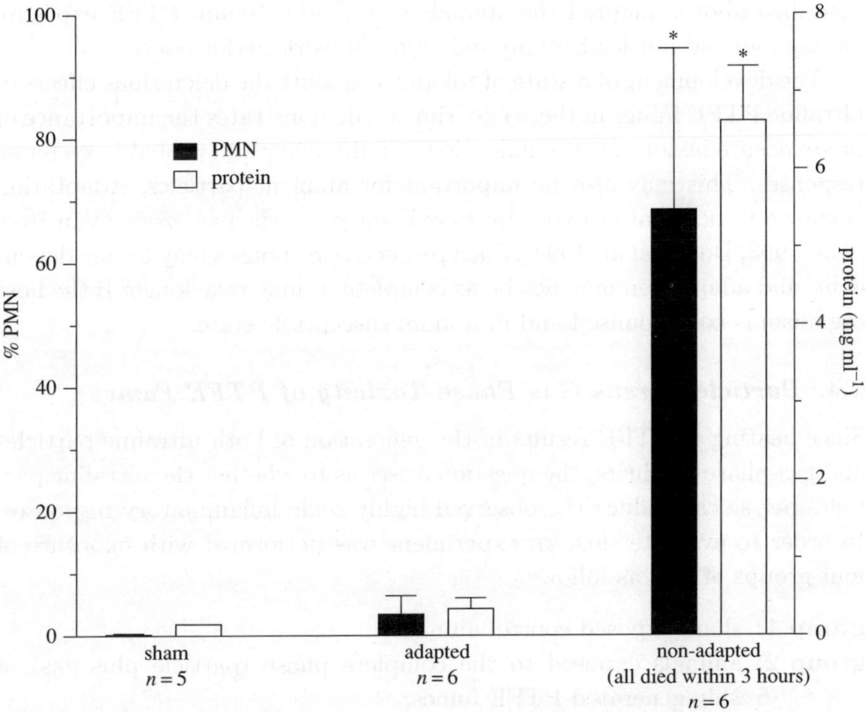

Fig. 3. Lavage parameters 4 h after a 15 min exposure to PTFE fumes in adapted and non-adapted F-344 rats. The percentage neutrophils and total lavage protein content are shown. ($N = 6$ per group, mean ±SD; ∗, significantly different from sham and adapted group (ANOVA, $p < 0.05$).)

affected by the 15 min PTFE fume exposure on day four: all of them had to be euthanized within 3 h post exposure because of severe dyspnea. Sham-exposed control animals, like the adapted animals, did not show any effects (figure 3).

The high inflammatory effects in the lung observed after PTFE fume exposure are likely to also lead to systemic responses. This in turn would affect the rats' behaviour, and work performance is expected to drop significantly. Indeed, exposure of rats to PTFE fumes for 10 min was found to reduce performance on a running wheel by up to 40%, the effect starting immediately after exposure and lasting several days. However, pre-exposure of rats on three consecutive days, as was done in the adaptation study described above, adapted the animals so that the 10 min PTFE exposure on day four did not lead to any reduction in work performance.

The development of a state of tolerance against the deleterious effects of ultrafine PTFE fumes in these experiments demonstrates the importance of pre-exposure history for the induction of pulmonary and probably systemic responses. This may also be important for ambient particles. Adaptation is known to occur after exposure to oxidant gases such as ozone (Van Bree *et al.* 1993; Dodge *et al.* 1994). Such protective responses may be age dependent, and adaptation may not be as complete or may take longer if the host organism is compromised and in a more susceptible state.

2.4. *Particle Versus Gas Phase Toxicity of PTFE Fumes*

Since heating of PTFE results in the generation of both ultrafine particles and gas-phase products, the question arises as to whether the ultrafine particles per se can induce the observed highly toxic inflammatory responses. In order to evaluate this, an experiment was performed with exposure of four groups of rats as follows:

group 1: sham-exposed control animals;
group 2: animals exposed to the complete phase (particle plus gas) of freshly generated PTFE fumes;
group 3: animals exposed to the gas phase only after filtering the ultrafine particles; and
group 4: animals exposed to ultrafine PTFE particles only.

For the latter group, the generation of PTFE fumes was changed from heating Teflon in air to heating in argon, which resulted in the generation of ultrafine particles with no detectable gas-phase compounds. The exposure was for 25 min in compartmentalized whole-body exposure chambers at an ultrafine particle number concentration of 5×10^5 particles cm^{-3}. Four hours after exposure, the animals were euthanized and inflammatory lung lavage parameters determined. As shown in figure 4, neither the group exposed to the gas phase alone nor the group exposed to the ultrafine particles alone showed significant responses with respect to lavage neutrophils or protein compared with control animals. In contrast, the group exposed to the complete fume, particle plus gas phase, confirmed the highly toxic nature of the PTFE fumes by responding with high increases in both lavage neutrophils and protein.

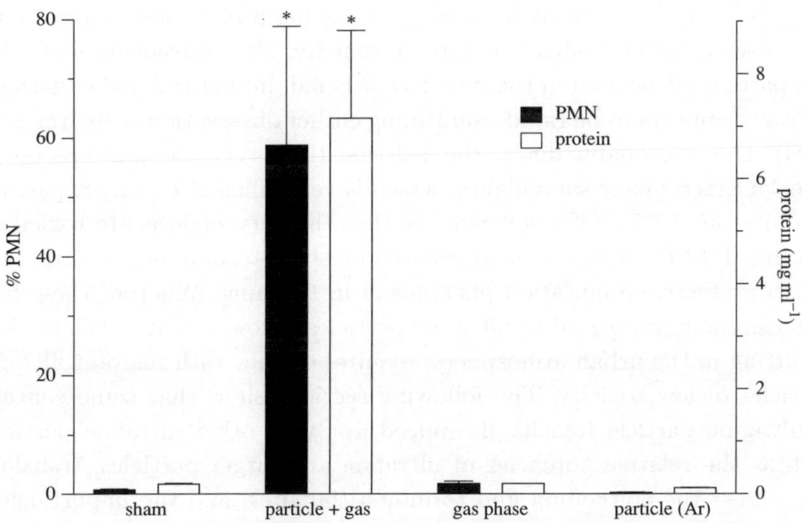

Fig. 4. Lung lavage parameters of rats 4 h after a 25 min exposure to the gas or particle phase of PTFE fumes or to the complete particle-plus-gas phase. The percentage neutrophils of total lavage cells and lavage protein content are shown. ($N = 5$ per group, mean ±SE; ∗, significantly different from all other groups (ANOVA, $p < 0.05$).)

This result indicates that 'clean' ultrafine PTFE particles (generated in argon) do not induce significant toxicity after short-term exposure. However, it is conceivable that the argon-generated PTFE particles may have different surface reactivity compared with the air-generated PTFE particles, which may contribute to their high toxicity. Clearly, the gas-phase constituents alone, when generated at below 500 °C, do not exert pulmonary toxicity, but their combination with the ultrafine particles does. Preliminary attempts to adsorb PTFE gas-phase constituents onto ultrafine carbon particles did not result in acute pulmonary, inflammatory responses. Earlier studies have also found that gas-phase compounds of PTFE fumes do not cause the high toxicity when generated below 500 °C (Waritz & Kwon 1968). However, additional studies may be needed to confirm that gas-phase compounds do not play a role in PTFE fume toxicity.

2.5. *Conclusions from PTFE Fume Studies*

The conclusions from these studies with PTFE fumes are that the ultrafine particles are, indeed, key to their high pulmonary toxicity. Furthermore, coagulation to larger particles during the ageing process results in a significant loss of the high ultrafine particle toxicity. The translocation of ultrafine particles deposited in the lung to epithelial, interstitial and endothelial sites also appears to be rapid, confirming earlier observations (Stearns *et al.* 1994). This is probably due to the fact that these very tiny particles escape alveolar macrophage surveillance, which is very efficient for larger particles (Hahn *et al.* 1977). Of significance is that the very high, acute toxicity of ultrafine PTFE fumes can be prevented by short-term pre-exposures, leading to protective adaptation phenomena in the lung. Whether these findings can be extrapolated to ultrafine particles of low toxicity, such as those occurring in the urban atmosphere, requires studies with relevant ultrafine particles of low toxicity. The following sections show that some concepts of ultrafine particle toxicity do indeed apply to other ultrafine particles, such as the relative potencies of ultrafine and larger particles, translocation across the epithelium and to interstitial sites, and the importance of combined exposure of ultrafine particles with an oxidant gas.

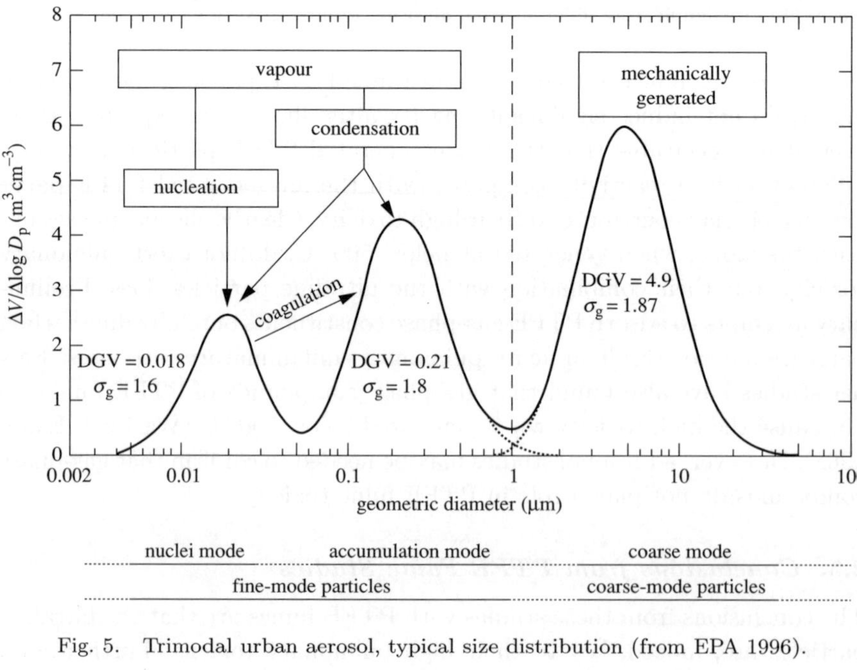

Fig. 5. Trimodal urban aerosol, typical size distribution (from EPA 1996).

Table 1. Urban atmosphere particle concentrations: ultrafines (below 100 nm). Nature of particles: carbonaceous; and others?.

condition	particles (cm^{-3})	($\mu g\ m^{-3}$)	reference
background average (Erfurt, Los Angeles)	1–5 × 10^4	0.8–1.6	Tuch *et al.* (1997), Hughes *et al.* (1998) (Cass: 10 t d^{-1} emitted in Southern California)
episodic event peak (Frankfurt)	3 × 10^5	*ca.* 50	Brand *et al.* (1992)
episodic events and traffic peak (Atlanta)	1 × 10^6	?	P. McMurry (2000, personal communication) (preliminary data)

3. Studies with Ultrafine Particles of Low Toxicity

3.1. *Dosimetric Aspects of Ultrafine Particle Toxicology*

As mentioned in § 1, ultrafine particles are one of three particle modes in the urban atmosphere (figure 5). Normal background levels of the ultrafine mode are below 2 $\mu g\ m^{-3}$, with number concentrations in the range 1–4 × 10^4 particles cm^{-3} (Hughes *et al.* 1998). Table 1 summarizes results of measurements which show that episodic events can give rise to much higher concentrations, approaching 1×10^6 particles cm^{-3} and mass concentrations of up to 50 $\mu g\ m^{-3}$. In addition to traffic-related increases, other sources seem to be responsible for these high concentrations as well, and further research is needed to determine the composition of these particles. Episodic increases in ultrafine particle concentrations occur frequently and can last for more than a day and it is probably during or after these events that health effects may occur. Therefore, these concentrations can serve as a guideline for designing toxicological studies using realistic exposure levels. Importantly, when performing studies with ultrafine particle exposures in experimental animals, one should attempt to estimate deposited lung doses of ultrafine particles with the goal of achieving lung burdens that would also be experienced by humans under a given urban exposure scenario. Exposure concentrations to achieve this may be different between humans and experimental animals, as is discussed below.

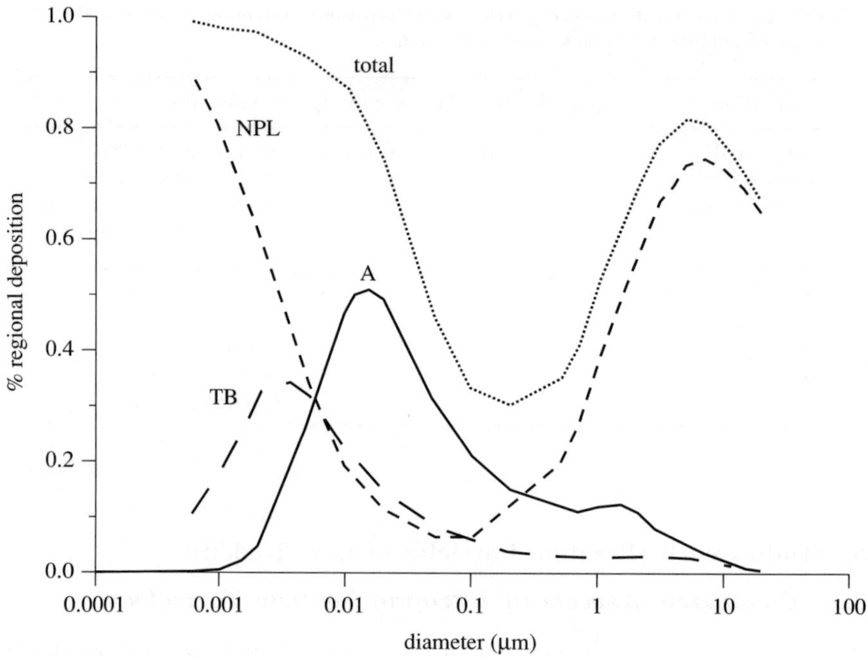

Fig. 6. Predicted deposition of inhaled particles of different sizes of unit density in the human respiratory tract during nose breathing, light exercise (ICRP 1994). 'NPL' denotes nasopharyngolaryngeal deposition; 'TB' denotes tracheobronchial deposition; 'A' denotes alveolar deposition; 'total' denotes the sum of particle depositions in the respiratory tract for all three compartments.

Deposition of inhaled ultrafine particles in the respiratory tract occurs by diffusional processes. Several studies have shown that the human nose is highly efficient in collecting inhaled ultrafine particles by this process in the nasal compartment (Cheng *et al.* 1991; Swift *et al.* 1992). However, these data also show that diffusional deposition in the nasopharyngeal compartment is highly dependent on the ultrafine particle size, such that ultrafine particles below *ca.* 10 nm deposit with high efficiency in the nose, whereas this deposition becomes less for particles between 10 and 100 nm. There is thus a misconception that all ultrafine particles deposit efficiently in the nose. However, predictive particle deposition models show that the probability of deposition of 20 nm particles is greatest in the alveolar region (figure 6). Likewise, the tracheobronchial region can also be a significant target for even smaller ultrafine particles, as indicated in figure 6. If the deposited dose is expressed per unit surface area of the epithelium, the tracheo-bronchially deposited dose of ultrafine particles can be up to 50-fold higher than that for the alveolar region. These are important dosimetric aspects in ultrafine particle toxicology, which need to be considered when studying these particles. In addition, an important factor is the fate of deposited ultrafine particles, that is their disposition, which seems to be different from larger particles, as was discussed in the PTFE fume experiments.

Table 2. Particle dosimetry. Human equivalent concentration for inhaled 20 nm particles (ICRP human model and Yeh & Schum rat model).

	alveolar deposition fraction	deposited dose over 6 h per m^2 alveolar surface at 10 μg m^{-3}	relative dose at same inhaled concentration
rat, nasal breathing at rest	0.27	365 ng	1
human, nasal breathing, light	0.5	692 ng	1.9

Another important aspect of particle toxicology in general relates to comparative dosimetry between experimental animals and humans. Rodents, as the most frequently used experimental animals, have different deposition efficiencies for particles in their respiratory tract compared with humans. Table 2 compares the deposition efficiency of inhaled 20 nm particles—which is the peak of the urban ultrafine particle mode—between rats and humans based on deposition models by Yeh & Schum (1980) for rats and the ICRP (1994) model for humans. As this table indicates, the human equivalent concentration for 20 nm ultrafine particles is about twice that for the rat. The aforementioned high episodic ultrafine particle concentration of 50 μg m^{-3} in urban air would, therefore, be equivalent to *ca.* 100 μg m^{-3} inhaled by rats. Since this comparison is based on predictive deposition models, and since these models have not been validated experimentally for ultrafine particles, there could be considerable uncertainties. For example, significant differences exist between the NCRP and the ICRP deposition model. Thus, there is an urgent need to validate these deposition models by experimental data in rodents as well as humans.

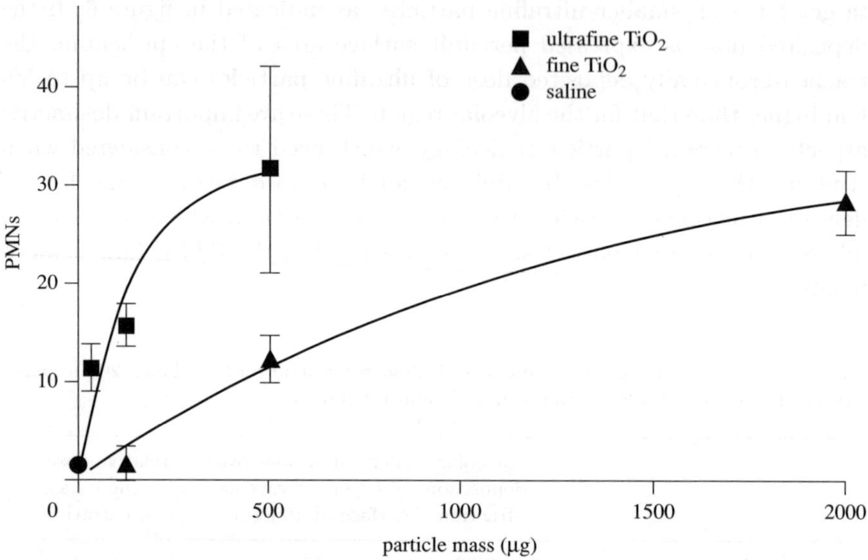

Fig. 7. Inflammatory response 24 h after instillation of different doses of ultrafine and fine TiO_2 in rats, expressed as percentage neutrophils in lung lavage as a function of instilled particle mass. Ultrafine TiO_2, *ca.* 20 nm; fine TiO_2, *ca.* 250 nm.

3.2. *Inflammatory Potential of Ultrafine Versus Fine Particles*

We postulate that deposited ultrafine particles induce a greater inflammatory response per given mass than larger particles of the size of the accumulation mode. Results of our studies with ultrafine freshly generated and larger, aged PTFE fumes described above are consistent with this hypothesis. We tested this hypothesis by dosing rats with two different particle types of a rather benign dust, TiO_2. Ultrafine TiO_2 with an average particle size of 20 nm and pigment grade (fine) TiO_2 with an average particle size of *ca.* 250 nm were used. Doses ranging from 30 to 2000 μg of TiO_2 were intratracheally instilled into groups of rats. The inflammatory response in their lungs was assessed by analysis of cellular and biochemical lung lavage parameters 24 h later. The result of this dose–response study is shown in figure 7. It is evident from this figure that ultrafine TiO_2 elicited a significantly greater inflammatory cell influx (neutrophils) for the same dose than larger sized TiO_2. However, when the deposited TiO_2 dose was

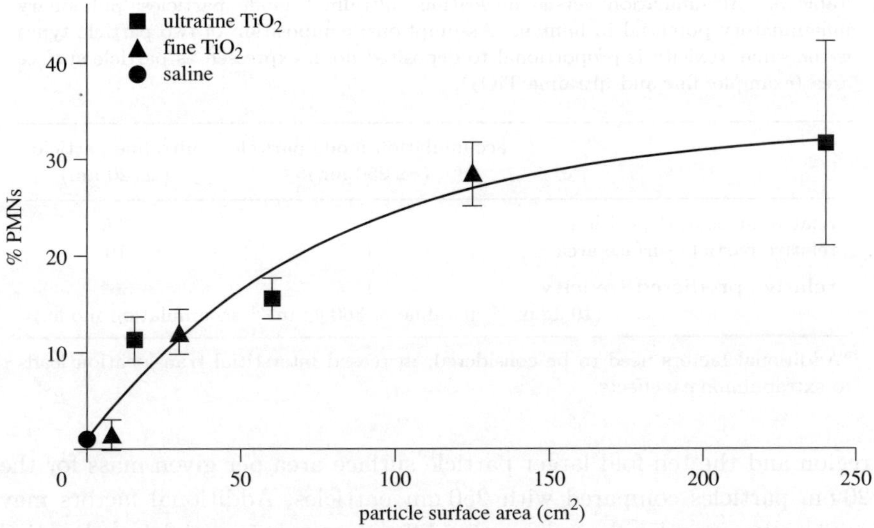

Fig. 8. Same data as shown in figure 7 with particle dose expressed as particle surface area in the lung.

expressed as particle surface area, the result was quite different, as shown in figure 8. Using the particle surface area as dosimetric resulted in virtually identical inflammatory responses of these two different sizes of TiO_2 particles. The importance of particle surface area for eliciting inflammatory responses in the lung has been confirmed by Li *et al.* (1996) with ultrafine and fine carbon-black particles. This concept of particle surface area as the appropriate dosimetric has been recognized as an important principle in particulate matter toxicology (Oberdörster 1996; Donaldson *et al.* 1998).

Considering differences in pulmonary deposition and the importance of dosimetric for characterizing the inflammatory potential of inhaled particles, one can deduce a relative potency ranking for the *in vivo* toxicity of inhaled ultrafine particles versus larger 250 nm particles of the accumulation mode. Assuming that the chemical composition of the two particle sizes is the same and that the toxicity is proportional to the deposited dose expressed as particle surface area, one can derive that the toxicity of ultrafine particles is about 36-fold greater than that of accumulation-mode particles in terms of the inhaled mass concentration. Table 3 shows that this factor is due to the 3.6-fold greater deposition efficiency in the alveolar

Table 3. Accumulation versus nucleation (ultrafine) mode particles: pulmonary inflammatory potential in humans. Assumptions: composition of two particle types is the same, toxicity is proportional to deposited dose, expressed as particle surface area (example: fine and ultrafine TiO_2).

	accumulation mode particle (*ca.* 250 nm)	ultrafine particle (*ca.* 20 nm)
relative alveolar deposition	1	3.6
relative particle surface area	1	10
relative predicted toxicity	1	36[a]
	($10\ \mu g\ m^{-3}$ ultrafine $\equiv$ $360\ \mu g\ m^{-3}$ accumulation mode.)	

[a]Additional factors need to be considered: increased interstitial translocation leads to extrapulmonary effects.

region and the ten-fold larger particle surface area per given mass for the 20 nm particles compared with 250 nm particles. Additional factors may need to be considered, such as the difference in interstitial translocation between the two particle sizes, and possibly also differences in translocation to extrapulmonary sites.

As mentioned above, TiO_2 particles are of a rather benign nature and have been used in the past in a number of studies as control particles of low toxic potency against which effects of other particle types have been compared. It is likely that the inflammatory response elicited by higher doses of ultrafine TiO_2 is based on a similar oxidative stress mechanism to that underlying the much greater pulmonary toxicity of PTFE fumes discussed above (Donaldson *et al.* 1998). One might, therefore, expect that adaptive responses observed in our PTFE experiments would also attenuate the inflammatory response of ultrafine TiO_2 based on the existence of cross-tolerance. The result of a study in rats indeed showed a significantly reduced inflammatory response in the lung to intratracheally instilled 100 μg of TiO_2 when the animals had been adapted to PTFE fumes for the previous three days (figure 9).

3.3. *Deposition Studies with Ultrafine Carbon and other Particles*

A major component of ambient particles generated by combustion processes is their carbonaceous core (Hughes *et al.* 1998). These particles consist of

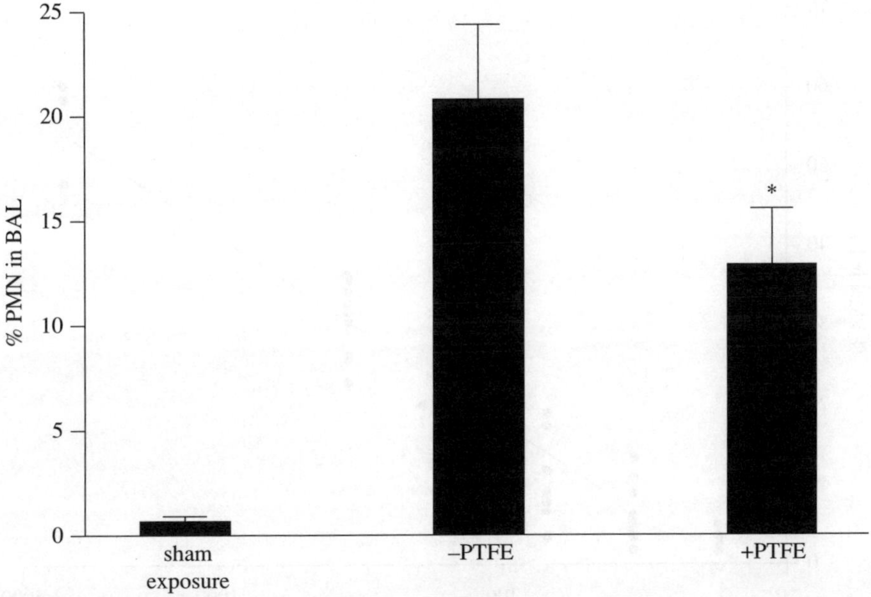

Fig. 9. Lavage neutrophil response in rats 24 h after intratracheal instillation of 100 μg of TiO_2 particles in PTFE-fume-adapted and non-adapted rats. ∗, significantly different groups without PTFE and sham exposure (ANOVA, $p < 0.05$)

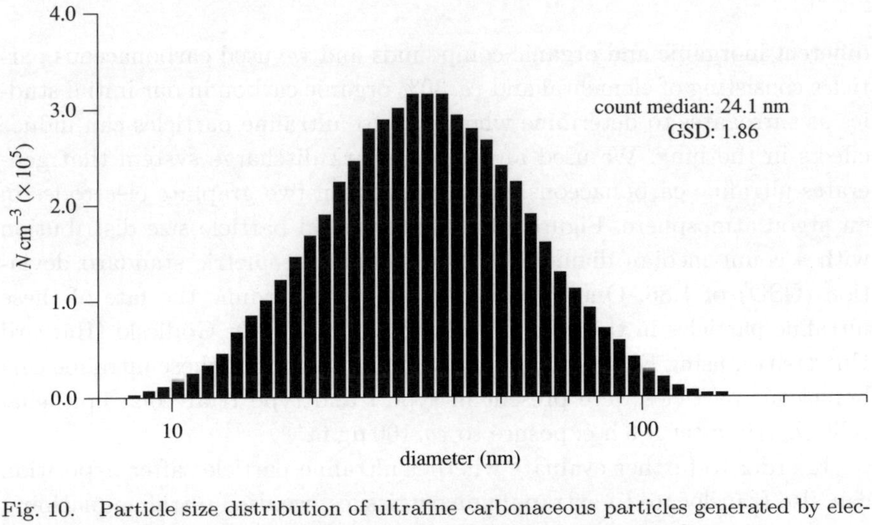

Fig. 10. Particle size distribution of ultrafine carbonaceous particles generated by electric spark discharge between graphite electrodes in an argon atmosphere.

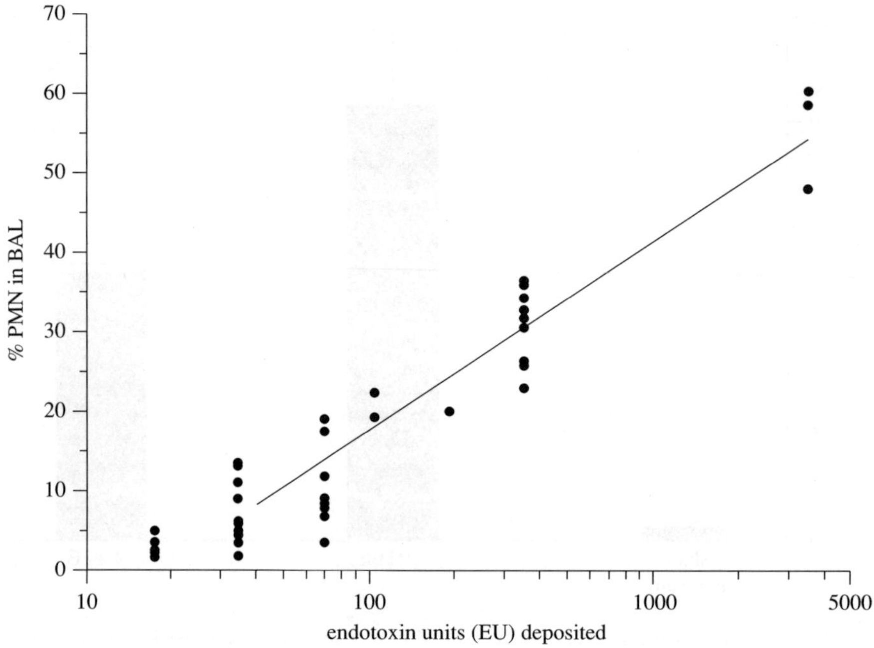

Fig. 11. Dose–response relationship of lung lavage neutrophils 24 h after different lung doses of inhaled endotoxin in young rats.

different inorganic and organic compounds and we used carbonaceous particles consisting of elemental and *ca.* 30% organic carbon in our initial studies as surrogates to determine whether these ultrafine particles can induce effects in the lung. We used an electric spark discharge system that generates ultrafine carbonaceous particles between two graphite electrodes in an argon atmosphere. Figure 10 shows a typical particle size distribution with a count median diameter of 24 nm and a geometric standard deviation (GSD) of 1.86. One of our goals was to determine the fate of these ultrafine particles in the lung. In collaboration with Dr Godleski (Harvard University), using EELS technology we could show that these ultrafine carbonaceous particles were present in type I and type II alveolar epithelial cells shortly after a 6 h exposure to *ca.* 100 $\mu g\ m^{-3}$.

In order to further evaluate whether ultrafine particles after deposition can also translocate to extrapulmonary tissues, we used ultrafine platinum particles (CMD 13 nm, GSD 1.7) and exposed a rat to these particles for 6 h

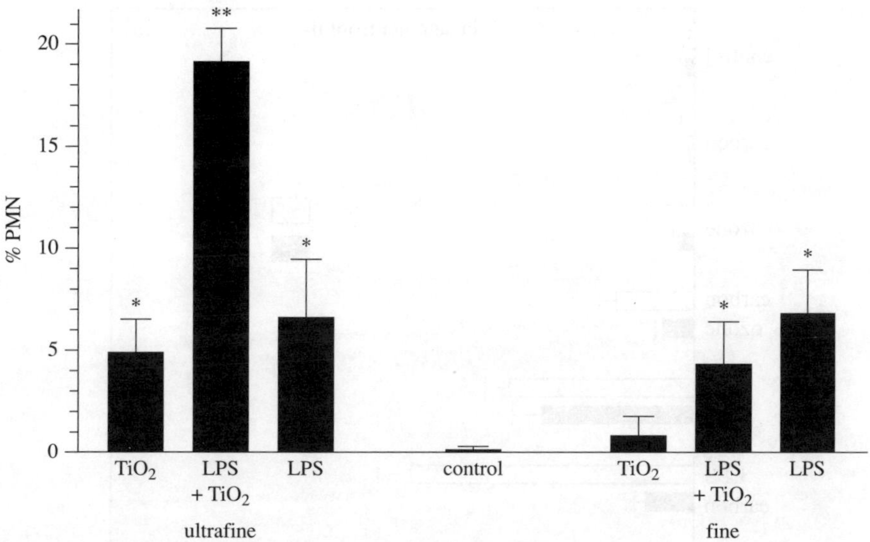

Fig. 12. Lung lavage neutrophils 24 h after intratracheal instillation of 50 μg of ultrafine (20 nm) or fine (250 nm) TiO_2 particles into rats with or without prior LPS priming compared with LPS alone and saline-instilled control rats (mean ±SE); *, significant difference to control rats; **, significant difference to LPS-primed group and to ultrafine particle only group ($P < 0.05$; one-way ANOVA).

at a concentration of *ca.* 100 μg m^{-3}. With the use of inductively coupled plasma mass spectroscopy, platinum levels were determined 30 min after the exposure in different lobes of the lung, the trachea and the liver. A total of 2.12 μg of platinum was found to be deposited in the lower respiratory tract, which corresponds to an estimated deposition efficiency of 20% of the inhaled ultrafine platinum particles. A significant finding was that platinum was also found in the liver, which amounted to *ca.* 7% of the lung platinum burden. However, it would be premature to conclude that this indicates translocation of the ultrafine particles from the lung, since it cannot be excluded that a small amount of the ultrafine platinum particles has been solubilized in the lung and may have reached the liver as soluble platinum. Thus, although platinum metal is considered to be very poorly soluble, additional studies with insoluble ultrafine particles need to be performed and are planned in our laboratory to determine more precisely the potential for ultrafine particles to reach extrapulmonary tissues.

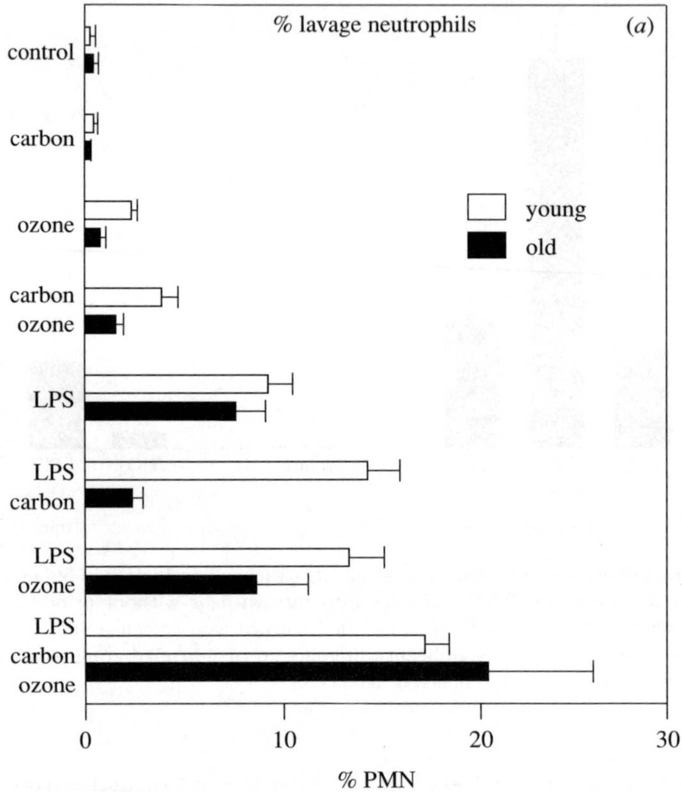

Fig. 13. Lung lavage inflammatory cell response of young (10 weeks) and old (22 month) rats following inhalation exposure to ultrafine carbonaceous particles (*ca.* 105 µg m^{-3}) ± O_3 (1 ppm) ± LPS priming. (*a*) Per cent neutrophils of total lavage cells.

3.4. *Animal Models of a Compromised Host to Study Ultrafine Particle Toxicity*

Associations between particulate air pollution and adverse health effects have only been observed in susceptible parts of the population and not in healthy people. An important aspect of studying potential causality of such effects includes, therefore, the use of animal models, which mimic a compromised respiratory or cardiovascular condition occurring in humans. Inhalation studies with particles in the past were typically performed in young, healthy animals using high exposure concentrations and doses in

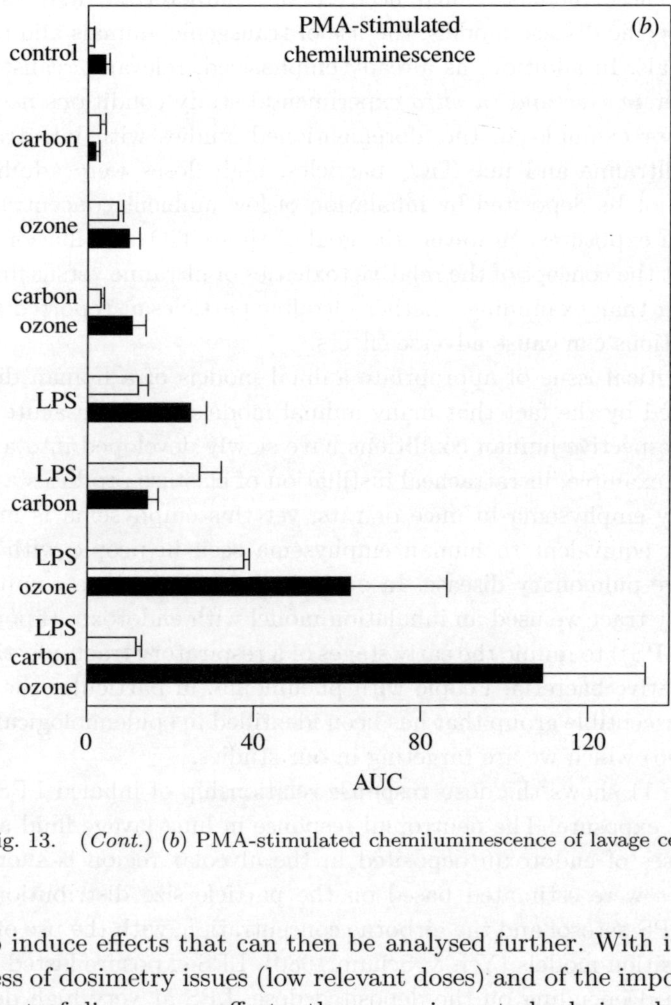

Fig. 13. (*Cont.*) (*b*) PMA-stimulated chemiluminescence of lavage cells.

order to induce effects that can then be analysed further. With increasing awareness of dosimetry issues (low relevant doses) and of the importance of the impact of diseases on effects, those previous study designs have to be questioned: are mechanisms underlying effects induced by high doses in the healthy mammalian organisms really the same as those of low doses in a compromised host? Or should we not rather assume that the dose/dose rate controls the mechanism, as has been concluded from a number of particle overload studies?

A change in particulate matter toxicology is taking place, switching from the use of healthy animals to that in animal models of compromised humans. These models, which need to be characterized and validated, include specific disease models, the use of transgenic animals and of senescent animals. In addition, as already emphasized, relevant, realistic doses both under *in vivo* and *in vitro* experimental study conditions need to be applied. For example, in the aforementioned studies with intratracheally instilled ultrafine and fine TiO_2 particles, high doses were administered that will not be deposited by inhalation of low ambient concentrations in short-term exposures. However, the goal of those TiO_2 instillation studies was to test the concept of the relative toxicities of ultrafine versus fine particles, rather than examining whether ultrafine particles at reported ambient concentrations can cause adverse effects.

The critical issue of appropriate animal models of a human disease is complicated by the fact that many animal models are of an acute nature, whereas respective human conditions have slowly developed into a chronic state. For example, intratracheal instillation of elastase produces a marked pulmonary emphysema in mice or rats, yet this emphysema is most certainly not equivalent to human emphysema seen in people with chronic obstructive pulmonary disease. In our initial studies with a compromised respiratory tract we used an inhalation model with endotoxin (lipopolysaccharide (LPS)) to mimic the early stages of a respiratory tract infection with gram negative bacteria. People with pneumonia, in particular the elderly, are one susceptible group that has been identified in epidemiological studies (EPA 1996) which we are targeting in our studies.

Figure 11 shows the dose–response relationship of inhaled LPS in rats 24 h after exposure. The neutrophil response in lung lavage fluid after different doses of endotoxin deposited in the alveolar region is shown. The LPS doses were estimated based on the particle size distribution of the inhaled LPS aerosol and the airborne concentration, with the use of predictive deposition models (Yeh & Schum 1980). LPS exposure lasted only for *ca.* 12 min. Depending on the deposited dose, LPS at very high doses can result in a severe ARDS-like pulmonary inflammation, with large amounts of neutrophils and protein in the lavage fluid. However, at lower doses, only a mild inflammatory response in terms of neutrophil influx and no increase in lavage protein occurs. We used this lower dose of *ca.* 70 endotoxin units (EUs) to prime the respiratory tract prior to exposure with ultrafine particles. The mild inflammatory response at this low deposited alveoles dose is characterized by a lavage neutrophil level of *ca.* 10% of the total cells 24 h post exposure.

Before using this LPS model to examine the response to inhaled low concentrations of spark discharge-generated ultrafine carbonaceous particles, we tested it using ultrafine and fine TiO_2 particles via intratracheal instillation. 50 μg of ultrafine and fine TiO_2 were intratracheally instilled in rats that had either received the LPS priming inhalation or a sham inhalation of NaCl aerosol. Instillation of the particles was performed within 30 min after the inhalation. Other rats received inhaled LPS alone or instilled saline (controls). As the result in figure 12 shows, only the ultrafine TiO_2 particles given after LPS induced a significantly greater neutrophil influx compared with TiO_2 alone and LPS alone, whereas the fine TiO_2 particles administered to the LPS-primed lung did not show a greater response than LPS alone. This result shows that priming of the respiratory tract with inhaled LPS can indeed amplify the response to a subsequent particulate stimulus, and it further confirms that for the same lung dose in terms of mass, ultrafine particles are significantly more potent than fine particles.

3.5. *Age and Ozone Co-exposure as Modulators of Ultrafine Carbon Particle Toxicity*

The senescent mammalian organisms may be more sensitive to inhaled toxicants than the younger organism. With respect to the potential effect of ambient particulate matter, it has been suggested that co-exposure to other pollutants, such as oxidant gases, may contribute to the adverse effects observed in the epidemiological studies (Burnett *et al.* 1997*a*, *b*; Samet *et al.* 1997). Our studies with ultrafine PTFE fumes, although inconclusive with respect to a contributory effect of gas-phase compounds, can be interpreted as showing some influence of gas-phase compounds. Our latest study with inhaled ultrafine carbonaceous particles was designed to evaluate their inflammatory effects in the lung of young and old rats with and without ozone co-exposure and with and without LPS inhalation priming. Eight groups of 10-week old and 22-month old rats were exposed to ultrafine carbonaceous particles (concentration *ca.* 105 $\mu g\, m^{-3}$), ozone (1 ppm) or inhaled LPS (*ca.* 70 EU estimated alveolar dose) and to combinations of these compounds. Sham-exposed control rats served as controls. Lung lavage parameters were determined 24 h later and lavaged inflammatory cells were subjected to a chemiluminescence assay *in vitro* to determine their unstimulated and phorbol ester (PMA)-stimulated oxidant release.

Figure 13 shows the result with respect to lavage neutrophils and the PMA-stimulated chemiluminescence. Three-way analyses of variance (ANOVAs) for the groups of young and old rats, as well as a four-way ANOVA for the two age groups combined, were performed; the three factors were ultrafine carbon, ozone and LPS and the fourth factor was age. These analyses showed that each of the three components (ultrafine carbon, ozone and LPS) induced significant effects independently. In addition, the ultrafine carbonaceous particle response in the aged rats was synergistic with the effects of ozone. In both old and young groups, the greatest inflammatory cell response was observed in the LPS-primed group with combined exposure to ultrafine carbonaceous particles and ozone (Elder *et al.* 2000*b*).

There was also a significant age effect, showing that the aged animals responded with greater oxidant release of the lavaged inflammatory cells compared with the young animals in the combined exposure groups. This greater release of reactive oxygen species implies a greater risk of oxidative lung injury in the aged organism under exposure conditions of ultrafine carbon particles in combination with ozone in the LPS-sensitized respiratory tract. These studies show that ultrafine carbonaceous particles can cause significant pulmonary inflammation. This occurs at inhaled concentrations leading to lung doses which are deposited in human lungs during episodic increases of urban ultrafine particles. Future studies will evaluate whether addition of transition metals to the carbon particles amplifies the inflammatory response.

4. Summary and Conclusions

Conclusions from these studies are as follows.

(1) Poorly soluble ultrafine particles cause a significantly greater pulmonary inflammation per given mass than larger particles. The appropriate dosimetric is their high specific surface area rather than the mass of these particles.
(2) Ultrafine carbonaceous particles at relevant inhaled concentrations can cause an inflammatory response in rodents.
(3) Ultrafine particles translocate readily to epithelial and interstitial sites. It is also conceivable that they may be transported to extrapulmonary organs; this needs to be confirmed in future studies.

(4) Specific modulating factors that increase ultrafine particle effects include age and a compromised/sensitized respiratory tract.

(5) Combined exposures with an oxidant gas can enhance ultrafine particle effects.

Acknowledgements

Studies with ultrafine particles mentioned in this manuscript have been supported by the Health Effects Institute (contract no. 95-11) and by the National Institute of Health Sciences grants (ES 04872 and ES 247). Detailed results of these studies were reported by Elder *et al.* (2000*a*–*c*), Johnston *et al.* (1996, 1998) and Oberdörster *et al.* (1995, 2000).

References

Brand, P., Ruob, K. & Gebhart, J. 1992 Performance of a mobile aerosol spectrometer for an *in situ* characterization of environmental aerosols in Frankfurt city. *Atmos. Environ.* A **26**, 2451–2457.

Burnett, R. T, Brook J. R., Yung, W. T., Dales, R. E. & Krewski, D. 1997*a* Association between ozone and hospitalization for respiratory diseases in 16 Canadian cities. *Environ. Res.* **72**, 24–31.

Burnett, R. T., Cakmak, S., Brook, J. R. & Krewski, D. 1997*b* The role of particulate size and chemistry in the association between summertime ambient air pollution and hospitalization for cardiorespiratory diseases. *Environ. Health Perspect.* **105**, 614–620.

Cheng, Y.-S., Yeh, H.-C. & Swift, D. L. 1991 Aerosol deposition in human nasal airway for particles 1 nm to 20 nm: a model study. *Radiation Protection Dosimetry* **38**, 41–47.

Dodge, D. E., Rucker, R. B. Pinkerton, K. E., Haselton, C. J. & Plopper, C. G. 1994 Dose-dependent tolerance to ozone. III. Elevation of intracellular clara cell 10-k protein in central acini of rats exposed for 20 months. *Toxicol. Appl. Pharmacol.* **127**, 109–123.

Donaldson, K., Li, X. Y. & MacNee, W. 1998 Ultrafine (nanometer) particle mediated lung injury. *J. Aerosol Sci.* **29**, 553–560.

Drinker, P., Thomson, R. M. & Finn, J. L. 1927 Metal fume fever. II. Resistance acquired by inhalation of zinc oxide on two successive days. *J. Ind. Hygiene Toxicol.* **9**, 98–105.

Elder, A. C. P., Johnston, C., Finkelstein, J. & Oberdörster, G. 2000*a* Induction of adaptation to inhaled lipopolysaccharide in young and old rats and mice. *Inhal. Toxicol.* **12**, 225–243.

Elder, A. C. P., Gelein, R., Finkelstein, J., Cox, C. & Oberdörster, G. 2000*b* Endotoxin priming affects the lung response to ultrafine particles and ozone in young and old rats. *Inhal. Toxicol.* (Suppl. I) **12**, 85–98.

Elder, A. C. P., Gelein, R., Finkelstein, J., Cox, C., Oberdörster, G. 2000*c* The pulmonary inflammatory response to inhaled ultrafine particles is modified by age, respiratory tract sensitization, and disease. *Inhal. Toxicol.* (In the press.)

EPA 1996 Air quality criteria for particulate matter, vol. III. EPA/600/P-95/001cF.

Goldstein, M., Weiss, H., Wade, K. *et al.* 1987 An outbreak of fume fever in an electronics instrument testing laboratory. *J. Occup. Med.* **29**, 746–749.

Gordon, T., Chen, L. C., Fine, J. M., Schlesinger, R. B., Su, W. Y., Kimmel, T. A. & Amdur, M. O. 1992 Pulmonary effects of inhaled zincoxide in human subjects, guinea pigs, rats and rabbits. *Am. Ind. Hygiene Ass. J.* **53**, 503–509.

Hahn, F. F., Newton, G. J. & Bryant, P. L. 1977 *In vitro* phagocytosis of respirable-sized monodisperse particles by alveolar macrophages. In *Pulmonary macrophages and epithelia cells* (ed. C. L. Sanders, R. P. Schneider, G. E. Dagle & H. A. Ragen), pp. 424–435. Energy Research and Development Administration Symposium Series.

Hart, B. A., Voss, G. W. & Willean, C. L. 1989 Pulmonary tolerance to cadmium following cadmium aerosol pretreatement. *Toxicol. Appl. Pharmacol.* **101**, 447–460.

Hinds, W. C. 1982 *Aerosol technology*, pp. 235–239. Wiley.

Hughes, L. S., Cass, G. R., Jones, J., Ames, M. & Olmec, L. 1998 Physical and chemical characterization of atmospheric ultrafine particles in the Los Angeles area. *Environ. Sci. Technol.* **32**, 1153–1161.

ICRP (International Commission on Radiological Protection) 1994 *Annals of the ICRP, human respiratory tract model for radiological protection.* ICRP publication 66. Oxford: Pergamon.

Johnston, C. J., Finkelstein, J. N., Gelein, R., Baggs, R. & Oberdörster, G. 1996 Characterization of the early pulmonary inflammatory response associated with PTFE fume exposure. *Toxicol. Appl. Pharmacol.* **140**, 154–163.

Johnston, C. J., Finkelstein, J. N., Gelein, R. M. & Oberdörster, G. 1998 Pulmonary inflammatory responses and cytokine and antioxidant mRNA levels in the lungs of young and old C57BL/6 mice after exposure to Teflon fumes. *Inhal. Toxicol.* **10**, 931–953.

Lee, K. P. & Seidel, W. C. 1991 Pulmonary response to perfluoropolymer fume and particles generated under various exposure conditions. *Fund. Appl. Toxicol.* **17**, 254–269.

Li, X. Y., Gilmour, P. S., Donaldson, K. & MacNee, W. 1996 Free radical activity and pro-inflammatory effects of particulate air pollution (PM_{10}) *in vivo* and *in vitro*. *Thorax* **51**, 1216–1222.

Makulova, I. D. 1965 The clinical picture in acute perfluoroisobutylen poisoning. *Gigiena Truda I Professionalnye Zabolevaniya* **9**, 20–23.

Oberdörster, G. 1996 Significance of particle parameters in the evaluation of exposure–dose–response relationships of inhaled particles. *Inhal. Toxicol.* (Suppl.) **8**, 73–89.

Oberdörster, G., Gelein, R. M., Ferin, J. & Weiss, B. 1995 Association of particulate air pollution and acute morality: involvment of ultrafine particles? *Inhal. Toxicol.* **7**, 111–124.

Oberdörster, G., Finkelstein, J. N., Johnston, C., Gelein, R., Cox, C., Baggs, R. & Elder, A. 2000 Investigator's report: acute pulmonary effects of ultrafine particles in rats and mice. Health Effects Institute final report. (In the press.)

Pryor, W. A., Nuggehalli, S. K., Scherer Jr, K. V. & Church, D. F. 1990 An electron spin resonance study of the particles produced in the pyrolysis of perfluoropolymers. *Chem. Res. Toxicol.* **3**, 2–7.

Rosenstock, L. & Cullen, M. R. 1986 *Clinical occupational medicine*, pp. 28, 232. Philadelphia, PA: Saunders.

Samet, J. M., Zeger, S. L., Kelsall, J. E., Xu, J. & Kalkstein, L. S. 1997 Particulate air pollution and daily mortality: analysis of the effects of weather and multiple air pollutants. The phase I.B. report of the particle epidemiology evaluation project. Cambridge, MA: Health Effects Institute report.

Seidel, W. C., Scherer Jr, K. V., Cline Jr, D., Olson, A. H., Bonesteel, J. K., Church, D. F., Nuggehalli, S. & Pryor, W. A. 1991 Chemical, physical, and toxicological characterization of fumes produced by heating tetrafluoroethene homopolymer and its copolymers with hexafluoropropene and perfluoro(propyl vinyl ether). *Chem. Res. Toxicol.* **4**, 229–236.

Stearns, R. C., Murthy, G. G. K., Skornik, W., Hatch, V., Katler, M. & Godleski, J. J. 1994 Detection of ultrafine copper oxide particles in the lungs of hamsters by electron spectroscopic imaging. In *Proc. Int. Conf. on Electron Microscopy, ICEM 13, Paris*, pp. 763–765.

Swift, D. L., Montassier, N., Hopke, P. K., Karpen-Hayes, K., Cheng, Y.-S., Su, Y. F., Yeh, H. C. & Strong, J. C. 1992 Inspiratory deposition of ultrafine particles in human nasal replicate cast. *J. Aerosol Sci.* **23**, 65–72.

Tuch, T. H., Brand, P., Wichmann, H. E. & Heyder, J. 1997 Variation of particle number and mass concentration in various size ranges of ambient aerosols in eastern Germany. *Atmos. Environ.* **31**, 4193–4197.

Van Bree, L., Koren, H. S., Devlin, R. B. & Rombout, P. J. A. 1993 Recovery from attenuated inflammation in lower airways of rats following repeated exposure to ozone. *Am. Rev. Respir. Dis.* **147**, A633.

Waritz, R. S. & Kwon, B. K. 1968 The inhalation toxicity of pyrolysis products of polytetrafluoroethylene heated below 500 degrees centigrade. *Am. Indust. Hygiene Ass. J.* **29**, 10–26.

Yeh, H. C. & Schum, M. 1980 Theoretical evaluation of aerosol deposition in anatomical models of mammalian lung airways. *Bull. Math. Biol.* **42**, 1–15.

Discussion

M. S. Bingley (*Cobham, Surrey, UK*). The European Community is about to remove lead (Pb) from solder. Replacement solders will have a higher melting point. Electronic circuitry makes extensive use of PTFE. In view of what you have said about the toxicity of ultrafine particles generated by heated PTFE, are electronic engineers to be put in danger by future EC legislation?

G. Oberdörster. The particles in fumes from solder seem to be bigger than ultrafines, and you can actually see those fumes in contrast to PTFE fumes. If PTFE is present on electronic circuitry and is heated at the same time by the solder, the emitted ultrafine PTFE particles will most likely coagulate onto the larger particles of the dense solder fume. Experiments adding PTFE fumes to diesel smoke or wood smoke have shown that these combined particles were 80 times less potent in their toxicity than PTFE fumes.

D. Costa (*US EPA, NC, USA*). Your last figure, showing the combined effects of ultrafines, ozone and LPS showed that 100 $\mu g\,m^{-3}$ of ultrafines alone had little impact on PMNs, and likewise, the ozone exposure also had little effect, where many published results show a significant effect. Yet, the interaction (combined) effect seems larger. Do you have an explanation for this?

G. Oberdörster. Indeed, the response of neutrophils in BAL was rather low after ozone alone and after ultrafine carbon alone. However, the combined exposure in the LPS-primed animals showed a large response, which emphasizes the importance of establishing animal models of increased susceptibility for evaluating otherwise subtle effects of inhaled ultrafine carbon. I think that the development and use of specific animal models—with a compromised pulmonary or cardiovascular system—is crucial for future progress in PM research.

CHAPTER 13

ULTRAFINE PARTICLES: MECHANISMS OF LUNG INJURY

K. Donaldson[1,2], V. Stone[1,2], P. S. Gilmour[2], D. M. Brown[1] and W. MacNee[2]
[1] *Biomedicine Research Group, School of Life Sciences, Napier University, 10 Colinton Rd, Edinburgh EH10 5DT, UK* (k.donaldson@napier.ac.uk)
[2] *ELEGI Colt Laboratory, University of Edinburgh Medical School, Teviot Place, Edinburgh EH8 9AG, UK*

Many ultrafine particles comprised classically of low-toxicity, low-solubility materials such as carbon black and titanium dioxide have been found to have greater toxicity than larger, respirable particles made of the same material. The basis of the increased toxicity of the ultrafine form is not well understood and a programme of research has been carried out in Edinburgh on the toxicology of ultrafines aimed at understanding the mechanism. We used fine and ultrafine carbon black, TiO_2 and latex and showed that there was an approximately 10-fold increase in inflammation with the same mass of ultrafine compared with fine particles. Using latex particles in three sizes—64, 202 and 535 nm—revealed that the smallest particles (64 nm) were profoundly inflammogenic but that the 202 and 535 nm particles had much less activity, suggesting that the cut-off for ultrafine toxicity lies somewhere between 64 and 202 nm. Increased oxidative activity of the ultrafine particle surface was shown using the fluorescent molecule dichlorofluorescein confirming that oxidative stress is a likely process by which the ultrafines have their effects. However, studies with transition-metal chelators and soluble extracts showed that the oxidative stress of ultrafine carbon black is not necessarily due to transition metals. Changes in intracellular Ca^{2+} levels in macrophage-like cells after ultrafine particle exposure suggested one way by which ultrafines might have their pro-inflammogenic effects.

Keywords: ultrafine; particulate matter; lung; PM_{10}; inflammation; air pollution

1. Introduction

This paper summarizes research carried out collaboratively in the ELEGI Colt Laboratory at Edinburgh University and the Biomedicine Research Group at Napier University, Edinburgh. The research is focused on the mechanisms of pathogenicity of ultrafine particles in the lungs. Many toxicological studies over the last 10 years have confirmed earlier research indicating that the particles in the ultrafine size range (less than 100 nm) pose special problems to the lungs (reviewed in Donaldson *et al.* (1998)). Typically, ultrafine particles cause more inflammation in experimental studies than respirable particles above the ultrafine size range made from the same material (Donaldson *et al.* 1998).

Attention has focused on ultrafine particles lately because of

(1) increased application and use in industry with concomitant potential for occupational exposure (Pui & Chen 1997); and
(2) research on particulate air pollution, $PM_{10}/PM_{2.5}$, has shown adverse effects at very low levels, resulting in a research thrust into which components of the PM_{10} particle mix might be responsible. The ultrafine particles have been hypothesized to be one component, amongst many, that could account for some of the adverse health effects of PM (MacNee & Donaldson 1999).

2. Materials and Methods

The particle types used are shown in table 1.

We used bronchoalveolar lavage (see, for example, Li *et al.* 1996) to quantify inflammation following instillation of particles into rat lungs with

Table 1. Characteristics of particles used in the studies.

name	abbreviation	origin	1^0 size (nm)	surface area ($m^2\ g^{-1}$)
normal carbon black	CB	Haeffner	260.2	7.9
ultrafine carbon black	ufCB	Degussa	14.3	253.9
normal TiO_2	TiO_2	Degussa	250	6.5
ultrafine TiO_2	$UfTiO_2$	Degussa	20	50
normal latex	latex	Polysciences	202	28.3
ultrafine latex	uflatex	Polysciences	64	89.3

or without the thiol antioxidant nacystelin, which was a kind gift from SMB Pharmaceuticals and which we have previously shown to be able to prevent pro-inflammatory responses by particles (Brown *et al.* 1999). The intracellular Ca^{2+} concentration was assessed using the dye Fura 2, which fluoresces in proportion to the amount of Ca^{2+} present in the cell, and thapsigargin as a stimulus for the release of endoplasmic reticulum stores of Ca^{2+} (Stone *et al.* 2000). mRNA for *IL-8* was measured by reverse transcriptase-polymerase chain reaction (RT-PCR), as described in Schins *et al.* (2000). The E1A positive A549 epithelial cell line was a gift from Professor J. C. Hogg, Vancouver, Canada.

3. Results and Discussion

3.1. *Increased Inflammation Caused by Ultrafine Particles Compared with Fine Particles of the Same Material*

As shown in figure 1, all three types of ultrafine particle were capable of causing more inflammation than their non-ultrafine counterparts. Note that there were differences in the dose used: 125 μg in the case of latex and TiO_2, and 500 μg in the case of CB. Fine or ultrafine particles were instilled into the rat lung at the same mass, and there are remarkably similar increases in inflammation compared with the non-ultrafine material in each case. There are differences in the proportional increase in polymorphonuclear neutrophils (PMN) between the particle types, that is, for about four times more particle by mass of ufCB at 125 μg, there is an approximately 10-fold increase in the extent of the inflammation. This suggests that the composition, or the surface area of the particles, is important (see table 1). We do not have data for CB and ufCB at the time of writing but these experiments are under way.

3.2. *Evidence for a Size Cut-off for Ultrafine Particle-Mediated Inflammation using Latex Particles*

The three different sizes of latex particle were used to shed light on the size at which the ultrafine particle effect appears (figure 2). The 64 nm latex caused much more inflammation than the 202 or the 535 nm latex at 125 mg. At the 500 μg dose, all of the particles caused more inflammation, but the 64 nm latex was markedly more inflammogenic than the other two and again there was little difference between the 202 and the 535 nm latex.

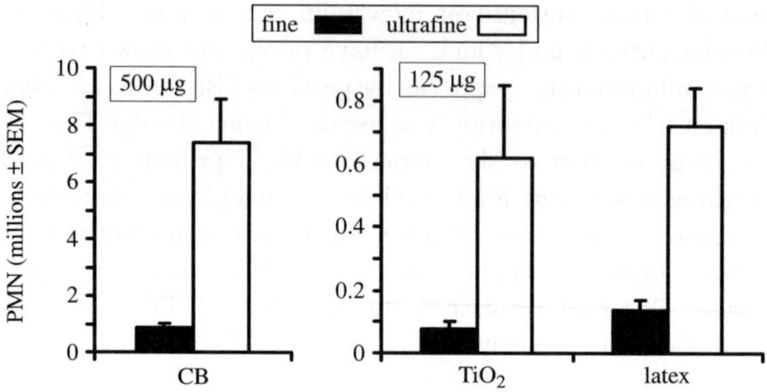

Fig. 1. Inflammation, measured as the number (mean ± SEM of three rats) of neutrophils (PMN) in the lavage of rats instilled with either 125 or 500 µg of fine or ultrafine carbon black (CB), titanium dioxide (TiO_2) or latex 24 h previously.

This suggests that 64 nm particles show the ultrafine effect of producing enhanced inflammation and suggests that the cut-off for considering particles to be 'ultrafine' (less than 100 nm) may be approximately correct. More research with particles above and just below the 100 nm size are required to clarify this question of a size cut-off.

3.3. *Role of Transition Metals in the Inflammation Caused by Ultrafine Carbon Black*

Since several types of carbon-based particle, such as residual oil fly ash, have their effects via transition metals, we examined the role of transition metals in the inflammation caused by ultrafine carbon black. We used two strategies as follows.

(1) We treated CB and ufCB with the transition-metal chelator desferal (desferioxamine) before instilling: this chelates any transition metal present on, or released by, the particles. These chelated particles are then washed and instilled into the lungs of rats and inflammation assessed.
(2) We incubated the ufCB and CB in saline to collect transition metals or any other soluble material and then instilled these soluble components into rat lungs and assessed the inflammation.

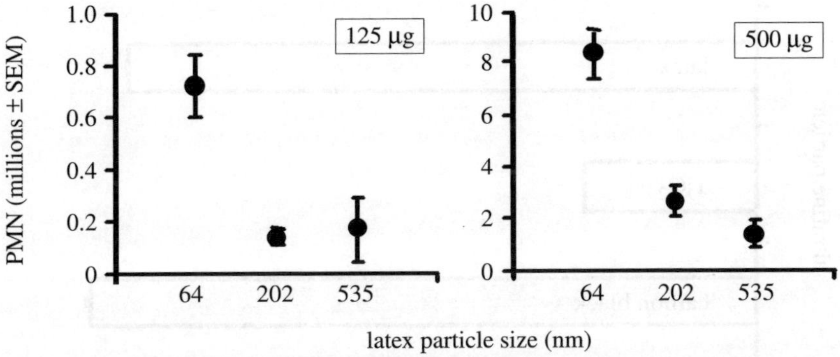

Fig. 2. Inflammation, measured as the number (mean ± SEM of three rats) of neutrophils (PMN) in lavage after instillation of 125 or 500 μg of latex particles of various sizes 24 h previously.

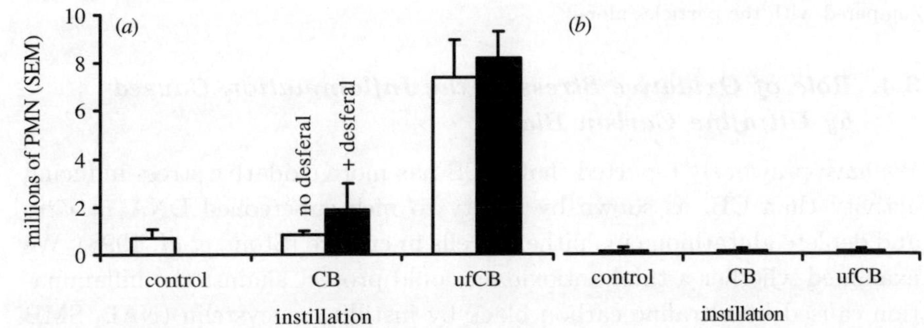

Fig. 3. (*a*) Particles treated with iron chelator. (*b*) Diffusable material from the surface of particles. Inflammation, measured as millions of neutrophils (mean ± SEM PMN in three rats) in lavage. Treatments were (*a*) instillation of 125 or 500 μg of CB or ufCB that had been incubated in saline (no desferal (open bars) or desferal solution (black bars) prior to instillation); (*b*) lungs of rats instilled with 0.5 ml of saline alone (control) or saline that had been incubated with 1 mg ml^{-1} of CB or ufCB and the particles centrifuged out to leave the diffusable components.

As shown in figure 3*a*, the chelated particles were no different from the unchelated particles in their ability to cause lung inflammation. Figure 3*b* shows that there was no inflammogenic activity in the saline wash of the particles, demonstrating that no soluble transition metals, or other soluble components, were mediating the increased inflammation of the ufCB.

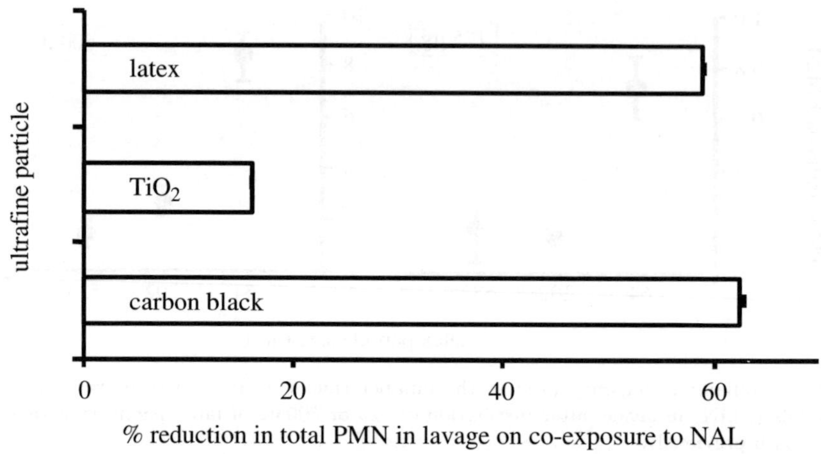

Fig. 4. Percentage reduction in the inflammatory response (mean ± SEM of lavage neutrophils in three rats) caused by ultrafine particles instilled along with nacystelin compared with the particles alone.

3.4. *Role of Oxidative Stress in the Inflammation Caused by Ultrafine Carbon Black*

We have previously reported that ufCB has more oxidative stress-inducing activity than CB, as shown by ability to nick supercoiled DNA *in vitro* and deplete glutathione in epithelial cells in culture (Stone *et al.* 1998). We examined whether a thiol antioxidant could protect against the inflammation caused by ultrafine carbon black by instilling nacystelin (NAL, SMB Pharmaceuticals, Belgium) along with ufCB, ufTiO_2 and uflatex. As shown in figure 4, co-instillation of particles with NAL caused significant amelioration of the inflammation caused by the different types of ultrafine particle on their own. The protective effect of NAL was most substantial with ufCB and uflatex and much less marked with ufTiO_2.

3.5. *Studies on the Mechanism of Lung Inflammation Caused by Ultrafine Particles*

We have examined the cellular and molecular basis of the increased inflammation caused by ultrafines. Calcium, as Ca^{2+}, is an important signalling mechanism for gene expression via activation of transcription factors (Dolmetsch *et al.* 1998). We have hypothesized that ultrafine particles may

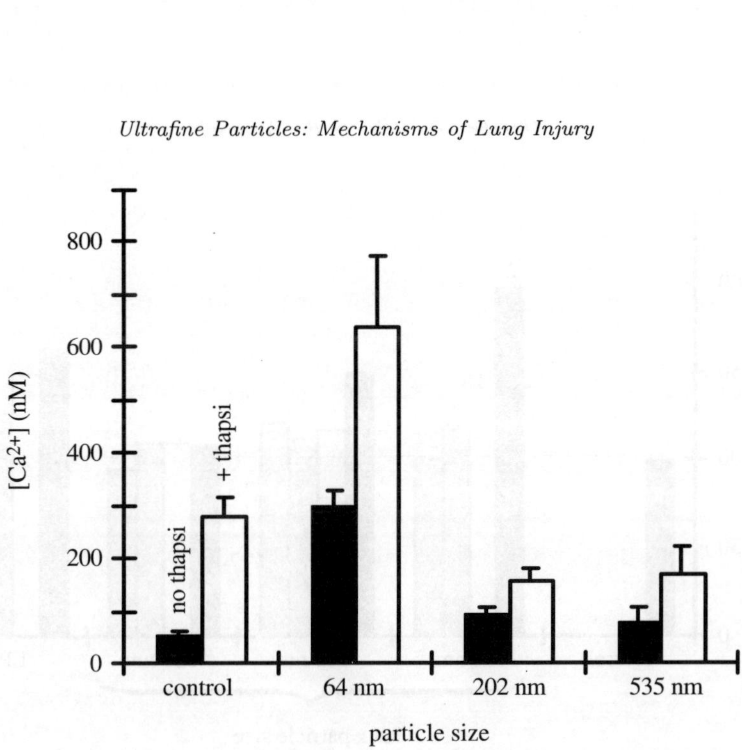

Fig. 5. Resting ('no thapsi') and thapsigargin-stimulated ('+thapsi') intracellular Ca^{2+} levels (mean ± SEM of three separate experiments) in macrophage-like cells Monomac 6 exposed to different sized latex particles.

induce Ca^{2+}-mediated signalling for activation of transcription of the chemokine *IL-8*, which is highly chemotactic for PMN and could explain the inflammation produced by these particles. We used the dye Fura 2 that fluoresces in the presence of Ca^{2+} to assess the levels of Ca^{2+} in macrophage-like cells. As shown in figure 5, the resting intracellular Ca^{2+} levels, and the thapsigargin-stimulated intracellular Ca^{2+} levels, are rapidly and significantly increased on treatment with the ultrafine latex but not with the larger sizes of latex particle. We reported this previously for ufCB and CB and consider this to be an important 'priming' effect for gene expression that results from a direct or indirect effect of ultrafine particles on the membrane Ca^{2+} channels (Stone *et al.* 2000). This effect can be inhibited by antioxidants such as NAL, and so a role for oxidative stress is suggested in the Ca^{2+} effect. Figure 6 supports these findings by showing a particle-size-related effect of the latex particles on levels of mRNA for *IL-8*, showing that transcription of this important chemokine is indeed increased more by uflatex particles than the other sizes of latex studied.

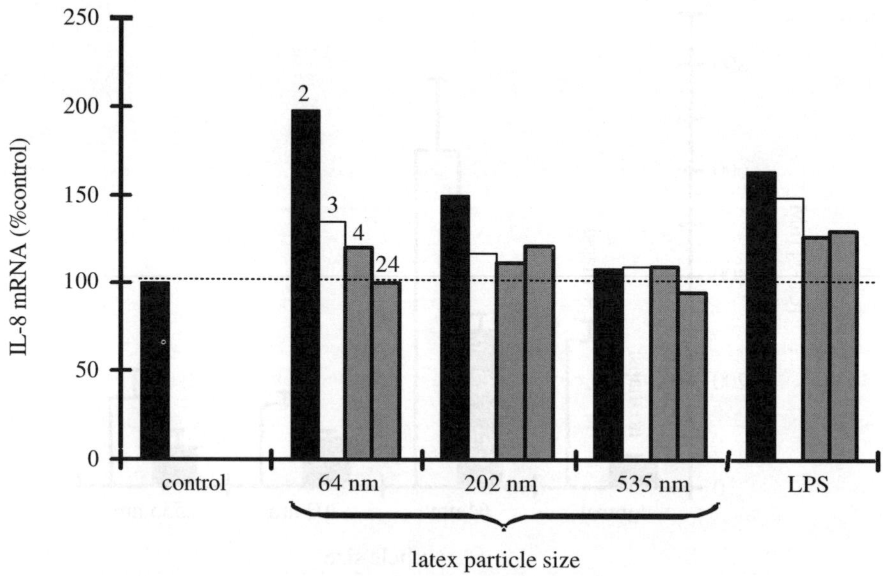

Fig. 6. *IL-8* mRNA levels in A549 epithelial cells exposed to different sizes of latex or lipopolysaccharide. Different shades of bar represent time after exposure in hours as indicated on the 64 nm group. Data from a single experiment.

3.6. *Pulmonary Adenoviral Infection as Susceptibility Factor in Inflammation Caused by Ultrafines*

The adverse health effects of PM_{10} are seen only in certain susceptible populations and little is understood of the factors that underlie this susceptibility. We hypothesized that cells transfected with the adenoviral gene E1A might show increased susceptibility to ultrafine particles in terms of the pro-inflammatory effects these particles induce. This is based on reports that cells expressing E1A showed hyper-responsivity of the NF-κB pathway, an oxidative stress-responsive transcription pathway that we have shown to be important to the pro-inflammatory effects of PM_{10} (K. Donaldson *et al.*, unpublished data). Figure 7 shows preliminary data from a single experiment showing the *IL-8* protein release from control A549 cells (E1A− (negative)) and A549 cells that have been stably transfected with the E1A gene (E1A+ (positive)), in response to exposure to CB and ufCB. The white columns show that normal A549 cells show no discrimination between CB and ufCB but that the E1A+ cells release approximately threefold more.

The data show that the ufCB causes more release of *IL-8* than CB on a mass basis and that the E1A+ cells release more *IL-8* in response to ufCB than the E1A− cells.

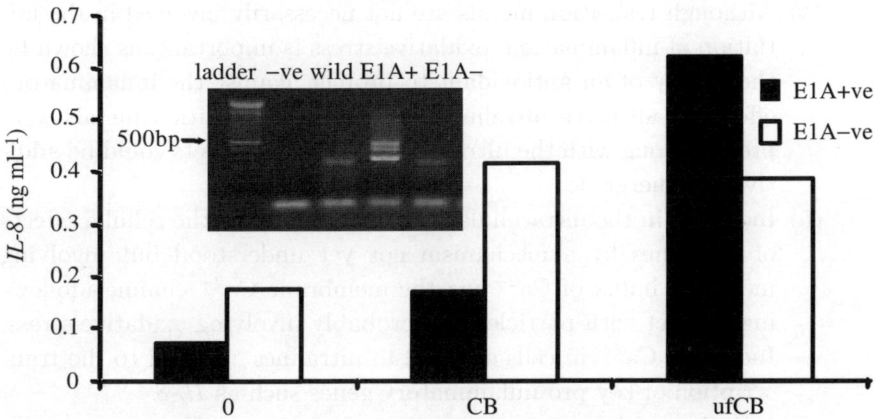

Fig. 7. Result of a single experiment showing release of *IL-8* protein by A549 epithelial cells that are stably transfected with the E1A gene (E1A+) or not (E1A−) in response to exposure to CB and ufCB. Inset shows RT-PCR product of the E1A gene.

4. Conclusions

The research described here is ongoing but suggests important new understandings of the likely mechanism of lung injury caused by ultrafine particles. By using a range of different ultrafine particles and their fine counterparts we have sought to learn about the generic effects of ultrafines. However, there are likely to be differences in toxicological effects of different materials presented as ultrafine particles, depending, for example, on the solubility, etc., of the particles (see Donaldson *et al.* 1998). However, based on the insoluble, low-toxicity ultrafine particle types used here, we suggest the following.

(1) Ultrafine particles are more inflammogenic than their fine but still respirable counterparts made from the same material.
(2) The cut-off size for this increased toxicity lies somewhere between 65 and 200 nm, although the cut-off may not be sharp.

(3) Ultrafine particles can cause inflammation via processes independent of the release of transition metals. The property that drives this toxicity is unknown but very likely relates to particle number or particle surface area and involves oxidative stress.
(4) Although transition metals are not necessarily involved in the initiation of inflammation, oxidative stress is important, as shown by the ability of an antioxidant to protect against the inflammatory effects of all three ultrafines used here. If transition metals were present along with the ultrafine particles, the effects could be additive or synergistic.
(5) Increases in the intracellular Ca^{2+} may underlie the cellular effects of ultrafines by a mechanism not yet understood but involving increased influx of Ca^{2+} via the membrane Ca^{2+} channels following contact with particles and probably involving oxidative stress. Increased Ca^{2+} in cells exposed to ultrafines can lead to the transcription of key pro-inflammatory genes such as *IL-8*.
(6) Infection with adenovirus, a virus that causes the common cold, may serve to render lung cells susceptible to the production of increased amounts of inflammatory mediators. This could occur via interaction of the E1A protein with oxidative stress-responsive transcription pathways rendering the cells more susceptible to the oxidative effects of particles and leading to enhanced expression of pro-inflammatory genes such as *IL-8*. There may also be a role for Ca^{2+} in this phenomenon.

Acknowledgements

This research was funded by the Medical Research Council, the British Lung Foundation, The Colt Foundation and the British Occupational Health Research Foundation. K.D. is the British Lung Foundation Transco Fellow in Air Pollution and Respiratory Health. We acknowledge the invaluable assistance of Dr Roel Schins, Dr Shizu Hayashi and Dr Jim Hogg. The authors thank Dr Robert Maynard, Department of Health, for his continued encouragement and support for this research and for making editing suggestions for this paper.

References

Brown, D. M., Beswick, P. H. & Donaldson, K. 1999 Induction of nuclear translocation of NF-κB in epithelial cells by respirable mineral fibres. *J. Pathol.* **189**, 258–264.

Dolmetsch, R. E., Xu, K. & Lewis, R. S. 1998 Calcium oscillations increase the efficiency and specificity of gene expression. *Nature* **392**, 933–936.

Donaldson, K., Li, X. Y. & MacNee, W. 1998 Ultrafine (nanometer) particle-mediated lung injury. *J. Aerosol Sci.* **29**, 553–560.

Li, X. Y., Gilmour, P. S., Donaldson, K. & MacNee, W. 1996 Free radical and pro-inflammatory activity of particulate air pollution (PM_{10}) *in vivo* and *in vitro*. *Thorax* **51**, 1216–1222.

MacNee, W. & Donaldson, K. 1999 Particulate air pollution: injurious and protective mechanisms. In *Air pollution and health* (ed. S. T. Holgate, J. M. Samet, H. S. Koren & R. L. Maynard), pp. 653–672. San Diego: Academic Press.

Pui, D. H. & Chen, D. R. 1997 Nanometer particles: a new frontier for multidisciplinary research. *J. Aerosol Sci.* **28**, 539–544.

Schins, R. P. F., McAlinden, A., MacNee, W., Jimenez, L. A., Ross, J. A., Guy, K., Faux, S. & Donaldson, K. 2000 Persistent depletion of I κ B α and *interleukin-8* expression in human pulmonary epithelial cells exposed to quartz. *Toxicol. Appl. Pharmacol.* (In the press.)

Stone, V., Shaw, J., Brown, D. M., MacNee, W., Faux, S. P. & Donaldson, K. 1998 The role of oxidative stress in the prolonged inhibitory effect of ultrafine carbon black on epithelial cell function. *Toxicol. In Vitro* **12**, 649–659.

Stone, V., Tuinman, M., Vamvakopoulos, J. E., Shaw, J., Brown, D., Petterson, S., Faux, S. P., Borm, P., MacNee, W., Michaelangeli, F. & Donaldson, K. 2000 Increased calcium influx in a monocytic cell line on exposure to ultrafine carbon black. *Eur. Respir. J.* **15**, 297–303.

Discussion

C. V. Howard (*Foetal Toxico-Pathology, University of Liverpool, UK*). Do you have the basis in your assays for comparing different substances at equal dosage and equal particle size ranges by inhalation, to construct a table of relative toxicities? It may be that this would be of use to policy makers to help them design strategies for trying to control those processes that produce the most toxic particles.

K. Donaldson. We agree that some measure of relative potency by inhalation is desirable but this would be costly.

M. Williams (*DETR, London, UK*). Professor Oberdörster plotted PMN response against *mass* and showed that ultrafine TiO_2 had a larger response than fine TiO_2, but, when plotted against *surface area*, all points fell on one curve. The graphs early on in your talk showed similar behaviour, and in many of your later histograms, the responses seemed to scale with surface area. Would one expect this if (cf. Dr Jefferson's paper) there was something special about ultrafines and it was not just a surface area effect?

K. DONALDSON. This is an important question that needs to be addressed by well-defined dose–response studies. However, our impression, from limited data, is that ultrafine particles have extra surface reactivity as well as extra surface, compared with non-uniformities.

A. D. MAYNARD (*NIOSH, Cincinatti, OH, USA*). There has been significant emphasis on the contribution that low-solubility particle surface area may make to the nature, magnitude and rate of biological interactions. However, characterization of 'biologically relevant' surface area will depend on the length-scale over which these interactions occur. Could you speculate on the order of magnitude of length-scale that is likely to be of greatest relevance in determining interaction mechanisms?

K. DONALDSON. The only information that I know about regarding the length or distance that cells can resolve in a paper by Wojciak-Stothard *et al.* (1996). This paper shows that macrophage-like cells align their cytoskeleton along grooves 44 nm in depth. This suggests that cells can discriminate, via one assumes surface receptors, well down into the ultrafine size range. The physiological responses that such a cytoskeletal reorganization might cause are of great potential interest.

Additional reference

Wojciak-Stothard, B., Curtis, A., Monaghan, W., MacDonald, K. & Wilkinson, C. 1996 Guidance and activation of murine macrophages by nanometric scale topography. *Exp. Cell Res.* **223**, 426–435.

CHAPTER 14

EPIDEMIOLOGICAL EVIDENCE OF THE EFFECTS OF ULTRAFINE PARTICLE EXPOSURE

H.-ErichWichmann[1,2] and Annette Peters[1]
[1] *GSF – Institute of Epidemiology,*
[2] *LMU – University of Munich,*
Ingolstadter Landstraße 1, D-85764 Neuherberg, Germany

In epidemiological studies associations have been observed consistently and coherently between ambient concentrations of particulate matter and morbidity and mortality. With improvement of measurement techniques, the effects became clearer when smaller particle sizes were considered. Therefore, it seems worthwhile to look at the smallest size fraction available today, namely ultrafine particles (UPs, diameter below 0.1 μm) and to compare their health effects with those of fine particles (FPs, diameter below 2.5 μm). However, there are only few studies available which allow such a comparison.

Four panel studies with asthma patients have been performed in Germany and Finland. A decrease of peak expiratory flow and an increase of daily symptoms and medication use was found for elevated daily particle concentrations, and in three of these studies it was strongest for UPs. One large study on daily mortality is available from Germany. It showed comparable effects of fine and ultrafine particles in all size classes considered. However, FPs showed more immediate effects while UPs showed more delayed effects with a lag of four days between particulate concentrations and mortality. Furthermore, immediate effects were clearer in respiratory cases, whereas delayed effects were clearer in cardiovascular cases.

In total, the limited body of studies suggests that there are health effects, due to both UPs and FPs, which might be independent from each other. If this is confirmed in further investigations, it might have important implications for monitoring and regulation, which until now does not exist for UPs. Data from Germany show that FPs cannot be used as indicator for UPs: the time trends for FPs decreased, while UPs was stable and the smallest size fraction of UPs has continually increased since 1991/92.

Keywords: ultrafine particles; fine particles; short-term effects; mortality; respiratory diseases; cardiovascular diseases

1. Introduction

The aim of this overview is the evaluation of the available epidemiological knowledge on health effects of ultrafine particles in ambient air. This is only possible in the context of particle epidemiology in general. Therefore, at the beginning a short summary of relevant studies is given, where the particle mass with a diameter below 2.5 or 10 μm ($PM_{2.5}$, PM_{10}) or total suspended particulates (TSP) have been measured.

The paper will be restricted to short-term effects, since until now no studies on long-term effects have been available where ultrafine particles have been measured. Furthermore, the role of copollutants will not be considered here, but we will address the question, if there are associations between ambient particles and morbidity or mortality, can they be attributed in part or totally to the ultrafine fraction?

We will use the following definitions:

ultrafine particles (UPs) have a diameter below 0.1 μm;
fine particles (FPs) have a diameter between 0.1 and 2.5 μm. They are mainly represented by $PM_{2.5}$;
coarse particles (CPs) have a diameter above 2.5 μm.

Furthermore we look at the following parameters:

number concentration (NC) is the concentration of the number of particles in 1 cm^3;
mass concentration (MC) is the mass of particles measured in $\mu g\, m^{-3}$.

As will be shown below, in a given volume the number of UPs is much higher than the number of FPs. Therefore, UPs are represented by the number concentration. In contrast, the mass of UPs is much smaller than the mass of FPs, and FPs are represented by the mass concentration.

1.1. *Epidemiological Knowledge on Particle Effects*

Epidemiological studies allover the world have consistently observed short-term effects of particulate matter on daily mortality (Dockery & Pope 1994; Schwartz 1994; Bascom *et al.* 1996; Katsouyanni *et al.* 1997; Pope & Dockery 1999). Often an immediate association was observed resulting in the largest effect estimates for the concurrent day or one day after. A recent review estimated that an increase of PM_{10} by 10 μg m^{-3} is associated with a 0.8% increase in mortality. The summary estimate for respiratory disease mortality was *ca.* 3% and for cardiovascular disease mortality *ca.* 1.3% (Pope & Dockery 1999). In studies where both PM_{10} and $PM_{2.5}$ were available to characterize the ambient concentrations of particles mass, there were indications that $PM_{2.5}$ was more strongly associated with mortality than PM_{10} (Dockery *et al.* 1992; Schwartz *et al.* 1996).

A pooled analysis based on data from four large, western European cities (London, Barcelona, Paris and Athens) as part of the APHEA project estimated that the risk of mortality increased in association with SO_2 and black smoke independently of each other (Katsouyanni *et al.* 1996, 1997). In the absence of more detailed air pollution measurements, black smoke might be regarded as a surrogate measure for ambient particles in urban air.

The impact of particulate matter on respiratory symptoms has been reinforced by studies on exacerbation of respiratory diseases from the 1960s to the 1990s (Dockery & Pope 1994; Bascom *et al.* 1996; Pope & Dockery 1999; Peters *et al.* 1997*b*, *c*). However, a biological mechanism linking the association between exacerbation of cardiovascular diseases and inhalation of ambient particulate matter had to be established. Seaton *et al.* (1995) have hypothesized that pulmonary inflammation may trigger systemic hypercoagulability. During the 1985 Europe-wide air pollution episode, the WHO MONICA survey (*Monitoring of trends and determinants in cardiovascular disease*) was conducted in Augsburg, Germany. Increases in plasma viscosity (Peters *et al.* 1999*a*–*c*) have been observed in randomly selected healthy adults in association with high particulate air pollution in both men and women from Augsburg. The odds of observing plasma viscosity levels above the 95th percentile tripled during the air pollution episode. Analyses of the C-reactive protein concentrations of healthy, middle-aged men (aged 45–64) based on data from the same study, showed an odds ratio of 3.5 for C-reactive protein concentrations above the 90th percentile. In addition, the TSP were associated independently from the episode with elevated CRP concentrations. Both CRP and plasma viscosity have been identified to be independent cardiovascular risk factors for subsequent myocardial infarctions (Danesh *et al.* 1998; Koenig *et al.* 1998). Plasma viscosity characterizes the physical properties of the blood. Elevated plasma viscosity increases the shear forces at an atherosclerotic lesion (Koenig & Ernst 1992). C-reactive protein is an acute phase reactant released as part of an inflammatory cascade. Its increase in association with particulate air pollution might point towards the inflammatory processes that particles elicit in the alveoli. A study published last year based on the Whitehall study also showed an increase of fibrinogen in association with nitrogen dioxide (Pekkanen *et al.* 1999*b*). Fibrinogen is also considered to be an acute phase reactant, and is one of the main determinants of plasma viscosity (Koenig & Ernst 1992). However, a study particularly designed to investigate the effects of ambient air pollution on blood in a panel of elderly subjects in Edinburgh was unable to confirm these associations (Seaton *et al.* 1999). Instead a decrease in red blood cells was observed in the blood samples repeatedly collected from panel members.

Toxicological studies conducted by Godleski and co-investigators suggested that concentrated ambient particles might alter the autonomic nervous system response (Stone & Godleski 1999). Epidemiological evidence was found for increased heart rate (Pope *et al.* 1999*a*; Peters *et al.* 1999*c*). The data collected in the MONICA study, Augsburg, in a random sample of the population (Peters *et al.* 1999*c*), as well as in a panel study in elderly subjects (Pope *et al.* 1999*a*), cohere. Three panel studies on the alteration of the autonomic control by ambient particles in elderly subjects have been reported on (Liao *et al.* 1997; Pope *et al.* 1999*b*; Gold *et al.* 2000). Heart rate variability was calculated based on either 24 holter EKG recording or 5–6 min intervals of EKG recording. An overall decrease in the standard deviation of all normal R–R intervals was observed (Liao *et al.* 1997; Pope *et al.* 1999*b*; Gold *et al.* 2000). However, the results differed with respect to measures that capture the sympathetic and parasympathetic portions of the nervous system control. Differences in the subjects, the EKG recordings and analyses or the different pollution mixtures and levels might account for these inconsistencies (Pope 2000). Additional evidence for the impact of particulate air pollution on arrhythmia was found in a follow-up study of patients with implanted cardioveter defibrillators (Peters *et al.* 2000). One hundred patients with a history of coronary artery disease and often syncope were enrolled into the study. Therapeutic interventions due to sustained tachycardia or defibrillation were analysed, and statistically significant odds ratios were noted in association with increased concentrations of $PM_{2.5}$ and NO_2.

Both mechanisms, the changes in coagulability of the blood and the alteration of the autonomic nervous control of the heart, might potentially increase the likelihood of ischemic events and arrhythmia, especially in persons with manifest atherosclerotic disease.

1.2. *Possible Role of Ultrafine Particles*

Ambient concentrations of particles are classically characterized by their mass concentrations. However, depending on their sizes, quite substantial differences in numbers or surfaces might constitute the same mass. While only one particle per cm^3 with a diameter of 2.5 μm is sufficient to result in a mass concentration of $10\,\mu g\,m^{-3}$, more than two million particles of a diameter of 0.02 μm are needed to obtain the same mass concentration (Oberdörster *et al.* 1995).

Ultrafine particles are deposited in the deep lung (ICRP 1994; US EPA 1996) and have been hypothesized to be responsible for the associations between particle matter and health outcomes at the current ambient concentrations (Oberdörster *et al.* 1995; Seaton *et al.* 1995). There are a number of potential mechanisms that can contribute to increased toxicity of UPs.

(i) For a given aerosol mass concentration, there is a much higher particle number and a much larger surface area when compared with larger sized particles. Since fine and ultrafine particles can act as a carrier to the deep lung for adsorbed reactive gases, radicals, transition metals or organic compounds, the larger surface area of ultrafines can transport more toxic surface adsorbed materials than larger particles.
(ii) Deposition of inhaled ultrafine particles is very high in the respiratory tract. Predicted deposition of inhaled 0.02 μm particles can be up to 50% in the alveolar region of the human lung and it is also very high in the lower tracheobronchial tree.

(iii) For particles not readily soluble in the epithelial lining fluid, the surface area provides the interface between the retained particles and cells, fluids, and tissues of the lungs; hence the dramatically increased surface area of ultrafine particles is likely to increase surface dependent reactions.

(iv) Protection resulting from the avid phagocytosis by alveolar macrophages is impaired since ultrafine particles are less well recognized by these cells, while there are many more ultrafine particles spread over the surface area of the alveolar epithelium less likely to be phagocytized when compared with larger particles.

(v) After deposition ultrafines penetrate more rapidly into interstitial sites. Preliminary evidence that ultrafine particles can be translocated to remote organs such as the liver and heart has been collected.

2. Epidemiological Studies on Ultrafine Particles

2.1. *Particle Measurements in these Studies*

Since 1991, daily measurements of UPs and, more general, of particle size distributions have been performed in the framework of epidemiological studies. The first equipment used was the mobile aerosol spectrometer (MAS). As described elsewhere (Brand *et al.* 1991, 1992; Tuch *et al.* 1997; Wichmann *et al.* 2000*a*), it consists of two instruments covering different size ranges. Particles in the size range 0.01 to 0.5 μm are measured using a differential mobility analyser (DMA) combined with a condensation particle counter (CPC). This set is termed differential mobility particle sizer (DMPS). Particles in the size range from 0.1 up to 2.5 μm are classified by an optical laser aerosol spectrometer (LAS-X). The DMA allows the segregation of particle fractions of uniform electrical mobility from a polydisperse aerosol. The number of particles selected by the DMA is counted by the CPC in 13 discrete size ranges. The LAS-X classifies particles according to their light scattering into 45 size-dependent channels. MAS yields a differential particle number concentration. Based on parallel measurements of $PM_{2.5}$, the mean density of ambient particles has been determined as $1530\ kg\ m^{-3}$, which is in excellent agreement with the literature value of $1500\ kg\ m^{-3}$. The differential mass distribution is calculated on this basis.

MAS measurements have been performed in Erfurt since 1991/92 in the framework of several epidemiological studies (Peters *et al.* 1997*a*; Von Klot

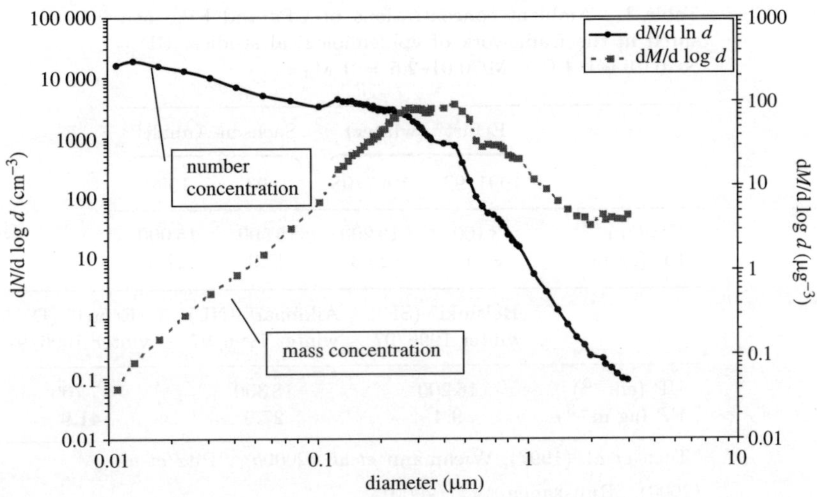

Fig. 1. Typical particle number and mass distribution averages from approximately 10 000 single measurements, Erfurt. From Wichmann *et al.* (2000*a*).

et al. 2000; Wichmann *et al.* 2000*a*, *b*) and also in three places in Sachsen-Anhalt, Eastern Germany (Pitz *et al.* 2000). A typical distribution of the particle number and the particle mass over the size range from 0.01 to 2.5 μm is shown in figure 1. In the years 1995–98 in Erfurt, 58% of the number concentration (NC) was found between 0.01 and 0.03 μm and 88% were UPs (between 0.01 and 0.1 μm). In contrast, only 3% of the mass was found below 0.1 μm, 78% between 0.1 and 0.5 μm and 95% below 1 μm. In other words, $PM_{0.5}$ equals $0.81PM_{2.5}$ and PM_1 equals $0.95PM_{2.5}$ in this study (Wichmann *et al.* 2000*a*).

The annual means of the number and mass concentrations are shown in table 1. UPs varied between 10 000 and 20 000 particles per cm^3 with a 24 h maximum of 50 000 particles per cm^3 and was stable over time. In contrast, FPs decreased substantially during the period of observation.

In the European ULTRA study, measurements of UPs have been performed in Finland, The Netherlands and Germany (Pekkanen *et al.* 1999*a*; Kreyling *et al.* 1999; Ruuskanen *et al.* 2000). In parallel to MAS in Erfurt, a similar device has been used in Alkmaar (denoted DAS) and a third spectrometer in Helsinki (denoted EAS), which measured the particle size distribution in the size range 0.01–10 μm by an electrical method alone. (In

Table 1. Ambient concentrations of UPs and FPs measured in the framework of epidemiological studies. UP = NC0.01–0.1, FP = MC0.01–2.5 = '$PM_{2.5}$'.

	Erfurt[a] (winter)		Sachsen-Anhalt[b]	
	1991/92	1997/98	1993	1999
UP (cm^{-3})	13 100	19 200	15 500	15 000
FP ($\mu g\ m^{-3}$)	82.1	25.3	47.9	22.6

	Helsinki[c] (SF) winter 1996/97	Alkmaar[c] (NL) winter 1996/97	Erfurt[c] (D) winter 1996/97
UP (cm^{-3})	16 200	18 300	17 700
FP ($\mu g\ m^{-3}$)	9.4	27.0	41.9

[a]Tuch *et al.* (1997), Wichmann *et al.* (2000*b*). [b]Pitz *et al.* (2000). [c]Ruuskanen*et al.* (2000).

an earlier side-by-side comparison, the different measurement principles had shown good agreement both in the number concentration of UPs and the total number concentration (Tuch *et al.* 2000*a*, *b*).) The concentrations of UPs in the three locations were comparable, whereas the concentrations of FPs differed substantially between the cities (table 1).

For source apportionment, in addition to the measurement of particle size distributions and gases, elemental composition in five size fractions has been determined. Particles have been collected with a Berner impactor in the size range between 0.05 and 1.4 μm and have been analysed by proton-induced X-ray emission (PIXE) spectrometry (technique described in Wichmann *et al.* (2000*b*)). In Erfurt, measurements have been performed every 10th day from September 1995 to August 1997 and every day from September 1997 to December 1998. Three sources have been considered, namely natural dust, domestic heating/fuel combustion of brown coal and oil, and motor vehicle exhausts. Using information based on crustal enrichment factors (enrichment of an element in the aerosol sample compared with the composition of the natural crust), correlations between the components, and patterns of the concentrations during the day, during the week and in summer and winter, the following associations have been found. In Erfurt, natural dust is especially represented by silicon, aluminium and titanium. Combustion of brown coal and oil is represented by sulphur, vanadium, nickel and sulphur dioxide. Motor vehicle exhausts are best characterized by the number concentration of the smallest available size fraction, namely NC0.01–0.03, followed by UPs, lead, NO, NO_2, CO and finally $PM_{2.5}$ (Wichmann *et al.* 2000*b*).

2.2. *Observed Health Effects*

Until now only a few epidemiological studies have been published which address the role of ultrafine particles. These deal with short-term effects in adults and children with asthma and daily mortality.

2.2.1. *Study on Adults with Asthma in Erfurt, Germany 1991/92*

In Erfurt, 27 non-smoking asthmatics recorded the peak expiratory flow (PEF) and respiratory symptoms daily during the winter season 1991/92 (Peters *et al.* 1997*a*). Most of the particles were in the ultrafine fraction, whereas most of the mass was attributable to particles in the size range 0.1–0.5 μm. Since these two fractions did not have similar time courses, comparison of their health effects was possible (correlation coefficient, $r = 0.51$). Both fractions were associated with a decrease of PEF and an increase in cough and feeling ill during the day. Health effects of the number of ultrafine particles were larger than those of the mass of the fine particles. The effects were strongest for the five days mean of the particle concentrations (tables 2 and 3, figure 2).

2.2.2. *Study on Adults with Asthma in Erfurt, Germany 1996/97*

Daily medication use was reported in 58 asthmatic adults in Erfurt from October 1996 to March 1997 (Von Klot *et al.* 2000). Number and mass concentrations in the size range of 0.01–2.5 μm diameter were determined concurrently. Overall prevalence of bronchodilator use and inhaled corticosteroid were analysed with a logistic regression model controlling for trend, temperature, weekend, holidays and autocorrelation. The results are shown in table 2. Corticosteroid use and bronchodilator use both increased in association with cumulative exposure over 14 days of UPs and FPs. A comparable effect was found for cumulative exposure over 5 days. The data suggest that asthma medication use increases with particulate air pollution.

Table 2. Effects of UPs and FPs on PEF of asthmatics in epidemiological studies. Δ is the interquartile range; $*$, $p < 0.05$.

	Δ	morning PEF coefficient (l min^{-3})	evening PEF coefficient (l min^{-3})
Adults with asthma Erfurt 1991/92[a]			
UP	9200 cm^{-3}	−2.55*	−3.58*
FP	50 µg m^{-3}	−1.42*	−2.18*
PM_{10}	50 µg m^{-3}	−1.51	−2.31*
Adults with asthma Helsinki 1996/97[b]			
PNC	7300 cm^{-3}	−1.16*	−1.66*
FP	6.6 µg m^{-3}	0.32	−0.41
PM_{10}	9.3 µg m^{-3}	1.68*	1.13*
Children with asthma symptoms Kuopio 1994[c]			
NC0.01–0.03	20 700 cm^{-3}	−0.73	0.35
NC0.03–0.1	13 100 cm^{-3}	−0.48	0.10
PM_{10}	13 µg m^{-3}	−2.24*	0.04

[a]Peters *et al.* (1979): 5 days mean, UP = NC0.01–0.1, FP = MC0.1–0.5.
[b]Penttinen *et al.* (2000): 5 days mean, PNC is the total particle number count, FP = MC0.1–0.5.
[c]Pekkanen *et al.* (1997): 4 days mean.

The effect might be more delayed but stronger on anti-inflammatory medication than on bronchodilators.

2.2.3. *Study on Adults with Asthma in Helsinki, Finland 1996/97*

Seventy-eight adult asthmatics were followed with daily peak-flow (PEF) measurements and symptoms and medication diaries for six months in the winter and spring season 1996/97 in Helsinki (Penttinen *et al.* 2000). The associations between daily health end-points and indicators of air pollution were examined by multivariate, autoregessive linear regression. Daily mean number concentration, but not particle mass (PM_{10}, $PM_{2.5}$), was negatively associated with daily PEF deviations. The strongest effects were seen for particles in the ultrafine range. No significant effect of particulate pollution on symptoms or bronchodilator use was seen (tables 2 and 3).

Table 3. Symptoms and medication used in asthmatics depending on UPs and FPs. Δ is the interquartile range; ∗, $p < 0.05$.

Adults with asthma Erfurt 1991/92[a]			
	Δ	feeling ill during the day OR [95% CI]	cough OR [95% CI]
UP	9200 cm^{-3}	1.44 [1.15, 1.81]*	1.26 [1.06, 1.50]*
FP	50 μg m^{-3}	1.21 [1.06, 1.38]*	1.02 [0.91, 1.15]
PM_{10}	50 μg m^{-3}	1.47 [1.16, 1.86]*	1.30 [1.09, 1.55]*
Adults with asthma Erfurt 1996/97[b]			
	Δ	corticosteroid use OR [95% CI]	bronchdilator use OR [95% CI]
UP	7700 cm^{-3}	1.34 [1.22, 1.47]*	1.09 [0.99, 1.21]
FP	20 μg m^{-3}	1.29 [1.21, 1.38]*	1.03 [0.96, 1.11]
Adults with asthma Helsinki 1996/97[b]			
	Δ	asthmatic symptoms % coefficient	cough % coefficient
PNC	7300 cm^{-3}	0.001	0.076*
FP	6.6 μg m^{-3}	−0.010*	−0.008
PM_{10}	9.3 μg m^{-3}	−0.010	−0.016

[a]Peters *et al.* (1979): 5 days mean, UP = NC0.01–0.1, FP = MC0.1–0.5.
[b]Von Klot *et al.* (2000): 14 days mean, UP = NC0.01–0.1, FP = MC0.01–0.5 = '$PM_{2.5}$'.
[c]Penttinen *et al.* (2000): 5 days mean, PNC is the total particle number count, FP = $PM_{2.5}$.

2.2.4. *Study on Children with Asthma Symptoms in Koupio, Finland 1994*

The effects of daily variations in particles of different sizes on peak expiratory flow (PEF) were investigated during a 57-day follow-up of 39 asthmatic children aged 7–12 years in 1994 in Koupio. In addition to PM_{10} and black smoke (BS) concentrations, an electrical aerosol spectrometer (EAS) was used to measure particle number concentrations in the size range of 0.01–10 μm. All pollutants tended to be associated with declines in morning PEF. In this study, the concentration of UPs was less strongly associated with variations in PEF than PM_{10} or BS (table 2).

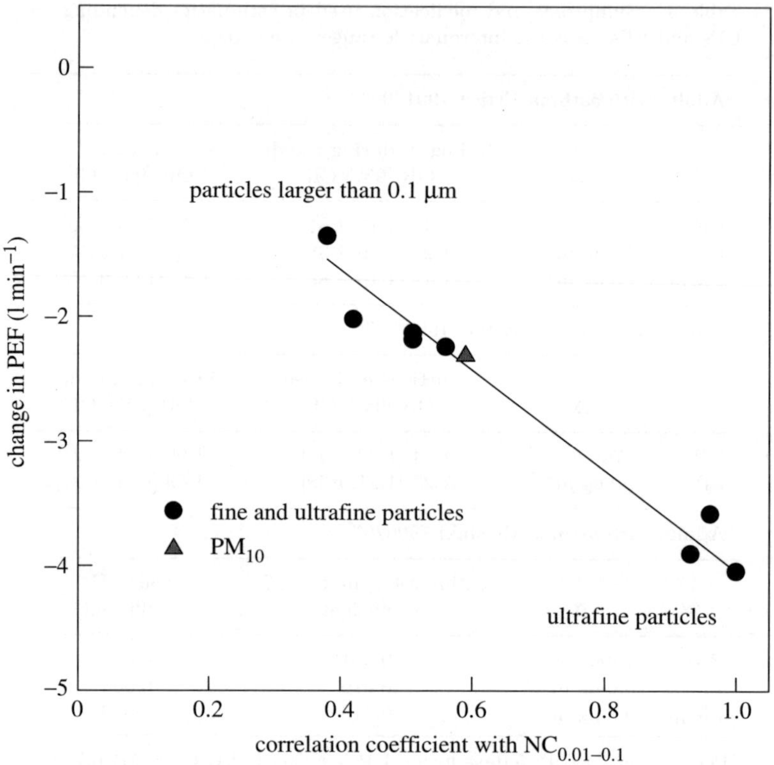

Fig. 2. Changes in evening peak expiratory flow (PEF) by correlation between all size fractions and the number concentration of ultrafine particles (NC0.01–0.1). From Peters *et al.* (1979).

2.2.5. *Mortality Study in Erfurt, Germany 1995–98*

Mortality data were collected prospectively over a 3.5 year period from August 1995 to December 1998. Death certificates were obtained from the local health authorities. The death certificates were aggregated to daily time-series of total counts or counts for subgroups. These were compared with particle data: besides $PM_{2.5}$ and PM_{10}, size specific number and mass concentration data in six size classes between 0.01 and 2.5 μm were derived from measurements with the MAS (Wichmann *et al.* 2000*a*). Furthermore, elemental composition was analysed by PIXE, as described above (Wichmann *et al.* 2000*b*).

Some of the UP and FP concentrations are given in table 1. All particulates had a strong seasonality with maximal concentrations in winter. The UP concentrations showed a strong day of the week effect with concentrations during the weekend 40% lower than during the week. This and a clear increase of the UP concentrations during the rush hours suggests that the main source for UPs was automobile traffic.

The association with daily mortality was analysed using Poisson regression techniques with generalized additive modelling (GAM) to allow nonparametric adjustment for the confounders. The pollutants were included either untransformed or log transformed, depending on goodness of fit. Mortality increased in association with ambient particulates after adjustment for season, influenza epidemics, day of week and meteorology. In a sensitivity analysis, the results proved stable against changes of the confounder model.

As shown in figure 3*a*, associations between particle number and particle mass concentrations have been observed in different size classes, and both immediate effects (lags 0 or 1 days) and delayed effects (lags 4 or 5 days) were found. There was a tendency for more immediate effects of the mass concentrations (i.e. in the larger size ranges) and for more delayed effects of the number concentrations (i.e. in the smaller size ranges). However, this pattern could not be separated clearly, and distributed lag models comprising the days 0 to 5 showed similar results.

The effects could be found for total mortality but also for respiratory and cardiovascular causes (figure 3*b*). There was a tendency for more immediate effects on respiratory causes and more delayed effects for cardiovascular causes. Again this could not be distinguished statistically.

2.3. *Ongoing Studies*

2.3.1. *Study on Adults with Cardiovascular Diseases in Three European Cities (EU-ULTRA) 1996–1999*

In the first part of this study, UP and FP measurements have been compared in Finland, The Netherlands and Germany (Pekkanen *et al.* 1999*a*; Ruuskanen *et al.* 2000). The results are shown in table 1. In the epidemiological part, a panel study of 150 elderly with cardiovascular diseases was performed in the winter season 1998/99, using symptom diaries and performing biweekly EKG and lung function measurements. The analysis is ongoing.

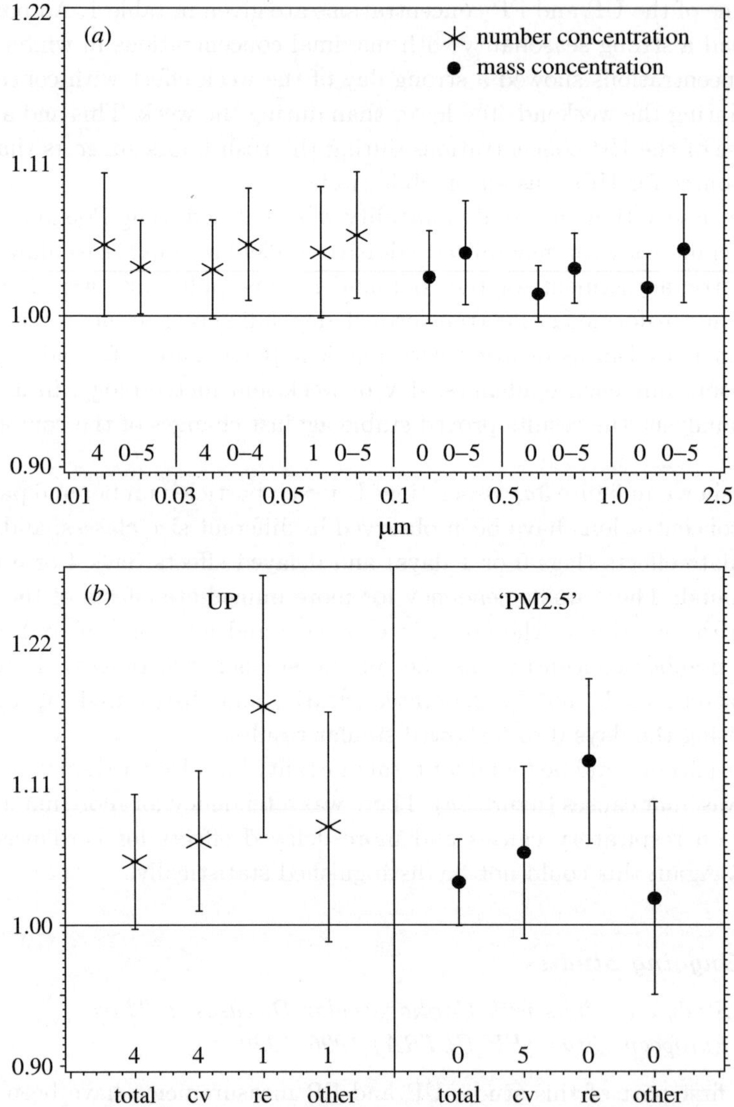

Fig. 3. (*a*) Effects of different size classes of UPs and FPs on daily mortality in Erfurt, 1995–1998. Left, best one-day lag; right, distributed lag model. The lags (days) are given at the bottom. (*b*) Effects of UPs and FPs on mortality for prevalent diseases (total, cardiovascular, respiratory, others). Best day-lag model. There seems to be a stronger immediate effect (lag 0 or 1 days) on respiratory causes and a stronger delayed effect (lag 4 or 5 days) on cardiovascular causes. Modified from Wichmann *et al.* (2000*a*).

2.3.2. *Study on Survivors of Acute Myocardial Infarction in Augsburg, Germany 1999–2001*

A case crossover study is performed based on the Coronary Event Registry in Augsburg. Cases are survivors of an acute myocardial infarction. Measurements of fine particle mass and total number concentration are performed on an hourly basis.

2.3.3. *Study on Cardiovascular Diseases and COPD in Erfurt (as part of the EPA Rochester Ultrafine Particle Centre) 2000–2004*

In the first part, a panel of 50 patients with adult cardiovascular patients is observed for six months. In the second phase a corresponding protocol is used for adult patients with chronic obstructive pulmonary disease (COPD). Daily respiratory and cardiovascular symptoms are recorded, biweekly EKGs and blood parameters are determined. Fine and ultrafine particles are measured using the MAS device, as well as $PM_{2.5}$, PM_{10} and PIXE. The aim of the study is the characterization of the association between ambient particle exposure and changes in biomarkers of inflammation of the cardiorespiratory system in patients with stable coronary artery disease and/or COPD.

3. Discussion

3.1. *Ultrafine Particles in Ambient Air*

The ambient aerosol is a dynamic system which may change its concentration and size distribution due to coagulation and chemical reactions. Because of their high diffusivity UPs coagulate with other aerosol particles depending on the ambient aerosol conditions such as concentration, size distribution, thermodynamic parameters, etc. (Fuchs 1964; Willeke & Baron 1993).

Measurements of ultrafine particles in the framework of epidemiological studies are only available for a limited number of places in Europe like Erfurt (Tuch *et al.* 1997; Wichmann *et al.* 2000*a*, *b*) as well as the ULTRA study in Germany, Finland and The Netherlands (Tuch *et al.* 2000*a*, *b*; Ruuskanen *et al.* 2000; Pekkanen *et al.* 1999*a*; Mirme *et al.* 2000) and three places in Sachsen-Anhalt (Pitz *et al.* 2000). These data show a surprisingly homogeneous picture, but the number of places is not sufficient to see which range exists within Europe. Furthermore, no data of the spatial distribution of UPs within a city are available. It is important to note that the correlation of UPs and FPs is surprisingly low, suggesting that different sources may be relevant and that the coagulation of UPs to FPs is a complex process.

3.2. *Health Effects of Fine and Ultrafine Particles*

3.2.1. *Lesson from the Asthma Panel Studies*

From the studies described above the following can be learned:

- there are clearer effects on adults with asthma than on children with asthma symptoms;
- effects of both UPs and FPs are observed, and the effects of UPs are slightly stronger;
- cumulative effects over 5 days (for medication use up to 14 days) are stronger than same-day effects;
- in two pollutant models, the effect on the same day is stronger for FPs, whereas the cumulative effect is stronger for UPs (Peters *et al.* 1997*a*).

3.2.2. *Lesson from the Mortality Study*

From the only available mortality study (Wichmann *et al.* 2000*a*, *b*) we learn:

- there are particle effects on total mortality as well as on respiratory and cardiovascular causes;
- effects of both UPs and FPs are observed;
- there are immediate effects (lag 0–1 day) and delayed effects (lag 4–5 days), which can be combined into cumulative effects (by distributed lag models);
- there is a tendency that FPs show slightly stronger immediate effects and that UPs show slightly stronger delayed effects;
- there is a tendency that mortality of respiratory cases is more immediately affected, whereas mortality of cardiovascular cases is more delayed;
- in two pollutant models, immediate (lag 0 day) and delayed effects (lag 4 days) are independent (Wichmann *et al.* 2000*a*).

3.2.3. *Which Pathophysiological Mechanisms are Plausible?*

Based on the knowledge from animal experiments and on the pathway of particles in the respiratory tract, the following mechanisms would be plausible.

- Since FPs are deposited in the small airways, one would expect to see effects there. These should be proportional to the volume (mass) deposited. One could think of soluble toxic agents. The larger a particle is, the more material can be dissolved from it. This would be directly available to the respiratory system and the dose would depend on the mass concentration. These soluble compounds could initiate inflammation and lead to an acute local inflammatory response in the lung and thereby may contribute to the exacerbation of pre-existing diseases (Bates 1992).

- UPs are deposited mainly in the alveolar region. Since the mass of UPs is negligible, mass-related effects are less probable. Therefore, not the soluble but the insoluble compounds are expected to be relevant. For this causal fraction, time would be required to translocate the particles to sites of reaction and/or initiation of chain of reactions. UPs are phagocytized less readily by alveolar macrophages and are found not only on the epithelium but in interstitial sites (Ferin *et al.* 1991; Stearns *et al.* 1994). At the same time inflammatory indicators may be upregulated, suggesting that the increased access of UPs to the interstitium triggered an inflammatory response. In other words, UPs may be translocated to reactive sites in and beyond the epithelium which may activate endothelial and circulating leukocytes and endothelial adhesion molecules in the blood, alter blood coagulability (Utell & Frampton 1999), and this process may need more time to become effective. These events could lead to an exacerbation of pre-existing cardiovascular disease.

3.2.4. *Do the Epidemiological Data Support the Described Mechanisms?*

The following observations are in favour of the mechanisms described in § 3 *b* (iii):

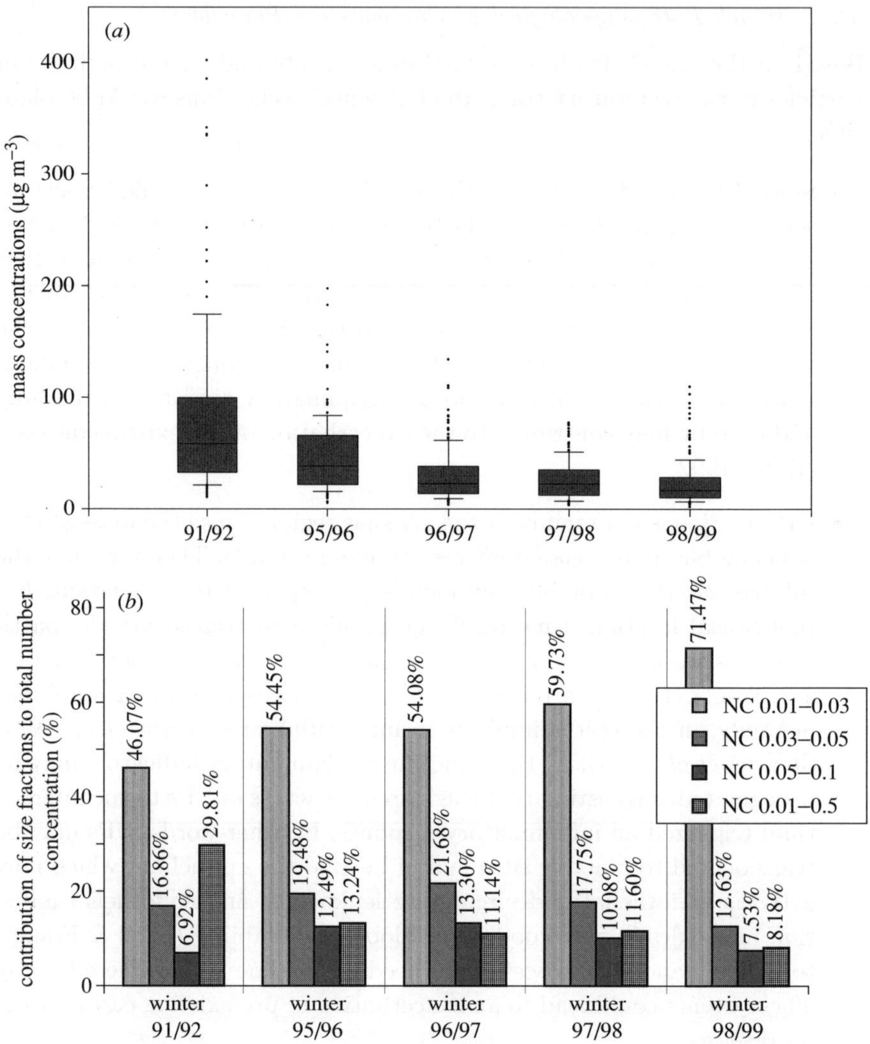

Fig. 4. (*a*) Seven years trend of the mass concentration (MC 0.01–2.5 = $PM_{2.5}$ of FPs in Erfurt, winters 1991/2 to 1998/9. From Wichmann *et al.* (2000*a*). (*b*) Seven years trend of the relative particle number concentration (in %); different size ranges (0.01–0.03, 0.03–0.05, 0.05–0.1. 0.1–0.5 µm diameter). The concentration of UPs is approximately constant (see table 1) and the fraction in the smallest size fraction increases steadily. From Wichmann *et al.* (2000*a*).

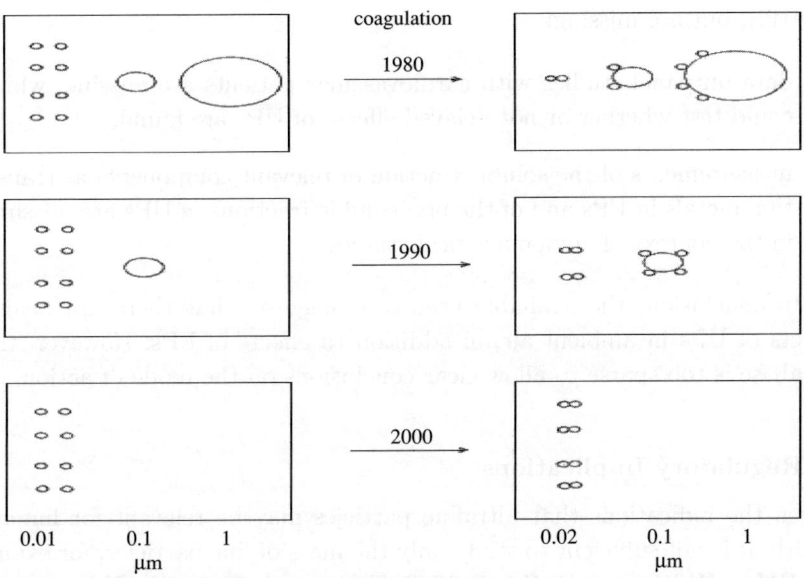

Fig. 5. Simplified model of the coagulation dynamics in Erfurt from 1980 to 2000. In 1980, large particles have been in the air, which effectively scavenged the ultrafine particles, leading to a short half-life of UPs. In 2000, mainly very small particles are in the air. They coagulate much slower and the coagulation products are still UPs. In total, in the year 2000, the half-life of UPs is clearly longer than in the year 1980, i.e. if the production rate is constant, the measure ambient concentration of UPs increases. From Wichmann *et al.* (2000*a*).

- the tendency of more direct effects of FPs in asthmatics and on mortality with respiratory causes;
- the tendency of more delayed effects of UPs on mortality with cardiovascular causes;
- the fact that these two mechanisms seem to be independent and show a positive interaction.

The following observations cannot be easily explained by these mechanisms:

- there are also delayed or cumulative effects of FPs (although weaker);
- the delayed or cumulative effects are not only seen in cardiovascular mortality but also in patients with asthma.

The following data would be very important to test the hypotheses in § 3*b* (iii), but are missing:

- data on panel studies with cardiovascular patients are missing, which could test whether or not delayed effects of UPs are found;
- measurements of the soluble fraction of relevant components as transition metals in FPs and of the non-soluble fractions in UPs are missing, in the context of epidemiological studies.

In conclusion, the available literature suggests that there are health effects of UPs in ambient air, in addition to effects of FPs. However, the database is too sparse to allow clear conclusions on the mode of action.

4. Regulatory Implications

Given the indications that ultrafine particles may be relevant for human health, it is not sufficient to study only the mass of fine particles, for example $PM_{2.5}$ (Wichmann & Peters 1999; Tuch *et al.* 2000*a*, *b*). This may be illustrated by the development in Erfurt as shown in figure 4. The mass of fine particles was clearly reduced since 1991/92. However, during the same period the number concentration of ultrafine particles was not decreased, and especially the fraction of very small particles between 0.01 and 0.03 μm diameter increased steadily over the seven years of observation. This makes clear that, with respect to regulation, the reduction of the fine mass does not automatically mean that the number of ultrafine particles is also reduced. Therefore, to identify the relevant particle fraction with respect to human health is crucial for sound regulatory activities.

The ambient aerosol is a dynamic system which may change its concentration and size distribution due to sources and due to coagulation and chemical reactions. Hence, specific pollution control measures to reduce fine particle mass concentration, which effectively reduces the FPs concentration, may paradoxically increase the persistence and thus number concentration of UPs. The drastic reduction of larger particles in the last 20 years in Erfurt may have reduced the scavenging of ultrafine particles and thus prolonged their half-life in the atmosphere. As a result, even if emissions of UPs were constant, their ambient concentration nevertheless may have increased. This is shown schematically in figure 5.

It is important to realize that technologies different from the ones currently used to reduce the mass emission are needed to reduce the particle number emission.

References

Bascom, R., Bromberg, P. A., Costa, D. A., Devlin, R., Dockery, D. W., Frampton, M. W., Lambert, W., Samet, J. M., Speizer, F. E. & Utell, M. 1996 Health effects of outdoor air pollution. *Am. J. Respir. Crit. Care Med.* **153**, 3–50.

Bates, D. V. 1992 Health indices of the adverse effects of air pollution. The question of coherence. *Environ. Res.* **59**, 336–349.

Brand, P., Gebhart, J., Below, M., Georgi, B. & Heyder, J. 1991 Characterization of environmental aerosol on Helgoland Island. *Atmos. Environ.* A **25**, 581–585.

Brand, P., Ruoß, K. & Gebhart, J. 1992 Technical note: performance of a mobile aerosol spectrometer for *in situ* characterization of an environmental aerosol in Frankfurt city. *Atmos. Environ.* A **26**, 2451–2457.

Danesh, J., Collins, R., Appleby, P. & Peto, R. 1998 Association of fibrinogen, C-reactive protein, albumin, or leukocyte count with coronary heart disease: meta-analyses of prospective studies. *J. Am. Med. Ass.* **279**, 1477–1482.

Dockery, D. W. & Pope C. A. 1994 Acute respiratory effects of particulate air pollution. *A. Rev. Public Health* **15**, 107–132.

Dockery, D. W., Schwartz, J. & Spengler, J. D. 1992 Air pollution and daily mortality: association with particulates and acid aerosols. *Environ. Res.* **59**, 362–373.

Ferin, J., Oberdörster, G., Soderholm, S. C. & Gelein, R. 1991 Pulmonary tissue access of ultrafine particles. *J. Aerosol Med.* **4**, 57–68.

Fuchs, N. A. 1964 *The mechanic of aerosols*, pp. 288–302. Oxford: Pergamon.

Gold, D. R., Litonjua, A., Schwartz, J., Lovett, E., Larson, A., Nearing, B. D. *et al.* 2000 The relationship between paticulate pollution and heart rate variability. *Circulation.* (In the press.)

ICRP (International Commission of Radiological Protection) 1994 Human respiratory tract model for radiological protection. (ICRP Publication no. 66.) *Ann. ICRP* **24**, 36–52.

Katsouyanni, K. (and 12 others) 1996 Short term effects of air pollution on health: a European approach using epidemiologic time series data: the APHEA protocol. *J. Epidemiol. Commun. Health* (Suppl.) **50**, 12–18.

Katsouyanni, K. (and 12 others) 1997 Short term effects of ambient sulfur dioxide and particulate matter on mortality in 12 European cities: results from time series data from the APHEA project. *Br. Med. J.* **314**, 1658–1663.

Koenig, W. & Ernst, E. 1992 The possible role of hemorheology in altherothrombogenesis. *Atheroclerosis* **94**, 93–107.

Koenig, W., Sund, M., Filipiak, B., Döring, A., Löwel, H. & Ernst, E. 1998 Plasma viscosity and the risk of coronary heart disease: results from the MONICA–

Augsburg cohort study, 1984 to 1992. *Arterioscler. Thromb. Vasc. Biol.* **18**, 768–772.

Kreyling, W. G., Khlystov, A., Mirme, A., Tuch, T., Ruuskanen, J., Vallius, M., Ten Brink, H., Roth, C., Kos, G. A. & Pekkanen, J. 1999 Exposure assessment for fine and ultrafine particles in ambient urban aerosoles. In *Proc. Third Colloquium on Particulate Air Pollution and Human Health in Durham, UC Irvine*, 4-80–4-91.

Liao, D., Cai, J., Rosamond, W. D., Barnes, R. W., Hutchinson, R. G., Whitsel, E. A. *et al.* 1997 Cardiac autonomic function and incident coronary heart disease: a population-based case-cohort study. The ARIC Study. Atherosclerosis Risk in Community Study. *Am. J. Epidemiol.* **145**, 696–706.

Mirme, A., Tuch, T., Khlystov, A., Kos, G., Ten Brink, H. M., Ruuskanen, J., Kreyling, W. G. & Pekkannen, J. 2000 Intercomparison of aerosol spectrometers for ambient air monitoring. *Atmos. Environ.* (Submitted.)

Oberdörster, G., Gelein, R. M., Ferin, J. & Weiss, B. 1995 Association of particulate air pollution and acute mortality: involvement of ultra-fine particles? *Inhal. Toxicol.* **7**, 111–124.

Pekkanen, J., Timonen, K. L., Ruuskanen, J., Reponen, A. & Mirme, A. 1997 Effects of ultrafine and fine particles in an urban air on peak expiratory flow among children with asthmatic symptoms. *Environ. Res.* **74**, 24–33.

Pekkanen, J., Brunekreef, B. & Wichmann, H. E. 1999*a* Exposure and risk assessment for fine and ultrafine particles in ambient air (ULTRA). Final report, EU Environment Programme Contract ENV4-CT95-0205, Brussels.

Pekkanen, J., Brunner, E., Anderson, H. R., Tittanen, P. & Atkinson, R. W. 1999*b* Air pollution and plasma fibrinogen. *Am. J. Respir. Crit. Care Med.* **54**, 1027–1032.

Penttinen, P., Timonen, K. L., Tiittanen, P., Mirme, A., Ruuskanen, J. & Pekkanen, J. 2000 Fine and ultrafine particulate matter in ambient air are associated with peak flow decreases in adult asthmatic subjects. *Am. J. Respir. Crit. Care Med.* (In the press.)

Peters, A., Wichmann, H. E., Tuch, T., Heinrich, J. & Heyder, J. 1997*a* Respiratory effects are associated with the number of ultra-fine particles. *Am. J. Respir. Crit. Care Med.* **155**, 1376–1383.

Peters, A., Dockery, D. W., Heinrich, J. & Wichmann, H. E. 1997*b* Medication use modifies the health effects of particulate sulfate pollution in children with asthma. *Environ. Health Perspect.* **105**, 430–435.

Peters, A., Dockery, D. W., Heinrich, J. & Wichmann, H. E. 1997*c* Short-term effects of particulate air pollution on respiratory morbidity in asthmatic children. *Eur. Respir. J.* **10**, 872–879.

Peters, A., Döring, A., Wichmann, H. E. & Koenig, W. 1997*d* Increased plasma viscosity during the 1985 air pollution episode: a link to mortality? *Lancet* **349**, 1582–1587.

Peters, A., Perz, S., Döring, A., Stieber, J., Koenig, W. & Wichmann, H. E. 1999*a* Activation of the autonomic nervous system and blood coagulation in association with an air pollution episode. In *Proc. Third Colloquium on Particulate Air Pollution and Human Health, 6–8 June 1999, Durham* (ed. R. Phalen & Y. Bell), 8-71–8-85.

Peters, A., Perz, S., Döring, A., Stieber, J., Koenig, W. & Wichmann, H. E. 1999*b* Increases in heart rate during an air pollution episode. *Am. J. Epidemiol.* **150**, 1094–1098.

Peters, A., Wichmann, H. E. & Koenig, W. 1999*c* Air pollution exposure influences cardiovascular risk factors: a link to mortality? In *Proc. Int. Inhal. Symp. Hanover, Germany.*

Peters, A., Liu, E., Verrier, R. L., Schwartz, J., Gold, D. R., Mittleman, M. *et al.* 2000 Air pollution and incidence of cardiac arrhythmia. *Epidemiology* **11**, 11–17.

Pitz, M., Heinrich, J., Tuch, T., Kreyling, W. G. & Wichmann, H. E. 2000 Change of particle size distribution in Sachsen-Anhalt between 1993 and 1999. (Submitted.)

Pope, C. A. 2000 Epidemiology of fine particulate air pollution and human health: biological mechanisms and who's at risk? *Environ. Health Perspect.* (In the press.)

Pope, C. A. & Dockery, D. W. 1999 Epidemiology of particle effects. In *Air pollution and health* (ed. S. T. Holgate, J. M. Samet, H. S. Koren & R. L. Maynard), pp. 673–705. San Diego: Academic Press.

Pope, C. A., Dockery, D. W., Kanner, R. E., Villegas, G. M. & Schwartz, J. 1999*a* Oxygen saturation, pulse rate, and particulate air pollution. *Am. J. Respir. Crit. Care Med.* **159**, 365–372.

Pope, C. A., Verrier, R. L., Lovett, E. G., Larson, A. C., Raizenne, M. E., Kanner, R. E. *et al.* 1999*b* Heart rate variability associated with particulate air pollution. *Am. Heart J.* **138**, 890–899.

Ruuskanen, J. (and 12 others) 2000 Concentrations of ultrafine, fine and $PM_{2.5}$ particles in three European cities. *Atmos. Environ.* (In the press.)

Schwartz, J. 1994 Air pollution and daily mortality: a review and meta analysis. *Environ. Res.* **64**, 36–52.

Schwartz, J., Dockery, D. W. & Neas, L. M. 1996 Is daily mortality associated specifically with fine particles? *J. Air. Waste. Management Ass.* **46**, 927–939.

Seaton, A., MacNee, W., Donaldson, K. & Godden, D. 1995 Particulate air pollution and acute health effects. *Lancet* **345**, 176–178.

Seaton, A., Soutar, A., Crawford, V., Elton, R., McNerlan, S., Cherrie, J. *et al.* 1999 Particulate air pollution and the blood. *Thorax* **54**, 1027–1032.

Stearns, R. C., Murthy, G. G. K., Skornik, W., Hatch, V., Katler, M. & Godleski, J. J. 1994 Detection of copper oxide particles in the lungs of hamsters by electron spectroscopic imaging. *ICEM* **13**, 763–764.

Stone, P. H. & Godleski, J. J. 1999 First steps toward understanding the pathophysiologic link between air pollution and cardiac mortality. *Am. Heart J.* **138**, 804–807.

Tuch, T., Brand, P., Wichmann, H. E. & Heyder, J. 1997 Variation of particle number and mass concentration in various size ranges of ambient aerosols in Eastern Germany. *Atmos. Environ.* **31**, 4193–4197.

Tuch, T., Mirme, A., Tamm, E., Heinrich, J., Heyder, J., Brand, P., Roth, C., Wichmann H. E., Pekkanen, J. & Kreyling, W. G. 2000*a* Comparison of two particle size spectrometers for ambient aerosol measurements. *Atmos. Environ.* **34**, 139–149.

Tuch, T., Kreyling, W. G., Peters, A., Heinrich, J., Heyder, J. & Wichmann, H. E. 2000*b* Reduction of particle mass parallels increase in particle number in the atmosphere. (Submitted.)

US EPA 1996 Air quality criteria for particulate matter research. Triangle Park Research: EPA.

Utell, M. J. & Frampton, M. W. 1999 Clinical relevance of particle related effects. *J. Aerosol Med.* **12**, 104 (Abstract 56).

Von Klot, S., Wölke, G., Tuch, T., Heinrich, J., Docker, D. W., Schwarz, J., Wichmann, H. E. & Peters, A. 2000 Short-term effects of ultrafine and fine particles on medication use in asthmatic adults. *Proc. Conf. American Thoracic Soc.* 2000 Toronto (Abstract).

Wichmann, H. E. & Peters, A. 1999 Epidemiological studies on health effects of fine and ultrafine particles in Germany. In *The health effects of fine particles: key questions and the 2003 Review Report of the Joint Meeting of the EC and HEI, 14–15 January 1999, Brussels, Belgium. HEI Commun.* **8**, II-163–172.

Wichmann, H. E., Spix, C., Tuch, T., Wölke, G., Peters, A., Heinrich, J., Kreyling, W. G. & Heyder J. 2000*a* Daily mortality and fine and ultrafine particles in Erfurt, Germany, Part A: Role of particle number and particle mass. HEI report.

Wichmann, H. E., Spix, C., Tuch, T., Wittmaack, K., Cyrys, J., Wölke, G., Peters, A., Heinrich, J., Kreyling, W. G. & Heyder, J. 2000*b* Daily mortality and fine and ultrafine particles in Erfurt, Germany, Part B: Role of sources, elemental composition and other pollutants. HEI Report.

Willeke, K. & Baron, P. A. (eds) 1993 *Aerosol measurements: principles, techniques and applications.* New York: Van Nostrand Reinhold.

Discussion

H. R. ANDERSON (*St George's Hospital Medical School, Cranmer Terrace, London, UK*). Your studies in Erfurt have found clear associations between health effects in adult asthmatics, but similar studies among children in Kuopio, Finland, have not been so conclusive. Studies of hospital admissions for asthma also tend to find associations between particles and admissions in adults but not in children. Can you think of any explanation for this difference between adults and children?

H.-E. WICHMANN. Indeed, the effects of ultrafine particles seem to be more pronounced in asthmatic adults than in asthmatic children. I have no explanation for this result.

D. COSTA (*US EPA, NC, USA*). The aerometric data from the first half of this issue suggest that the ultrafine PM is quite variable in concentration and time over the course of the day. Yet your early data showed correlations with five-day averages. Does this suggest that the ultrafine effects on impact is cumulative?

H.-E. WICHMANN. Our data suggest cumulative effects on daily mortality. The influence of cumulative exposure over five days seems stronger than the influence of every single day. This is true both for fine and ultrafine particles. However, if one considers single-day effects, these seem to be more immediate for fine particles (lag 0 days) and more delayed for ultrafine particles (lag 5 days).

CHAPTER 15

DIFFERENTIAL EPIDEMIOLOGY OF AMBIENT AEROSOLS

H. R. Anderson
St George's Hospital Medical School, Cranmer Terrace, London SW17 0RE, UK
(r.anderson@sghms.ac.uk)

There is now a large body of epidemiological evidence associating exposure to ambient particles with short- and long-term effects on health. Most authorities consider that at least some of these associations represent a causal relationship with particles. The size fraction of particles that could potentially harm health is PM_{10}, since only particles less than this size can plausibly reach the small airways and alveoli. Studies of mechanisms and theoretical considerations suggest that the fine ($PM_{2.5}$) and ultrafine ($PM_{0.1}$) particles are probably more important than larger particles, because of their relatively greater numbers and deeper penetration of the lung. Because of limited population exposure data, there is little direct epidemiological evidence about the effects of ultrafine particles. Indirect evidence falls into three groups. The first comes from studies that have directly compared the coarse ($PM_{2.5-10}$) with the fine ($PM_{2.5}$) fractions; the findings of these few studies have not been consistent. The second comes from studies of chemical species or measures of particles (sulphates, acid aerosol and black smoke) that reside mainly in the fine fraction; many of these have found associations with adverse health effects. The third group are those few studies that have compared the effects of size/number concentrations with size/mass concentrations; the findings of these have either been inconclusive or have suggested that numbers may be more important than mass.

Inference about the toxic component of particles will depend on all the evidence, especially from toxicology, as well as epidemiology. At present, epidemiological evidence points towards the fine fraction being important, but an effect of the coarse fraction cannot be excluded.

Because of a lack of data, epidemiology has little to say about the relative importance of the ultrafine fraction. This is an urgent research need.

Keywords: air pollution; epidemiology; particles; PM_{10}; $PM_{2.5}$

1. Introduction

The development of our knowledge about the health effects of ambient air pollution has depended on two very different scientific disciplines. One may broadly be described as toxicology, which is laboratory based and experimental in concept. The other is epidemiology, which is population based and observational in concept. While toxicology is important for telling us whether an environmental agent might be important, and possible mechanisms of effect, epidemiology is important for telling us whether effects actually occur in the real-life situation. The earliest evidence of adverse effects came from simple epidemiological analyses of major air pollution episodes, notably the 1952 London air pollution episode (Ministry of Health 1954). Around the same time it was observed that ill health and mortality tended to be higher in polluted areas (Gardner *et al*. 1969; Lave & Seskin 1970). As pollution improved in western developed countries, evidence for health effects, using the crude epidemiological techniques available at the time, became marginal, and this was interpreted as indicating that there was no longer a problem. The resurgence of concern about air pollution is due in part to the development and application of more sensitive statistical tools for the epidemiological analysis of time-series and cohort data, which have identified associations at levels of pollution well under guideline values. Toxicology has, until recently, contributed mainly to the understanding of the mechanism of effects of pollutant gases, such as ozone, or selected chemically pure particles such as sulphuric acid, or titanium dioxide. More recently, experimental techniques have been developed to study ambient particles themselves. At present there is intense interest in identifying the important toxic components of the particle mixture and the field is becoming driven by mechanistic theories relating to aspects such as the chemistry, size and number concentration of particles. This, in turn, presents new challenges to epidemiology to raise and test hypotheses in exposed populations. In this paper I shall first review briefly the development of epidemiological knowledge of the health effects of inhalable particles, and then focus on epidemiological evidence concerning the responsible fraction, in terms of size, chemistry and numbers.

2. Evidence that Ambient Particles have Health Effects

2.1. *Short-term Associations: Air Pollution Episodes*

One method of epidemiological enquiry is to analyse data arranged as a time-series to look for short-term associations between air pollution and health outcomes. The earliest form of this approach is seen in reports of air pollution episodes, where a simple graphical display alone may be sufficient to show a convincing increase in daily mortality or some other outcome coinciding with a major increase in air pollution, such as the 1952 London episode. This is not a very sensitive way of detecting smaller effects and it is not always possible to exclude other explanations, such as a coincidental respiratory epidemic, or the effects of the weather conditions which predisposed to the episode in the first place: these will include cold in the case of winter episodes and heat in the case of summer episodes. The majority of major episodes comprise elevated concentrations of both particles and gases and there is no way of satisfactorily separating out the effects of the various components in a single episode analysis. In special situations such as certain types of volcanic eruption, where the population is exposed mainly to particles, adverse health effects have been found, which suggests that particles alone are sufficient to have effects (Baxter *et al.* 1983).

2.2. *Short-term Associations: Ecological Time-Series Analyses*

These are regression analyses that use aggregated data such as daily counts of mortality or hospital admissions from a large population, usually a city, obtained from routine health data systems. The method is statistically powerful and enables a range of potential confounding factors to be controlled for. Confounding factors are those that may be related to both air pollution and the outcome of interest, and failure to control for them could lead to spurious associations. They include time trends, seasonal variations, weather, day of the week, and epidemics of respiratory disease. It is the development of appropriate statistical methodology that has brought to light associations between daily mortality and air pollution at low (i.e. below guidelines) levels. These studies identify short-term associations in a statistical sense but from the point of view of the individual, the health effects may be either transient, e.g. a stay in hospital, or permanent (as in the case of mortality). It is likely that the increase in events such as admissions and mortality is due to air pollution acting as an added factor in a situation already loaded with other risk factors.

Typically, the exposure data for such analyses come from stations which routinely monitor background pollution. It is likely that this will lead to misclassification of individual exposure. In most circumstances, this will bias the effect estimate towards the null. This has implications for comparing different fractions of the particle mixture. Use of a community monitor also carries the risk of bias in the estimate of community exposure, the effect of which on the estimate of health risk of air pollution could be up or down.

A large number of such studies have now been reported from cities throughout the world (figure 1). There is a remarkably consistent tendency to positive effects, not only for mortality but for hospital admissions and emergency room visits (American Thoracic Society 1996; Dockery & Pope 1994; Pope *et al.* 1995*a*; USEPA 1996; Department of Health Committee on the Medical Effects of Air Pollutants 1995*a*). When looked for, it has also been common to observe similar associations with pollutant gases such as SO_2, O_3 and, to a lesser extent perhaps, CO and NO_2. In most cases the associations with particles are more or less maintained after controlling for covarying pollutants.

2.3. *Short-term Associations: Panel Studies*

The other time-series technique is to study relationships at an individual level by following a panel of subjects over time and monitoring such outcomes as lung function, symptoms and medication use. These have also been found to be associated with air pollution (see references above), though not so consistently. For example, a very large and carefully conducted European study of children, the Pollution Effects on Asthmatic Children in Europe (PEACE) study, did not observe an association between particles and health effects (Roemer *et al.* 1998). Pollutant gases, especially ozone, have also been found to be associated with such outcomes. The method only identifies short-term associations. It is likely that most of the outcomes recorded are short-term physiopathological adaptations or represent the functioning of defence mechanisms, but longer-term effects cannot be excluded, especially if exposure to pollutants is associated with other pathogenic factors.

The causality of associations identified by ecological and panel time-series studies have been questioned, mainly on the basis of inadequate con-

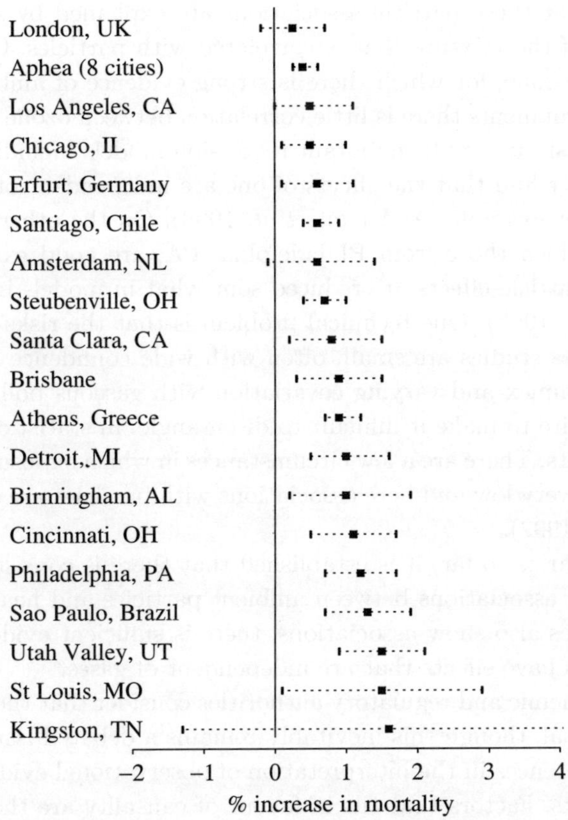

Fig. 1. Particulate matter with aerodynamic diameter less than 10 μm (PM_{10}) and daily mortality from cities around the world. Expressed as a percentage change in daily mortality associated with a 10 μg m^{-3} increase in PM_{10}.

trol for confounders, or failure to separate particle effects from those of other pollutants in the mixture (Gamble & Lewis 1996). Most authorities believe that this is not the case (Department of Health Committee on the Medical Effects of Air Pollutants 1995*a*), and for the purposes of this paper I shall accept that we are discussing a real effect on health, and not a spurious and, therefore, non-causal association.

I have already referred to some evidence that particles alone may be associated with health effects. In most situations, however, populations are exposed to particle–gas mixtures and it is important to consider the possibility that these particle associations are explained by some gaseous component of the mixture that is correlated with particles. One potential candidate is ozone, for which there is strong evidence of ambient toxicity. In most environments there is little correlation between ozone and particles on a daily basis and multi-pollutant regression models including both pollutants usually find that the effects of one are independent of those of the other (see, for example, Anderson *et al.* 1996). On the other hand, many studies, of which those from Philadelphia, PA, are good examples, have found that particle effects are reduced somewhat in models including SO_2 (Kelsall *et al.* 1997). One technical problem is that the risks identified by the time-series studies are small, often with wide confidence intervals and there is a complex and varying covariation with gaseous pollutants; these factors conspire to make it difficult to disentangle, in statistical terms, the separate effects. There are a few circumstances in which exposure to gaseous pollutants is very low and here associations with particles are still observed (Pope *et al.* 1992).

To summarize so far, it is established that there is consistent evidence of short-term associations between ambient particles and health, and that although gases also show associations, there is sufficient evidence to show that particles have effects that are independent of gases.

Most academic and regulatory authorities consider that the associations could be causal, though this inevitably remains a debated issue because of personal differences in the interpretation of observational evidence of small increased risks. Factors that tilt in favour of causality are the consistency of findings across many cities, climates, pollution sources and investigators, a specificity for cardiorespiratory diseases, the exposure–response relationship, and the growing toxicological evidence for biological plausibility.

2.4. *Associations with Chronic Disease*

The other epidemiological strategy for studying air pollution compares the health of populations exposed, long term, to different levels of pollution. Comparisons can only be made at a group level because the exposure is at a group level. This approach addresses more important health outcomes, such as mortality rates and chronic illness, but is hampered by the potential for confounding by factors common to both the outcome and pollution level (social class for example). The most satisfactory techniques are those that compare the prevalence or incidence of disease according to different levels of exposure while controlling for confounding factors at an individual level (smoking, household environment, occupation, etc.). These studies have found associations between air pollution and premature mortality, the incidence and prevalence of chronic respiratory disease, respiratory symptoms and reduced lung function (Dockery *et al.* 1989, 1993; Raizenne *et al.* 1996; Pope *et al.* 1995*b*; Abbey *et al.* 1995; Ackermann-Liebrich *et al.* 1997). Interestingly, however, there is very little evidence to suggest that air pollution affects the incidence of asthma, though it does play a role in exacerbations (Department of Health Committee on the Medical Effects of Air Pollutants 1995*b*). The majority of prevalence and cohort studies have identified particles as important, though a role for gases cannot be excluded. As for the time-series studies, the causality of associations is open to different interpretations, but most authorities accept at least the possibility of causality.

3. Which Component of the Particle Mixture is Important?

Having concluded that ambient particles in low concentrations may be toxic to humans, the next step is to consider the relative importance of different components of this very complex mixture of air pollutants. In considering the effects of particles on the lung, heart and blood, the first consideration is that of size. Particles of diameter greater than 10 µm have a low probability of reaching the intra-thoracic airways. This is the reason for the widespread adoption of PM_{10} as a measure of particles. The proportion of particles delivered to the air-exchanging parts of the lung increases with decreasing size of particle. Although very fine particles account for a small part of the total mass, either in the ambient aerosol or deposited in the lung, they account for the greatest number. Theoretical reasons now backed by some experimental evidence support the idea that large numbers of ultrafine particles may present the most risk (Seaton *et al.* 1995).

Apart from the size and number of particles, the chemical composition of particles must be considered. PM_{10} comprises particles from two very different sources and this is reflected in a bimodal size/mass distribution with a dip at *ca.* 2–3 µm. The so-called coarse mode or fraction ($PM_{2.5-10}$) is mainly from the degradation of crustal material and is composed of chemicals such as carbonates, silicates, etc. The fine mode ($PM_{2.5}$) is, in contrast, the result of condensations and aggregations of the gaseous products of combustion (APEG 1999). These may be divided into the primary fraction, which is derived directly from combustion (elemental carbon), and the secondary fraction, which is due to photochemistry and other processes acting on gaseous emissions. Relevant examples of the secondary generation of particles are sulphuric acid and ammonium sulphate from the oxidization of sulphur dioxide, and nitric acid and nitrate from the oxidization of oxides of nitrogen.

If this were not complicated enough, there are a host of trace metals and organic compounds in the mixture, and individual particles are not chemically homogeneous. There may be adsorption of other particles or gaseous pollutants onto a central particle core.

Table 1. Main measures of particles available for epidemiological investigation.

particle measure	comments
black smoke	Reflectance principle. Used in Europe for many years. Measures primary black carbonaceous particles under 4.5 µm in diameter. Resembles coefficient of haze in North America. Gradually going out of use in favour of PM_{10}.
TSP[a]	Gravimetric. No size cut-off, includes particles greater than PM_{10}. Common in North America and parts of Europe. Now being phased out by PM_{10}.
acid	Secondary pollutant mainly from oxidization of SO_2 to H_2SO_4.
sulphate	Results from ammonia reacting with H_2SO_4.
PM	Particulate matter of specified mean aerodynamic diameter. Gravimetric. Most commonly PM_{10}, but increasing information on $PM_{2.5}$.
particle numbers	Still essentially a research application.
other measures	These include chemical constituents such as nitrates, metals. Little epidemiological data.

[a]Total suspended particles.

The epidemiology of ambient particles is limited by the availability of appropriate measures. This in itself begs the question of what is appropriate in health terms, since we have little prior information from toxicology about the likely components. The measures most commonly encountered in epidemiological studies are listed in table 1.

Table 2. Selected time-series studies that have compared the effects of the fine particle fraction ($PM_{2.5}$) with those of the coarse fraction ($PM_{2.5-10}$) 'RR' denotes relative risk; 'OR' denotes odds ratio; 'MLR' denotes mean least-squares regression; and 'CI' denotes confidence interval.

study description	reference	outcome and pollution increment (interpercentile range) for which the relative risk is estimated	type of estimate (95% CI)	$PM_{2.5}$	$PM_{2.5-10}$
six eastern US cities	Schwartz *et al.* (1996)	daily all-cause mortality (5–95)	RR	1.058 (1.042, 1.074)	1.02 (0.992, 1.049)
Birmingham, UK 1994–1996	Anderson *et al.* (unpublished data)	daily all-cause mortality (10–90)	RR	1.006 (0.985, 1.027)	0.993 (0.958, 1.029)
		daily cardiovascular mortality (10–90)	RR	1.009 (0.979, 1.04)	0.992 (0.955, 1.035)
		daily respiratory mortality (10–90)	RR	0.999 (0.946, 1.055)	0.924 (0.861, 0.991)
Toronto 1992–1994 (summers)	Burnett *et al.* (1997)	daily respiratory admissions (25–75)	RR	1.037 (1.015, 1.059)	1.023 (1.010, 1.036)
		daily cardiac admissions (25–75)		1.031 (0.997, 1.066)	1.036 (1.015, 1.057)
Birmingham, UK 1994–1996	Anderson *et al.* (unpublished data)	daily respiratory admissions (10–90)	RR	1.012 (0.991, 1.034)	1.002 (0.975, 1.031)
		daily cardiovascular admissions (10–90)	RR	1.005 (0.974, 1.016)	1.007 (0.963, 1.023)
Harvard Six Cities	Schwartz & Neas (2000)	lower respiratory symptoms (25–75)	OR	1.33 (1.11, 1.58)	1.14 (0.98, 1.66)
Diary Study		coughwithout other respiratory symptoms (25–75)	OR	1.16 (0.99, 1.36)	1.20 (1.07, 1.35)
Philadelphia	Schwartz & Neas (2000)	peak flow (ml) (25–75)	MLR	−0.68 (−1.14, −0.22)	0.30 (−1.18, 1.78)
Philadelphia	Neas *et al.* (1999)	afternoon peak flow ($l\ min^{-1}$) (25–75)	MLR	−0.51 (−2.26, 1.24)	0.30 (−0.76, 1.35)
		next morning peak flow ($l\ min^{-1}$) (25–75)	MLR	−1.84 (−3.72, 0.04)	−0.69 (−1.83, 0.44)
Kuopio	Tiittanen *et al.* (1999)	morning peak flow ($l\ min^{-1}$) (lag 1) (25–75)	MLR	−1.06 (−2.08, −0.41)	−0.86 (−1.82, 0.10)
		cough (lag 2) (25–75)	OR	1.13 (1.01, 1.26)	1.15 (1.04, 1.27)

4. Fine or Coarse Particles?

4.1. *Daily Mortality and Hospital Admissions*

Overall, the results from studies that have directly compared the fine and coarse fractions have been mixed (table 2). A meta analysis of six eastern US cities that had taken part in a planned study of air pollution and health found that the associations between fine particles and daily mortality were larger and more significant than those of coarse particles (Schwartz *et al.* 1996). In two-pollutant models, the effects of $PM_{2.5}$ tended not to be affected when $PM_{2.5-10}$ was added to the model, whereas those of $PM_{2.5-10}$ were reduced to near zero when PM_{10} was included in the model. On the other hand, in an earlier report from one of these cities (St Louis; see Dockery *et al.* (1992)), it was noted that the associations between effects of both coarse and fine particles on daily mortality were similar when considered simultaneously in the model. The only other data on mortality are unpublished, from Mexico City and Birmingham, UK. In Mexico City, it was found that coarse particles were associated with daily mortality from all causes, and from respiratory and cardiovascular diseases more strongly than fine particles. When the two fractions were considered together, coarse particles were dominant (Loomis 2000). In Birmingham, UK, neither the fine nor coarse fractions were positively associated with all-cause or disease-specific mortality, and two pollutant models did not further clarify their relative importance. There were, however, hints of differences in the behaviour of the two modalities, the most notable being that the coarse fraction showed a significant *negative* association with respiratory mortality (H. R. Anderson *et al.*, unpublished data; see also table 2).

In an attempt to look at this question in a different way, Schwartz *et al.* (1999) studied the effect of periodic dust storms on mortality in Spokane, WA. These produce high levels of PM_{10}, but, being of crustal origin, are likely to be of coarse rather than fine mode particles. No effect on mortality was found and Schwartz concluded that this indicates that coarse particles are not the toxic component of PM_{10}. The relevance of these findings to the coarse mode found in more usual urban situations is unclear.

Results from the few hospital-admissions studies that have addressed this question tend not to show a clear difference between the coarse and fine fractions. In an analysis of summer hospital admissions to Toronto hospitals in 1992–1994 (Burnett *et al.* 1997), the fine and coarse fraction both showed statistically significant associations of a similar size with respiratory admissions. In the case of cardiac admissions, coarse particles had a slightly larger effect, which was significant, while the effect of fine particles fell below significance. The confidence intervals of the fine and coarse particle estimates overlapped considerably (table 2). These results conflict somewhat with an earlier study from Toronto for the years 1986–1988, in which fine but not coarse particles were significantly associated with cardiorespiratory admissions, though the overlapping of the confidence intervals indicates that this could be a chance difference (Thurston *et al.* 1994).

Table 2 shows the results from Birmingham, UK (H. R. Anderson *et al.*, unpublished data). Here it was found that neither the fine nor the coarse fraction had a significant association with either respiratory or cardiovascular outcomes, and in the case of cardiovascular admissions, the estimates were very similar in size. Lastly, in a study of asthma admissions in Seattle, WA, it was found that both the fine and coarse particle fractions had significant positive effects and that it was not possible to distinguish between them (Sheppard *et al.* 1999).

4.2. *Panel Studies*

Schwartz & Neas (2000) have recently reported a reanalysis of three panel studies, all carried out in the eastern US. The largest of these is of 1844 children in six cities, who kept a diary of respiratory symptoms (the Harvard Six City Diary Study). The investigators measured various particle indicators at a central monitor placed in a residential area of each community. There was a low correlation between the coarse and fine fractions. When all lower respiratory symptoms were considered, $PM_{2.5}$ showed the larger and significant effect, and had the most stability in two pollutant models (table 2). For the symptom of cough without other symptoms, the strongest effect was with nephelometry (a light-scattering method of measuring mainly sub-micronic particles), followed by a significant effect of coarse particles; the effect of $PM_{2.5}$ was similar to that of the coarse fraction but was not statistically significant. It is not clear why this particular single respiratory question was selected for analysis.

In two separate panel studies ($n = 83, 104$) conducted in Pennsylvania, also reported in Schwartz & Neas (2000), peak expiratory flow rates (PEFRs) for the evening and next morning were analysed in relation to

$PM_{2.1}$, $PM_{2.5-10}$ and sulphate. In the combined estimate for both panels, the effect of the fine fraction was negative and significant, whereas that of the coarse fraction was positive and non-significant, though the respective 95% confidence intervals overlapped (table 2). Further panel studies from Philadelphia, PA, during the summer period strengthen the impression that it is difficult to show a clear difference between the effects of coarse and fine fractions (Neas *et al.* 1999). In this study, the effects of fine particles, while larger than those of coarse particles, were non-significant, and clearly not statistically significantly different from those of the coarse particles (table 2).

Similar results were found in the very different environment of Kuopio, Finland, in which a panel of 49 children with chronic respiratory symptoms were studied (Tiittanen *et al.* 1999). In this case, the correlations between $PM_{2.5}$ and $PM_{2.5-10}$ were quite high (above 0.9). There were significant associations between cough symptom for both $PM_{2.5}$ and $PM_{2.5-10}$ after a lag of two days. The authors observed, more generally, that there were inconsistent associations at a variety of lags with all of the fractions studied. Different conclusions were drawn from a panel study in Mexico City (Gold *et al.* 1999), where effects were found with $PM_{2.5}$ but not with $PM_{2.5-10}$.

4.3. *Numbers or Mass?*

There is considerable current interest in the idea that high numbers of ultrafine particles are the most potentially toxic component of the ambient aerosol. Methods of counting particles do so within size categories and this gives an opportunity to compare the effects of particles in different size ranges using numbers or mass. An influential early report that addresses this question is that by Peters *et al.* (1997) among a panel of adults in the city of Erfurt. They measured the number concentrations and mass concentrations within size categories 0.01–2.5 (fine particles), 0.01–0.1 (ultrafine particles) and 0.5–2.5 μm, along with PM_{10} measured with a Harvard Impacter. These measures were analysed in relation to the symptoms and lung function of 27 non-smoking adults with chronic respiratory disease. Ultrafine particles made up 73% of particles but contributed only 1% to the mass of fine particles. Most of the mass provided by particles was between 0.5 and 2.5 μm in diameter. The time courses of changes in the number and mass concentrations were only moderately correlated, allowing their separate contributions to health effects to be analysed. The health effects of the number of ultrafine particles tended to be greater than that of the mass effects of fine particles and of PM_{10}. The study did not directly address the effects of coarse particles.

In the study of Tiittanen *et al.* (1999), particle numbers were also studied, but no coherent pattern of results emerged to give substantial support for any particular metric over another. In another panel study in Kuopio, Pekkanen *et al.* (1997) concluded that the number concentration of ultrafine particles was no more associated with variations in lung function than was PM_{10}, or black smoke. Taken together, these three studies provide only modest epidemiological support for the hypothesis that it is the number concentration of ultrafine particles, rather than the mass concentration of the aerosol, that is important in driving the health effects.

4.4. *Chronic Effects*

4.4.1. *Cohort Studies*

The results from three major cohort studies, all from the US, have all provided evidence for associations between fine particles and health effects. All have allowed for confounding at an individual level. Abbey *et al.* (1995) have followed a cohort of non-smoking Seventh Day Adventists to examine the association between $PM_{2.5}$ (estimated from an airport visibility index) and PM_{10}, and the incidence of chronic respiratory disease. While associations were reported for both particle indices, no direct comparison of fine with coarse particles was made (Abbey *et al.* 1995). The Six Cities Study examined the association between air pollution and mortality in a cohort of 8111 adults over 14–16 years. Significant associations were found with $PM_{10/15}$, $PM_{2.5}$ and sulphate, with similar rate ratios and confidence intervals. These associations were greater than those with total particles or acid aerosol. There was a specificity for deaths from lung cancer and cardiorespiratory causes (Dockery *et al.* 1993). Finally, in the largest cohort study, over half a million adults living in 151 metropolitan areas were followed from 1982 to 1989, using annual concentrations of sulphate (151 areas) and fine particles (50 areas) as indicators of air pollution exposure. Mortality was increased in association with both measures to a similar extent, but the effect of sulphates on cancer was greater (Pope *et al.* 1995*b*).

4.4.2. *Prevalence Studies*

Chronic respiratory symptoms and lung function measures are conveniently measured by prevalence surveys and allow the opportunity to compare areas with different air pollution exposures. The best modern studies control for confounding factors such as smoking, passive smoking, gas cooking, dampness, etc. A number of studies have used measures of fine particles and most have found either associations with symptoms or decrements in lung function, or both (Dockery *et al.* 1989). However, there is only one study, that by Raizenne *et al.* (1996), that directly compares the coarse and fine fraction. This was a study of children in 26 cities of the eastern US and Canada. There was a clear association between fine particles and lung function, but none was found for the coarse fraction; this is fairly convincing evidence that fine particles are more important.

5. Other Measures of Fine Particles

Apart from measures of $PM_{2.5}$ or other size fractions, we can also deduce something about the effect of fine particles from measures of sulphate, acid aerosol and black smoke, all of which reflect particles found mainly in the fine fraction, and for which there are sufficient epidemiological data.

5.1. *Sulphate and Acid Aerosols*

These are largely the result of the oxidation of SO_2 to sulphuric acid with subsequent reactions with ammonia, in particular, to form sulphates of various types. Nitric acid and nitrates also occur in the UK, but generally in lower concentrations than sulphate (APEG 1999). These are secondary pollutants and tend to have a regional distribution. The amount of associated acid varies according to the opportunities for neutralization and is generally higher in the eastern US, where most studies have been done, than in Europe, where farming activity produces enough ammonia to neutralize the acid.

Particle-associated acidity was a feature of the major air pollution episodes of the past, and it has been postulated that acid aerosol is harmful to health (Lippmann 1989). Without reviewing the evidence here, it is sufficient to say that a number of time-series studies support this view. One example is the study of hospital admissions in Toronto (referred to above;

see Thurston *et al.* (1994)), in which the ranking of effects of particles was

$$H^+ > \text{sulphate} > PM_{2.5} > PM_{10} > \text{TSP}$$

(though the effect of ozone was ten times greater). This was also found in later studies of Toronto (Burnett *et al.* 1997). On the other hand, the study of mortality in six eastern cities, also referred to earlier (Schwartz *et al.* 1996), found much lower and non-significant associations with H^+ than with sulphate and $PM_{2.5}$. In the Harvard Six Cities cohort study, fine particles, sulphates and inhalable particles were more strongly associated with mortality than acid aerosol. The results of panel studies in the eastern US show variable results. Little evidence is available from Europe.

There is probably stronger evidence to relate ambient sulphate to health effects, but it must be borne in mind that sulphate and acidity are closely associated in some atmospheres. Sulphate has been associated with daily mortality and hospital admissions in daily time-series studies (Schwartz *et al.* 1996; Thurston *et al.* 1994) and lung function in some panel studies in the US (Neas *et al.* 1995; Schwartz & Neas 2000) and Europe (Peters *et al.* 1996). The results tend to be less substantial and robust than those for $PM_{2.5}$. In cross-sectional studies, sulphate has been associated with mortality (Ozkaynak & Thurston 1987). More substantially, sulphate was associated with increased mortality in the American Cancer Society cohort study, with similar relative risks as for $PM_{2.5}$ (Pope *et al.* 1995*b*). In the recent studies from Birmingham, UK, sulphates showed inconsistent associations with mortality, but with a notable seasonal interaction, with larger effects in the warm season. There were weak effects on hospital admissions (H. R. Anderson *et al.*, unpublished data).

5.2. *Black Smoke*

The Black Smoke method, which uses a reflectance technique, has been the standby for particle measurement for many years in the UK and some other European countries. Unlike sulphate and acid, the method measures primary pollution from black carbonaceous particles. In cities such as London it is mainly measuring diesel exhaust particles. The inlet cut-off is at 4.5 μm, but most of the particles are probably in the fine fraction. It is, therefore, a measure of fine black primary particles. Most daily time-series studies have observed associations between black smoke and daily mortality and hospital admissions (Katsouyanni *et al.* 1997; Spix *et al.* 1998). In recent studies of London daily mortality, it was found that the effects of black smoke were more robust than those of PM_{10} (Bremner *et al.* 1999). In Amsterdam, the effects of black smoke and PM_{10} were almost identical (Verhoeff *et al.* 1996). This suggests that black smoke represents an important component of the toxic material included in PM_{10}. The results from panel studies have been more mixed. Many studies have found associations with lung function decrement and symptoms, whereas the European PEACE study of 14 centres found little evidence of associations between PM_{10} or black smoke and lung function and symptoms in panels of children with respiratory disease (Roemer *et al.* 1998).

6. Interpretation and Conclusion

The epidemiology of particle fractions is very patchy, due to a lack of appropriate measures and some inherent limitations of the epidemiological approach. Virtually nothing is known about ultrafines, apart from information now emerging from studies of particle number concentrations.

There is abundant evidence of short-term associations between ambient particles and mortality on hospital admissions and emergency-room visits. The evidence concerning short-term associations with lung function and respiratory symptoms is less consistent but generally persuasive. All of the major cohort studies have found associations between exposure to particles and mortality or disease incidence. Most authorities regard these associations as at least partly causal, and there is emerging mechanistic evidence from experimental studies which supports this view.

Epidemiological studies have made a contribution to understanding which component or components of the mixture are important. Firstly, this has been through studies using measurements of size-fractionated particles, mainly $PM_{2.5}$ (fine fraction). These have found that the associations with the fine fraction are similar to those of PM_{10}, which suggests that the $PM_{2.5}$ fraction is toxic. What these studies rarely address is whether the coarse fraction is also important, and there is not enough evidence at present to be sure that it is not. The second strand of evidence is that related to species of particles that are mainly fine; those that have been amenable to epidemiological study are mainly sulphate, acid and black smoke. Associations have been reported for all of these measures. Finally, the emerging evidence on particle numbers suggests that numbers of fine particles may be more important than mass.

It is important to mention a potential statistical problem with comparing different particle measures. This starts with the different behaviour of fine versus coarse particles in the atmosphere (Wilson & Suh 1997). The former have a low settling velocity and penetrate indoors quite effectively, whereas the latter settle out quite quickly. This means that concentrations of fine particles are more uniformly spread over large areas such as cities, and it follows from this that the community monitor or monitors used for epidemiological studies probably represent the population exposure more accurately than do monitors of coarse particles. Misclassification of exposure will, in most circumstances, bias the effect estimates of air pollution to the null. This argument has been used to postulate that the larger estimates for fine compared with coarse fractions reflect differential exposure misclassification rather than differences in toxicity (Lipfert & Wyzga 1997). Schwartz *et al.* (1996) have disputed this and the issue remains unresolved.

It is concluded that fine particles are associated with health effects, and that both secondary and primary particles may be important. It has not been shown that coarse particles are not important. The epidemiological evidence concerning the ultrafine fraction is meagre and will remain so until adequate series of data are available for epidemiological analysis.

References

Abbey, D. E., Ostro, B. D., Petersen, F. & Burchette, R. J. 1995 Chronic respiratory symptoms associated with estimated long-term ambient concentrations of fine particulates less than 2.5 microns in aerodynamic diameter ($PM_{2.5}$) and other air pollutants. *J. Expo. Analysis Environ. Epidemiol.* **5**, 137–159.

Ackermann-Liebrich, U. (and 23 others) 1997 Lung function and long term exposure to air pollutants in Switzerland. Study on Air Pollution and Lung Diseases in Adults (SAPALDIA) team. *Am. J. Respir. Crit. Care Med.* **155**, 122–129.

American Thoracic Society 1996 Health effects of outdoor air pollution. *Am. J. Respir. Crit. Care Med.* **153**, 3–50.

Anderson, H. R., Ponce de Leon, A., Bland, J. M., Bower, J. S. & Strachan, D. P. 1996 Air pollution and daily mortality in London: 1987–92. *Br. Med. J.* **312**, 665–669.

APEG (Airborne Particles Expert Group) 1999 *Source apportionment of airborne particulate matter in the United Kingdom.* London: Department of the Environment Transport and the Regions.

Baxter, P. J., Ing, R., Falk, H. & Plikaytis, M. S. 1983 Mount St Helens eruptions: the acute respiratory effects of volcanic ash in a North American community. *Arch. Environ. Health* **38**, 138–143.

Bremner, S. A., Anderson, H. R., Atkinson, R. W., McMichael, A. J., Bland, J. M., Strachan, D. P. & Bower, J. 1999 Short term associations between outdoor air pollution and mortality in London 1992–94. *Occup. Environ. Med.* **56**, 237–244.

Burnett, R. T., Cakmak, S., Brook, J. R. & Krewski, D. 1997 The role of particulate size and chemistry in the association between summertime ambient air pollution and hospitalization for cardiorespiratory diseases. *Environ. Health Perspect.* **105**, 614–620.

Department of Health Committee on the Medical Effects of Air Pollutants 1995*a* *Non-biological particles and health.* London: HMSO.

Department of Health Committee on the Medical Effects of Air Pollutants 1995*b* *Asthma and outdoor air pollution.* London: HMSO.

Dockery, D. W. & Pope, C. A. 1994 Acute respiratory effects of particulate air pollution. *Ann. Rev. Public Health* **15**, 107–132.

Dockery, D. W., Speizer, F. E., Stram, D. O., Ware, J. H., Spengler, J. D. & Ferris Jr, B. G. 1989 Effects of inhalable particles on respiratory health of children. *Am. Rev. Respir. Dis.* **139**, 587–594.

Dockery, D. W., Schwartz, J. & Spengler, J. D. 1992 Air pollution and daily mortality: associations with particulates and acid aerosols. *Environ. Res.* **59**, 362–373.

Dockery, D. W., Pope, C. A., Xu, X., Spengler, J. D., Ware, J. H., Fay, M. E., Ferris B. G. & Speizer, F. E. 1993 An association between air pollution and mortality in six US cities. *New Engl. J. Med.* **329**, 1753–1759.

Gamble, J. F. & Lewis, R. J. 1996 Health and respirable particulate (PM_{10}) air pollution: a causal or statistical association? *Environ. Health Perspect.* **104**, 838–850.

Gardner, M. J., Crawford, M. D. & Morris, J. N. 1969 Patterns of mortality in middle and early old age in the county boroughs of England and Wales. *Br. J. Prev. Soc. Med.* **23**, 133–140.

Gold, D. R., Damokosh, A. I., Pope, C. A., Dockery, D. W., McDonnell, W. F., Serrano P., Retama, A. & Castillejos, M. 1999 Particulate and ozone pollutant effects on the respiratory function of children in southwest Mexico City. *Epidemiol.* **10**, 8–16.

Katsouyanni, K. (and 12 others) 1997 Short-term effects of ambient sulphur dioxide and particulate matter on mortality in 12 European cities: results from time series data from the APHEA project. Air Pollution and Health: a European Approach. *Br. Med. J.* **314**, 1658–1663.

Kelsall, J. E., Samet, J. M., Zeger, S. L. & Xu, J. 1997 Air pollution and mortality in Philadelphia, 1974–1988. *Am. J. Epidemiol.* **146**, 750–762.

Lave, L. B. & Seskin, E. P. 1970 Air pollution and human health. *Science* **169**, 723–733.

Lipfert, F. W. & Wyzga, R. E. 1997 Air pollution and mortality: the implications of uncertainties in regression modeling and exposure measurement. *J. Air Waste Management Ass.* **47**, 517–523.

Lippmann, M. 1989 Progress, prospects, and research needs on the health effects of acid aerosols. *Environ. Health Perspect.* **79**, 203–205.

Loomis, D. 2000 Sizing up air pollution research. *Epidemiol.* **11**, 2–4.

Ministry of Health 1954 *Mortality and morbidity during the London fog of December 1952.* Reports on Public Health and Medical Subjects, no. 95. London: HMSO.

Neas, L. M., Dockery, D. W., Koutrakis, P., Tollerud, D. J. & Speizer, F. E. 1995 The association of ambient air pollution with twice daily peak expiratory flow rate measurements in children. *Am. J. Epidemiol.* **141**, 111–122.

Neas, L. M., Dockery, D. W., Koutrakis, P. & Speizer, F. E. 1999 Fine particles and peak flow in children: acidity versus mass. *Epidemiol.* **10**, 550–553.

Ozkaynak, H. & Thurston, G. D. 1987 Associations between 1980 US mortality rates and alternative measures of airborne particle concentration. *Risk Analysis* **7**, 449–461.

Pekkanen, J., Timonen, K. L., Ruuskanen, J., Reponen, A. & Mirme, A. 1997 Effects of ultrafine and fine particles in urban air on peak expiratory flow among children with asthmatic symptoms. *Environ. Res.* **74**, 24–33.

Peters, A., Goldstein, I. F., Beyer, U., Franke, K., Heinrich, J., Dockery, D. W., Spengler, J. D. & Wichmann, H. E. 1996 Acute health effects of exposure to high levels of air pollution in eastern Europe. *Am. J. Epidemiol.* **144**, 570–581.

Peters, A., Wichmann, H. E., Tuch, T., Heinrich, J. & Heyder, J. 1997 Respiratory effects are associated with the number of ultrafine particles. *Am. J. Respir. Crit. Care Med.* **155**, 1376–1383.

Pope, C. A., Schwartz, J. & Ransom, M. R. 1992 Daily mortality and PM_{10} pollution in Utah Valley. *Arch. Environ. Health* **47**, 211–217.

Pope, C. A., Dockery, D. W. & Schwartz, J. 1995*a* Review of epidemiological evidence of health effects of particulate pollution. *Inhal. Toxicol.* **7**, 1–18.

Pope, C. A., Thun, M. J., Namboodiri, M. M., Dockery, D. W., Evans, J. S., Speizer, F. E. & Heath Jr, C. W. 1995*b* Particulate air pollution as a predictor of mortality in a prospective study of US adults. *Am. J. Respir. Crit. Care Med.* **151**, 669–674.

Raizenne, M., Neas, L. M., Damokosh, A. I., Dockery, D. W., Spengler, J. D., Koutrakis, P., Ware, J. H. & Speizer, F. E. 1996 Health effects of acid aerosols on North American children: pulmonary function. *Environ. Health Perspect.* **104**, 506–514.

Roemer, W., Hoek, G., Brunekreef, B., Haluszka, J., Kalandidi, A. & Pekkanen, J. 1998 Daily variations in air pollution and respiratory health in a multicentre study: the PEACE project. Pollution Effects on Asthmatic Children in Europe. *Eur. Respir. J.* **12**, 1354–1361.

Schwartz, J. & Neas, L. M. 2000 Fine particles are more strongly associated than coarse particles with acute respiratory health effects in school children. *Epidemiol.* **11**, 6–10.

Schwartz, J., Dockery, D. W. & Neas, L. M. 1996 Is daily mortality associated specifically with fine particles? *J. Air Waste Manag. Ass.* **46**, 927–939.

Schwartz, J., Norris, G., Larson, T., Sheppard, L., Claiborne, C. & Koenig, J. 1999 Episodes of high coarse particle concentrations are not associated with increased mortality. *Environ. Health Persp.* **107**, 339–342.

Seaton, A., MacNee, W., Donaldson, K. & Godden, D. 1995 Particulate air pollution and acute health effects. *Lancet* **345**, 176–178.

Sheppard, L., Levy, D., Norris, G., Larson, T. V. & Koenig, J. Q. 1999 Effects of ambient air pollution on nonelderly asthma hospital admissions in Seattle, Washington, 1987–1994. *Epidemiol.* **10**, 23–30.

Spix, C. (and 12 others) 1998 Short-term effects of air pollution on hospital admissions of respiratory diseases in Europe: a quantitative summary of APHEA study results. Air Pollution and Health: a European Approach. *Arch. Environ. Health* **53**, 54–64.

Thurston, G. D., Ito, K., Hayes, C. G., Bates, D. V. & Lippmann, M. 1994 Respiratory hospital admissions and summertime haze air pollution in Toronto, Ontario: consideration of the role of acid aerosols. *Environ. Res.* **65**, 271–290.

Tiittanen, P., Timonen, K. L., Ruuskanen, J., Mirme, A. & Pekkanen, J. 1999 Fine particulate air pollution, resuspended road dust and respiratory health among symptomatic children. *Eur. Respir. J.* **13**, 266–273.

USEPA (United States Environmental Protection Agency) 1996 *Air quality criteria for particulate matter.* Research Triangle Park, NC: USEPA.

Verhoeff, A. P., Hoek, G., Schwartz, J. & van Wijnen, J. H. 1996 Air pollution and daily mortality in Amsterdam. *Epidemiol.* **7**, 225–230.

Wilson, W. E. & Suh, H. H. 1997 Fine particles and coarse particles: concentration relationships relevant to epidemiologic studies. *J. Air Waste Manag. Ass.* **47**, 1238–1249

Discussion

H.-E. Wichmann (*GSF – Institute of Epidemiology, Neuherberg, Germany*). You mentioned that in the APHEA study, effects of black smoke on daily mortality have been observed in western Europe, but not in eastern Europe. I wonder whether the new insights into ultrafine particles could help us understand this. If we look at the atmosphere in eastern Europe at the time of APHEA study, there were many larger particles in the air which might have scavenged the ultrafine particles, leaving no room for health effects of the ultrafines. In contrast, in western Europe, there were probably many more ultrafines in the air, which might have contributed to daily mortality, and, since they are correlated to black smoke, this might have been attributed to black smoke.

CHAPTER 16

CONTRIBUTIONS THAT EPIDEMIOLOGICAL STUDIES CAN MAKE TO THE SEARCH FOR A MECHANISTIC BASIS FOR THE HEALTH EFFECTS OF ULTRAFINE AND LARGER PARTICLES

Morton Lippmann and Kazuhiko Ito

New York University School of Medicine, Nelson Institute of Environmental Medicine, 57 Old Forge Road, Tuxedo, NY 10987, USA

Epidemiology is a rather blunt tool for elucidating biological mechanisms that can account for the increased mortality and morbidity associated with population exposures to ambient air particulate matter (PM). However, it has an essential role to play. Recent studies indicate that three readily measurable ambient air PM concentration indices can be significantly associated with one or more elevations of rates of specific disease or dysfunction categories. These three indices, i.e. ultrafine particle number, fine particle mass ($PM_{2.5}$) and thoracic coarse mass ($PM_{10-2.5}$) differ not only in size range, but also in terms of their sources, deposition patterns, and chemical reactivities, factors that may account for their different associations with human health effects. Further epidemiological studies employing a wider array of air quality and health effects variables should enable us to resolve some of the outstanding questions related to causal relationships for PM components or, at the minimum, to pose some better questions.

Keywords: ultrafine particles; fine particles; thoracic coarse particles; epidemiology; air pollution; lung deposition

1. Introduction

Hypothesis-driven epidemiological studies will be needed to clarify the role(s) that ultrafine particles may play in the causation of the various health effects that have been associated with community air pollution. While it is generally acknowledged that typical ambient air pollutant mixtures in economically developed countries contribute to excess daily mortality, greater usage of clinical and medical facilities and services, reductions in school and work attendance, increased rates of cardiopulmonary symptoms and abnormal function, and reduced longevity, there is much less agreement on which pollutant components, or mixtures of components, are most influential on the health-related responses.

In terms of strength of association for one or more of the health effects, there have been positive epidemiological findings reported for each of the common pollutant gases, i.e. ozone (O_3), nitrogen dioxide (NO_2), sulphur dioxide (SO_2), and carbon monoxide (CO), as well as for various indices of particulate matter (PM) concentrations. These PM indices include black smoke (BS) and coefficient of haze (CoH), both of which are closely related to the elemental carbon content of the PM, as well as various size-selective gravimetric concentrations. These gravimetric PM indices include: so-called total suspended particulate matter (TSP), which had an effective upper cut-size that varied from 20 to 50 μm in aerodynamic diameter, dependent on wind speed and direction; thoracic PM (PM_{10}), which approximates the PM fraction inhaled through the larynx; fine particles in the accumulation mode ($PM_{2.5}$); and thoracic coarse particles ($PM_{10-2.5}$). This Discussion Meeting has been focused on another component of PM_{10}, i.e. ultrafine particles (UPs), which are mostly smaller than 0.1 μm in diameter and which contribute very little mass to the aforementioned gravimetric concentration indices. In fact, the implicit assumption is that the health effects that can be produced by UPs are more closely influenced by their number concentration than by their mass concentration. In almost all cases, the number concentration of UPs is nearly equal to the total number concentration of particles of all sizes in ambient air, and the underlying hypothesis is that the net cardiopulmonary responses are related to the summation of the individual responses caused or initiated by each ultrafine particle that deposits on respiratory and/or conductive lung airways. In this model of response, the size of the particle is not important, since each individual particle can initiate a local cellular response that contributes to the aggregate change in function, symptoms, and/or disability.

Support for a causal role of UPs comes from the results of recent studies involving measurements of both the number concentration of UPs and gravimetric concentrations, which reported closer associations of some health-related responses for UPs than for some simultaneously measured gravimetric indices (Peters *et al.* 1997; Ostro & Lipsett 2000).

Table 1. Mechanistic plausibility: coherence between PM-exposure associated health effects from epidemiological and toxicological studies (Schlesinger 2000). Mø, macrophage; UF, ultrafine; WBC, white blood cells; ROI, reactive oxygen intermediates; ROFA, residual oil fly ash; BALT, bronchus associated lymphoid tissue; COPD, chronic obstructive pulmonary disease; Δ, change in parameter noted.

	toxicological health endpoints	
epidemiological health endpoints	concentrated ambient PM	specific PM components
↑ hypertension/↑ stroke	Δ PM homeostasis (e.g. peripheral blood differential cell counts)	Δ Blood coagulation factor: UF Carbon ↑ platelets, WBC; diesel exhaust (whole)
↑ ischemic heart disease/ ↑ heart attack	Δ heart-rate variability Δ EKG waveform segments	↑ arrhythmia incidence; ROFA
↑ acute respiratory infection (e.g. acute bronchitis, pneumonia)	↓ Mø ROI production ↓ BALT Δ pulmonary cytokine profile	↓ Mø ROI production; ammonium sulphate Δ pulmonary cytokines; metals
exacerbation of COPD, asthma	—	↑ airway reactivity: H^+ Δ mucociliary function: H^+
↑ respiratory symptoms Δ lung function indices	—	pulmonary inflammation: UF, metals Δ pulmonary cytokines: metals

1.1. *Identifying PM Components as Possible Causal Factors for Health Effects*

Recent toxicological and clinical exposure studies using concentrated accumulation mode ambient aerosols have produced health-related responses that correspond to effects indices found in epidemiological studies, demonstrating that effects of concern can be produced by $PM_{2.5}$ of ambient air origin alone. In these concentrated ambient air PM studies, the ambient $PM_{10-2.5}$ component had been removed by inertial separators, and the ambient air UPs and pollutant gas components were not concentrated. A comprehensive summary of current mechanistic knowledge for the health effects of PM was recently prepared by Schlesinger (2000) and is presented in tables 1 and 2.

Table 2. Currently hypothesized PM physiochemical properties related to biological responses (Schlesinger 2000). FP, fine particulates; CP, coarse particulate; UF, ultrafine particulate; ROFA, residual oil fly ash.

	response	
PM characteristic	epidemiology	toxicology
mass concentration	associated with health outcomes	associated with biological responses
particle size	relative association with health outcomes often related to size mode (FP, CP, UF, etc.)	different biological responses noted with different size modes
metals	Utah Valley: effects from steel mill related to metals	ROFA: effects related to metals
acidity	some evidence for H^+ association with health outcomes	various biological responses
organics	association of PM with lung cancer possibly due to carcinogenicity of organic fraction	known mutagens/carcinogens
biogenic PM	possible association with health outcomes	generally allergenic
sulphate/nitrate salts	association with some health outcomes (markers for H^+)	generally not very toxic at low concentrations
peroxides	?	high levels may produce biological effects
elemental C (soot)	?	mutagenic/carcinogenic/ irritant

A number of recent epidemiological studies have examined the relative roles of $PM_{10-2.5}$, $PM_{2.5}$, and pollutant gases, and have concluded that $PM_{10-2.5}$ can also have a significant influence on short-term health responses. These findings are summarized in table 3.

1.2. *Differing Characteristics of Ambient Air* $PM_{10-2.5}$, $PM_{2.5}$ *and UPs*

$PM_{10-2.5}$, $PM_{2.5}$ and UPs can be considered to be different pollutants in terms of their particle size ranges, compositions, and potentials for causing

Table 3. Recent epidemiological studies reporting short-term health-related associations with $PM_{10-2.5}$, $PM_{2.5}$ and/or UPs. +, significantly associated; (+), approaching statistical significance; nr, not reported; −, not significantly associated; *, summer only; nm, not measured.

study	location	effects studied	pollutant associations					
			PM_{10}	$PM_{10-2.5}$	$PM_{2.5}$	O_3	CO	UPs
1. Castillejos *et al.* (2000)	Mexico City	mortality (acute)	+	+	+	nr	nr	nm
2. Cifuentes & Vega (2000)	Santiago, Chile	mortality (acute)						
		total and respiratory	nr	−	+	−	+	nm
3. Ostro & Lipsett (2000)	Coachella Valley, CA	mortality (acute)						
		cardiovascular	+	+	−	−	−	+
		respiratory	−	−	−	−	−	−
		all causes	−	−	+	−	−	+
4. Simpson *et al.* (2000)	Melbourne, Australia	mortality (acute)*						
		cardiovascular	−	nr	−	−	nr	nm
		respiratory	+	nr	+	+	nr	nm
		all causes	+	−	−	+	nr	nm
5. McDonnell *et al.* (2000)	California (non-smoking)	mortality (reduced longevity)						
		all causes (male)	+	−	+	nr	nr	nm
		all causes (female)	−	+	−	nr	nr	nm
6. Van Den Eeeden *et al.* (1999)	Los Angeles, CA	hospital admissions						
		cardiovascular	+	+	+	+	nr	nm
		chronic respiratory	−	−	−	−	nr	nm
		acute respiratory	−	−	−	−	nr	nm
7. Burnett *et al.* (1997)	Toronto, ON	hospital admissions*						
		cardiac	+	+ (+)	+	+	nm	
		respiratory	+	+	+	+	−	nm
8. Sheppard *et al.* (1999)	Seattle, WA	hospital admissions						
		asthma	+	+	+	+	+	nm
9. Neas *et al.* (1999)	Philadelphia, PA	peak expiratory flow (in children)	+	(+)	(+)	(+)	nm	nm
10. Creason *et al* (2000); Liao *et al.* (1999)	Baltimore, MD	heart rate variability	−	−	+	−	−	nm

adverse health effects. $PM_{10-2.5}$ is largely mineral dust derived from windblown soil, mineral ash, and resuspended road dust. $PM_{2.5}$ is dominated by the accumulation mode in the ambient air. It is largely derived from combustion sources and its mass is composed largely of aged reaction products such as the ammonium salts of sulphuric and nitric acids, aggregates of ultrafine carbon emitted from engine exhaust, and of semi-volatile organic droplets formed in engine and boiler exhaust streams and in complex photochemical reaction sequences. In the absence of rainfall, $PM_{2.5}$ concentrations have much less temporal variation than either $PM_{10-2.5}$ particles, which settle out fairly rapidly under low widespread conditions, or UPs, which coagulate rapidly, becoming part of the $PM_{2.5}$. It should also be noted that as the UPs age, coagulate, and react chemically with gaseous air pollutants, they tend to become less biologically and chemically reactive and/or biologically potent.

The three size ranges also differ in respiratory tract deposition efficiencies and locations. $PM_{10-2.5}$ particles have preferential deposition by impaction at branch points in the larger conductive airways, producing concentrated deposition hot-spots on a small fraction of the conducting airway surface. Particles less than 2.5 μm in aerodynamic diameter down to *ca.* 0.1 μm have very little impaction or diffusional deposition in large airways and penetrate, with the tidal convective flow, to the respiratory acinus, where some of them mix with residual lung air. The expiratory tidal flow carries most of the inhaled particles back out into the atmosphere. The remaining 10–30% of the particles left behind in the residual lung air deposit by sedimentation and/or diffusion in terminal bronchioles, respiratory bronchioles, and alveolar ducts, especially at or near small airway bifurcations, where the insoluble particles can accumulate as centrilobular deposits (Lippmann *et al.* 1994).

UPs have greater diffusional mobility, and in the low nanometre end of the range, where the number concentrations are highest, they can deposit by diffusion in both large conductive airways, as well as penetrate more deeply than their penetration depth by convective tidal transport into the lung acinus. They can deposit by diffusion to a greater extent in the gas exchanging alveoli of the lung than can the larger particles having more limited diffusional mobilities.

Thus, it is not unreasonable to expect that there would be different effects produced by concentrations of coarse-mode mineral particles

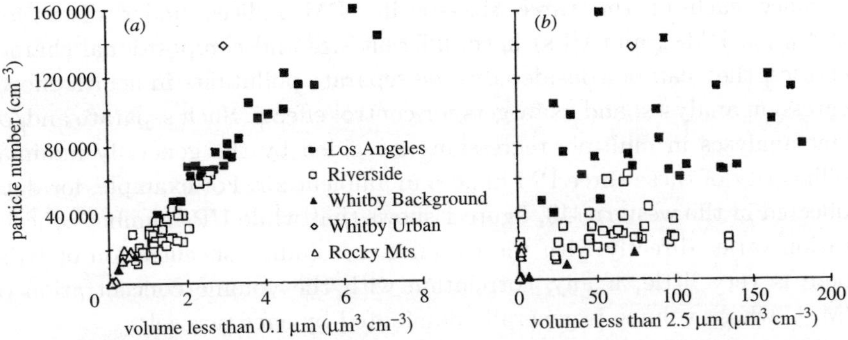

Fig. 1. Relationship between particle number and particle volume (volume at diameters of less than (*a*) 0.1 and (*b*) 2.5 μm). From *Air quality criteria for particulate matter*, vol. 1. EPA/600/P-95/001aF. US Environmental Protection Agency, Washington, DC, April 1996.

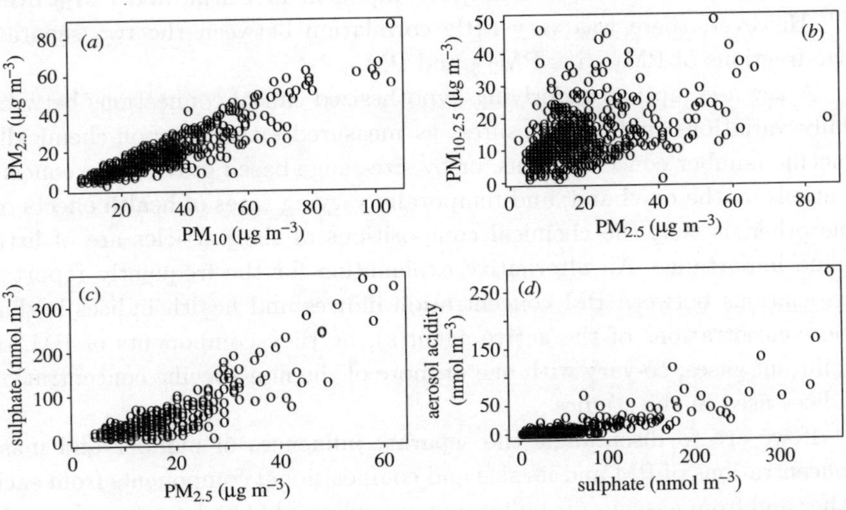

Fig. 2. Scatter plots among key PM components in Detroit for summer days in 1992–1994. (*a*) $PM_{2.5}$ versus PM_{10}; (*b*) $PM_{2.5}$ versus $PM_{10-2.5}$; (*c*) $PM_{2.5}$ versus sulphate; (*d*) sulphate versus acidity.

deposited on large conductive airway bifurcations, by aged accumulation mode acidic and organic particles depositing at limited surface areas in centrilobular lung regions, and by relatively freshly formed ultrafine singlets deposited relatively uniformly on the alveolar epithelium.

Since each of the three size-specific PM indices under discussion ($PM_{10-2.5}$, $PM_{2.5}$ and UPs) have different size and compositional characteristics, they can be considered to be separate pollutants in health effects regression analyses, and as targets for control effects. Such separate and/or joint analyses in multiple regressions are aided by the generally minimal collinearity of these three PM indices in ambient air. For example, for data collected in the western US, figure 1 shows that while UPs number concentration varies directly with the cumulative volume concentration of UPs, there is very little, if any, correlation with the volume concentration of $PM_{2.5}$, whose volume is generally dominated by aged aerosol.

In terms of the collinearity among gravimetric PM concentrations, figure 2, showing data from multiple monitoring stations in Detroit, shows that there was a reasonably high degree of correlation between sulphate, a major component of $PM_{2.5}$, and $PM_{2.5}$ overall. Likewise, there was fairly good correlation of $PM_{2.5}$, a major component of PM_{10}, and PM_{10} overall. However, there was very little correlation between the two separate size-fractions of PM_{10}, i.e. $PM_{2.5}$ and $PM_{10-2.5}$.

A key assumption underlying hypothesized causal connections between daily variations in PM exposures, as measured either by non-chemically specific number concentrations, or by size-range based gravimetric concentrations on the one hand, and temporally varying rates of health effects on the other, is that the chemical compositions of the particles are of little or no importance. An alternative explanation for the frequently reported associations between PM concentration indices and health indices is that the concentrations of the active agent(s), be they components of PM, or pollutant gases, co-vary with one or more of the non-specific concentration indices used in the studies.

If we are to disentangle the separate influences of number and mass concentrations of PM and its size and compositional components from each other and from gaseous air pollutants, we will need (1) a better mechanistic understanding of the physiologic and toxic responses to the inhaled agents, and (2) an increased number of more comprehensive, short-term, exposure–response studies in human populations that incorporate daily measurements of more of the components of the air pollution mixtures. Such studies would facilitate multipollutant regression analyses that could better define active components.

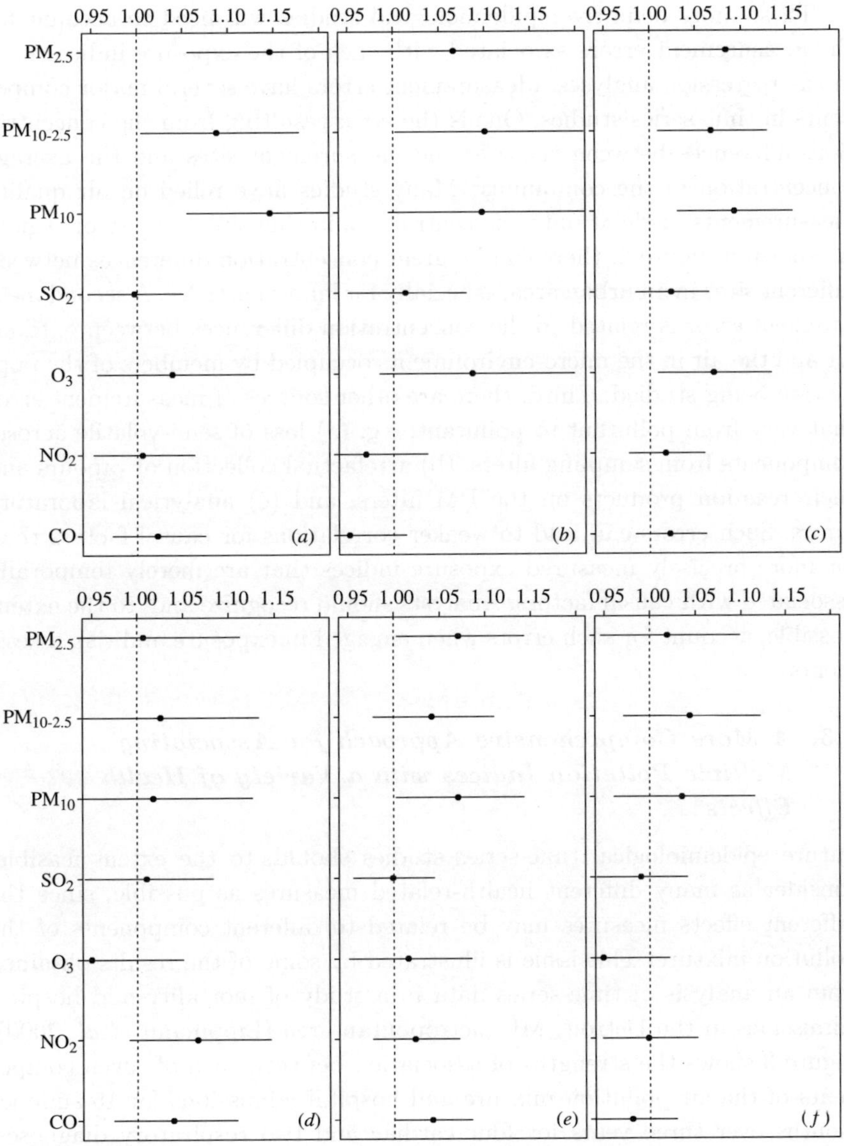

Fig. 3. Relative risks per 5–95th percentile of pollution levels for hospital admissions for Medicare patients in Detroit by disease diagnosis for 490 warm weather days in 1992, 1993 and 1994. (*a*) Pneumonia; (*b*) COPD; (*c*) ischemic heart disease; (*d*) dysrythmia; (*e*) heart failure; (*f*) stroke.

These more definitive epidemiological studies will need to account for the measurement errors associated with each of the exposure indices used in the regression analyses. Measurement errors have several major components in time-series studies. One is the error resulting from the concentration differences between those at the measurement sites and the average concentration in the community. Many studies have relied on air quality measurements made at only one central monitoring site in a city or region. As shown in figure 2, there can be great concentration differences between different sites in an urban area, especially for larger particles. A second measurement error is related to the concentration differences between outdoor air and the air in the micro-environments occupied by members of the population being studied. Third, there are other sources of measurement error that vary from pollutant to pollutant, e.g. (a) loss of semi-volatile aerosol components from sampling filters, (b) artefactual collection of vapours and their reaction products on the PM filters, and (c) analytical laboratory errors. Such errors can lead to weaker correlations for causal factors than for more precisely measured exposure indices that are merely temporally associated with causal factors. Analysts should recognize and, to the extent possible, account for such errors when engaged in exposure and risk assessments.

1.3. *A More Comprehensive Approach for Associating Multiple Pollution Indices with a Variety of Health Effects*

Future epidemiological time-series studies should, to the extent feasible, consider as many different health-related measures as possible, since the different effects measures may be related to different components of the pollution mixture. This issue is illustrated by some of the results obtained from an analysis of time-series data in a study of mortality and hospital admissions in the Detroit, MI, metropolitan area (Lippmann *et al.* 2000). Figure 3 shows the strengths of association between each of seven components of the air pollution mixture and hospital admissions for 15 summer months over three years for four cardiac and two respiratory diagnoses. Statistically significant associations were seen for certain pollutants and certain diagnoses, suggesting that different components within the mixture may have influenced the different health outcomes. For these Detroit data, combining all of the cardiovascular effects would have the effect of burying effects that may be both interesting and statistically significant.

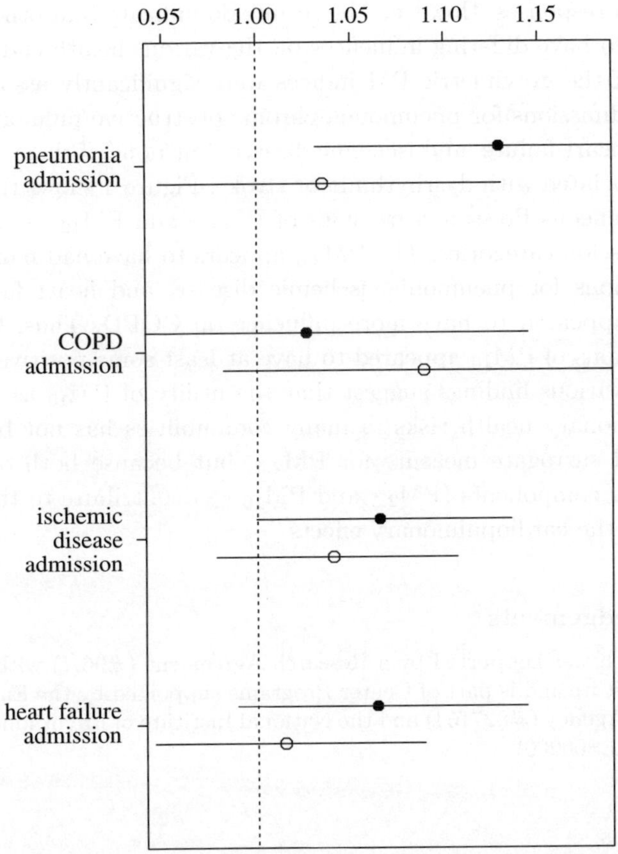

Fig. 4. Relative risks per 5–95th% increment of $PM_{2.5}$ (•) and $PM_{10-2.5}$ (○) simultaneously included in Poisson regressions, adjusting for seasonal cycles, temperature (locally estimated smoothing splines (LOESS) of same day and LOESS of average of 1–3 lags, and hot-and-humid indicator), and day-of-week, for 490 warm weather days in the Detroit metropolitan area in 1992, 1993 and 1994.

There is some biological plausibility for the results depicted in figure 3. For example, the indication that O_3 has its greatest apparent influence on ischemic disease is consistent with Seaton's alveolar inflammation hypothesis (Seaton *et al.* 1995), while the indication of CO having its greatest apparent influence on heart failure is consistent with CO's known effects on oxygen transport (Morris & Naumova 1998).

The Detroit study, lacking measurement data on UPs, cannot be used to help identify the role of ambient UPs concentrations on health-related population responses. However, the results do indicate that other PM components can have differing influences on the various health endpoints. One or more of the gravimetric PM indices were significantly associated with hospital admissions for pneumonia, chronic obstructive pulmonary disease (COPD), heart failure, and ischemic disease, but none of them were significantly associated with dysrhythmia or stroke. Figure 4 shows the results of the simultaneous Poisson regressions of $PM_{2/5}$ and $PM_{10-2.5}$ on four hospital admission categories. The $PM_{2.5}$ appears to have had more influence on admissions for pneumonia, ischemic disease, and heart failure, while $PM_{10-2.5}$ appeared to have more influence on COPD. Thus, both of the mass fractions of PM_{10} appeared to have at least some positive influence.

These various findings suggest that the utility of PM_{10} as an index of cardiopulmonary health risks in many communities has not been merely as a useful surrogate measure for $PM_{2.5}$, but because both of its major gravimetric components ($PM_{2.5}$ and $PM_{10-2.5}$) contribute to the elevation of some of the cardiopulmonary effects.

Acknowledgements

This research was supported by a Research Agreement (#96-1) with the Health Effects Institute and is part of Center Programs supported by the Environmental Protection Agency (#827164) and the National Institute of Environmental Health Sciences (#ES00260).

References

Burnett, R. T., Cakmak, S., Brook, J. R. & Krewski, D. 1997 The role of particulate size and chemistry in the association between summertime ambient air pollution and hospitalization for cardiorespiratory diseases. *Environ. Health Perspect.* **105**, 614–620.

Castillejos, M., Borja-Abarto, V. H., Dockery, D. W., Gold, D. R. & Loomis, D. 2000 Airborne coarse particles and mortality. *Inhal. Toxicol.* **12** (Suppl. 1), 61–72.

Cifuentes, L. A. & Vega, J. 2000 Effect of fine fraction vs. coarse mass in Santiago, Chile. In *Proc. of Particulate Matter and Health Conf., 24–28 January 2000, Charleston, SC* (Abstract).

Creason, J., *et al.* 2000 Effects of particulate matter on the heart rate variability of elderly residents in an East Coast retirement community: the Baltimore 1998 PM study. In *Proc. of Particulate Matter and Health Conf., 24–28 January 2000, Charleston, SC* (Abstract).

Liao, D., Creason, J., Shy, C., Williams, R., Watts, R. & Zweidinger, R. 1999 Daily variation of particulate air pollution and poor cardiac autonomous control in the elderly. *Environ. Health Perspect.* **107**, 521–525.

Lippmann, M., Briant, J. K. & Fang, C. P. 1994 The influence of axial core flow on particle penetration in lung airways. *Ann. Occup. Hyg.* **38** (Suppl. 1), 39–45.

Lippmann, M., Ito, K. & Burnett, R. T. 2000 Association of PM components with daily mortality and morbidity in urban populations. HEI Report. Cambridge, MA: Health Effects Institute. (In the press.)

McDonnell, W. F., *et al.* 2000 Association of mortality with the fine and coarse fractions of long-term ambient PM_{10} concentrations in nonsmokers. In *Proc. of Particulate Matter and Health Conf., 24–28 January 2000, Charleston, SC* (Abstract).

Morris, R. D. & Naumova, E. N. 1998 Carbon monoxide and hospital admissions for congestive heart failure: evidence of an increased effect at low temperature. *Environ. Health Perspect.* **106**, 649–653.

Neas, L. M., *et al.* 1999 Fine particles and peak flow in children: acidity versus mass. *Epidemiol.* **10**, 550–553.

Ostro, B. D. & Lipsett, M. J. 2000 Coarse and fine particles and daily mortality in the Coachella Valley, California: a follow-up study. In *Proc. of Particulate Matter and Health Conf., 24–28 January 2000, Charleston, SC* (Abstract).

Peters, A., Wichmann, E., Tuch, T., Heinrich, J. & Heyder, J. 1997 Respiratory effects are associated with the number of ultrafine particles. *Am. J. Respir. Crit. Care Med.* **54**, 1376–1383.

Schlesinger, R. B. 2000 Properties of ambient PM responsible for human health effects: coherence between epidemiology and toxicology. *Inhal. Toxicol.* **12** (Suppl. 1), 23–25.

Seaton, A., MacNee, W., Donaldson, K. & Godden, D. 1995 Particulate air pollution and acute health effects. *Lancet* **345**, 176–178.

Sheppard, L., Levy, D., Norris, G., Larson, T. V. & Koenig, J. Q. 1999 Effects of ambient air pollution on nonelderly asthma hospital admissions in Seattle, Washington, 1987–1994. *Epidemiology* **10**, 23–30.

Simpson, R. *et al.* 2000 Effects of ambient particle pollution on daily mortality in Melbourne, 1991–1996. In *Proc. of Particulate Matter and Health Conf., 24–28 January 2000, Charleston, SC* (Abstract).

Van Den Eeden, S. K., Quesenberry, C. P., Shan, J., Lurman, F., Lugg, M. & Segal, M. 1999 A parallel time-series study of air pollution in the Los Angeles Air Basin. In *Proc. 3rd Colloquium on Particulate Air Pollution and Human Health* (ed. R. Phalen & Y. Bell), Session 2. Relevant PM Properties Related, pp. 11–15, Abstract 057. Irvine: Air Pollution Health Effects Laboratory, University of California.

INDEX